The Age of Analogy

The Age of Analogy
Science and Literature between the Darwins

Devin Griffiths

Johns Hopkins University Press
Baltimore

Johns Hopkins University Press
2715 North Charles Street
Baltimore, Maryland 21218-4363
www.press.jhu.edu

Library of Congress Cataloging-in-Publication Data

Names: Griffiths, Devin, 1978– author.
Title: The age of analogy : science and literature between the Darwins /
Devin Griffiths.
Description: Baltimore : Johns Hopkins University Press, 2016. |
Includes bibliographical references and index.
Identifiers: LCCN 2015051184 | ISBN 9781421420769 (hardcover :
alk. paper) | ISBN 9781421420776 (electronic) | ISBN 1421420767
(hardcover : alk. paper) | ISBN 1421420775 (electronic)
Subjects: LCSH: Literature and science—England—History—19th
century. | English literature—19th century—History and criticism. |
Science in literature. | Nature in literature. | Darwin, Charles,
1809–1882—Criticism and interpretation. | Darwin, Erasmus,
1731–1802—Knowledge—Natural history. | Darwin, Erasmus,
1731–1802—Influence.
Classification: LCC PR468.S34 G75 2016 | DDC 820.9/36—dc23
LC record available at http://lccn.loc.gov/2015051184

A catalog record for this book is available from the British Library.

Special discounts are available for bulk purchases of this book.
For more information, please contact Special Sales at 410-516-6936 or
specialsales@press.jhu.edu.

Johns Hopkins University Press uses environmentally friendly book materials,
including recycled text paper that is composed of at least 30 percent
post-consumer waste, whenever possible.

CONTENTS

This book is about collaboration. Any work, whether a poem or a scientific treatise, is the product of many hands: authors, printers, publishers, colleagues,
family, friends—the larger world of people and things that help a work succeed. All are, to some degree, coauthors. Working on this book has made me
intensely aware of that larger contribution, and it's a pleasure to now perma-
nently acknowledge that help.

First, I want to thank my wife and partner, Meg. From my first confused
attempts to explain my intuitions about analogy, to the years of research trips
and rewrites that followed, she's been my closest confidant and, when needed,
critic. Neither of us realized what she was signing up for. She's had to live with
this book for more than a decade; our house is still littered with overdue books
and scratched-over papers. Not only has she made room for it all, she helped
make this the best book we could write. It's just one of the many reasons I love
her. Her name should be on the cover.

I've also been lucky to have the support of a wide community of scholars, colleagues, and mentors who've taught me what I know about the nineteenth century
and, even more daunting, how to write a book. George Levine has been a constant
source of guidance and inspiration; he taught me how to wed my love of biology to
my love of literature, and how to write clearly about that complex relationship. I
feel lucky to count Kate Flint as a mentor, colleague, and friend. Kate has been a
wellspring of advice and has helped me navigate the challenges of turning a messy
dissertation into a slightly less messy book. And I'm tremendously grateful to the
wider community of Rutgers University, where I first met George and Kate. Caroline Williams, Colin Jager, and Michael McKeon were strong supporters of this
work and helped me tease out some of its thornier problems. Like many, I was
fortunate to have the limitless support of Barry Qualls, who also mentored me in
my first attempts to teach the Victorian novel. At Rutgers I made friends I still
look for at conferences whenever I need a good laugh, a quick confab, or a hoppy
beer. I also want to thank the Center for Critical Analysis, whose curriculum
and support provide such a vibrant intellectual center. And what would I have
done without Cheryl Robinson, who, for so many of us, made Rutgers a home
away from home?

My two years at the University of Pennsylvania were thrilling; the conversations I had with colleagues, through venues like the Penn Humanities Forum
and the Benjamin Franklin Scholars Program, and in the various saloons,
coffee shops, and restos of Philly, provided the space I needed to turn a fresh

critical perspective on my work and leap from dissertation to book. Michael Gamer has been a close friend and constant support, from reading the manuscript in full to trading cocktail tips and the recipe for his marvelous braciole. I also picked up valuable pointers working with graduate students in our seminar together, and I'm thankful. The Penn community was endlessly supportive; I want to thank Jim English (always ready to trade pizza secrets) for reading the introduction and helping me leaven the argument. My thanks also to Rita Barnard, Toni Bowers, Max Cavitch, Jed Esty, Tsitsi Jaji, Paul Saint-Amour, and Peter Struck for their generous help and advice. My warm regard goes also to the graduate students at Penn, who were a consistent source of inspiration and support.

In moving to California, I've had the great fortune of finding a welcoming new community of friends and colleagues at the University of Southern California. I'm lucky to call it home. Hilary Schor is not only a close friend and confidante—her enthusiasm and her exhilarating range of interests are an inspiration. I'm totally reliant on her advice, whether about Victorian literature or the complexities of institutional protocol. Emily Anderson is a good friend and really too generous with her time, whether that has meant hosting me during a campus visit, going over proposal materials and drafts, or answering yet another dumb question about the graduate curriculum. I'm also thankful to Joe Boone, Rebecca Lemon, and Meg Russett both for their friendship and for their generous help with portions of the manuscript. I also have found an energizing band of colleagues within the Comparative Literature faculty and the race and empire study group. My gratitude goes to Neetu Khanna (the group's presiding genius), Olivia Harrison, Pani Norindr, Julie Van Dam, and the rest of the group for their advice and for their help with the introduction. My colleagues in history, Daniela Bleichmar and Dean Peter Mancall, have given valuable help in framing and marketing this book. Finally, I want to thank the graduate students at USC, particularly the participants in my spring 2015 seminar on science and literature, for their canny insights and their help with my introduction.

There is a much wider community of scholars here in California and beyond who've made this book possible through their generous reading and advice. Jay Clayton read this book in its earliest incarnation and gave me helpful advice and a boost of confidence in its possibilities. Nathan Hensley and Henry Turner helped me reshape the introduction and punch up its argument. Greta La Fleur and Alexander Regier helped me to sharpen my account of Erasmus Darwin's Romantic science in the first chapter. Ian Duncan, really the ideal reader for this book, was invaluable in helping to fill out my chapter on Scott, and I can't thank him enough. Jason Rudy and Jesse Hoffman (go Rutgers alums!) were sensitive readers of my Tennyson chapter, and I was thrilled to receive Jim Kinkaid's help in handling *In Memoriam*. Aaron Matz, Catherine Gallagher, Dehn Gilmore, and Jonathan Grossman both read the Eliot chapter and provided crucial help tightening its central thesis. I want to thank Dehn additionally for her generous dinners and her sage advice; she makes Los Angeles a warm place to study Victorian literature. And the nineteenth-century reading group that Jonathan organizes, along with his shrewd crew of graduate students at UCLA, constitutes both a center of gravity for Victorian studies here and also provided astute advice about my introduction and its argument. Finally, Darwin. Beyond the endless help and inspiration I received from George Levine, I was fortunate to have the help of John Plotz and Lynn Voskuil in tracing a clearer argumentative line for my final chapter. And I would hate to forget to thank Jesse Oak Taylor and the other members of the NAVSA Eco-caucus, who I've only recently come to know well, but whose conversations have inspired the late turn in my thinking toward Darwin's ecological imagination and its implications for the epistemology of life.

Since it's a first book (and, as I keep telling myself, you only get to do this once), I also want to thank the friends and family who've supported the literary and scientific tracks in my imagination. These include many gifted teachers over the years, especially Carol Mackay, who introduced me to *Middlemarch* at UT Austin, and Brent Iverson, who welcomed me into his lab there. It wasn't until I hit grad school that I realized that being an English professor means you never have to leave an interest behind, and I feel lucky to have found a career where I can continue to pursue my fascination with both science and literature.

Even though I've been trying hard to keep track, I'm sure there are many I've forgotten to thank personally for their generous help and support, and that includes the marvelous administrative staff and archivists who've helped me at the various institutions where I've pursued my work. The staffs at the Huntington Library in San Marino, the National Library of Scotland, and the British Library continue to inspire me with their mastery of their collections and seemingly endless generosity. I also want to thank Andrew Ascherl and Patrick Hruby, as well as Matt McAdam, Catherine Goldstead, Brian MacDonald, and the tireless staff at Johns Hopkins University Press. They've shown tremendous care and generosity to my firstborn. I'm grateful.

Finally, I've saved this ultimate place to thank the two people who have supported my work the longest, and who encapsulate between them my fascination with both literature and science: my parents, Therese and Scott. They have been a constant source of strength, inspiration, and understanding. Not least, they've been tirelessly devoted to supporting my studies (with cash when needed) and giving me a shoulder to lean on when things seemed bleak. It felt like a long haul, and I'm lucky they were not only along for the ride but willing to help steer or fix a tire when necessary. My love and thanks.

Work on this book was generously supported by grants and fellowships, including a year-long graduate fellowship from the Center for Cultural Analysis at Rutgers, a two-year Mellon Postdoctoral Teaching Fellowship from the University of Pennsylvania, and a generous grant from the Office of the Dean of Dornsife College, here at USC.

My gratitude to *SEL: Studies in English Literature* for allowing me to reprint portions of my article "The Intuitions of Analogy in Erasmus Darwin's Poetics" in the first chapter. My thanks also to *Nineteenth-Century Contexts* for allowing me to republish portions of my article "Flattening the World: Natural Theology and the Ecology of Darwin's *Orchids*" as part of my fifth chapter.

The Age of Analogy

Analogy under a Different Form

In the summer of 1837, Charles Darwin unlocked the clasp of a new brown-backed journal, the first of a series of notebooks in which he scratched away at a radical new approach to the mutability of species. Though most scientists agreed that species were fixed, Darwin had been mulling over a theory of species transformation for at least two years, ever since visiting the Galápagos Islands as the naturalist of the HMS *Beagle*. The new notebook was a fresh start. Darwin filled the top of the page with one word, underlined and scored in dark ink: "Zoonomia" (figure 1).

This was a reference to the masterwork of his grandfather, Erasmus Darwin's *Zoonomia, or the Laws of Organic Life* (1794). Charles used this citation to inaugurate the jumbled series of notes, quotes, musings, and diagrams that culminated in *On the Origin of Species* (1859). Respectively, these were the two most important evolutionary treatises of eighteenth-century and nineteenth-century Britain. Yet Charles's radically different approach to the problem of evolution, and the profound contrast between his cautious prose and his grandfather's soaring experiments in verse and natural philosophy, have frustrated attempts to answer a long-standing biographical question about the Darwins: What is the relationship between these two naturalists and their works?

Surprisingly, the answer until now has been: very little. For most of his life, Charles denied taking his grandfather's work seriously. Erasmus was notorious for his speculative and poetic fancy; until his grandson's success, to "Darwinize" in Britain meant to substitute wild speculation for sober scientific study. Alongside Erasmus's wild and racy botanical epics—which included elaborate orgies between the personified sex organs of plants—Charles's lengthy studies of barnacles, botany, and speciation seemed soberly Victorian. Privately, Charles insisted that his grandfather had not had "any effect" on his work. When a short historical sketch of previous evolutionary thinkers was added to later editions of the *Origin*, Erasmus was tucked into a brief footnote. It was only much later, as Charles was writing a biography of his grandfather, and approaching the end of his own life, that he admitted what had seemed obvious to many: "Hearing rather early in life such views maintained and praised may have favoured my upholding them under a different form."[1]

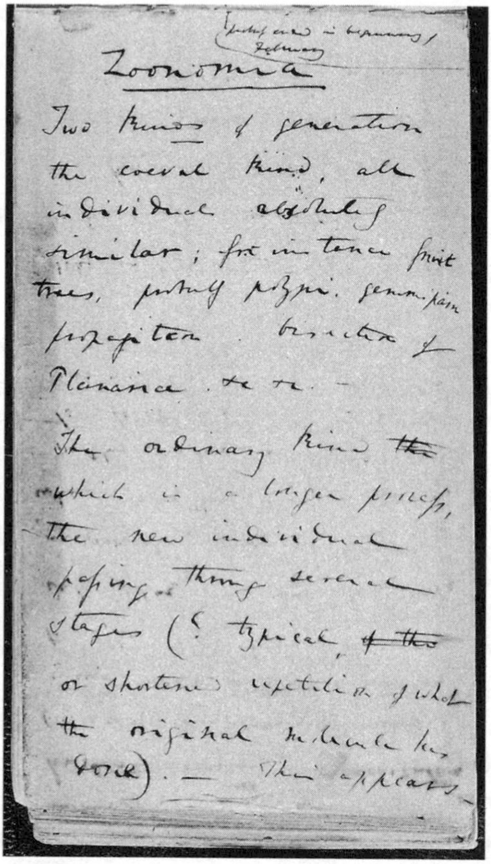

Figure 1. Manuscript page. From Charles Darwin, "Notebook B" (1837), 1r. Courtesy of Cambridge University Library, Cambridge

To address the relation between Erasmus and Charles Darwin, we must reconsider the intimate relation between scientific practice and literary writing, both because Erasmus foregrounded their relation in far-reaching ways and because his grandson's success was predicated on finding a new way to imagine and describe scientific study. This book understands Charles's "different form" as a new mode of historical description furnished by contemporary historical fiction and other imaginative genres. Emerging after Erasmus Darwin's death, these new literary modes and historical procedures, which took definitive form in the historical novel, constituted a new technology for writing about past events and thinking about their complex relations to present experience. These techniques shared a commitment to analogy, using it as a tool that brings the relation between previous ages and the present into focus, seeking the origin of contemporary social and natural order within the patterns of past events. This comparative turn has gone unnoticed by historiographers because it developed outside of traditional historical genres in a host of associated fields. These alternative theorists of the past, including philologists, anatomists, mythologists, and antiquarians, as well as literary authors, used comparative analysis to address the past in terms of contingent patterns of relation and differentiation. As Thomas Carlyle would put this, in an influential analysis of this new regime of historical description: "These Historical novels have taught all men this truth, which looks like a truism *but was as good as unknown to writers of history* . . . that all the bygone ages of the world were actually filled with living men, not by protocols, state papers, controversies and abstractions" (emphasis added).[2] Carlyle is describing a new sense of history's relation to "living" experience that ultimately extended well beyond the novel, and he notes that its most startling feature was that it emerged outside of traditional histories. Studies of nineteenth-century historical methods have emphasized the growing importance of empirical verification, source criticism, and primary documents; yet as Carlyle observes here, the new sense of "living" history rested on standards of truth and methods that were alien to the study of "protocols, state papers," and other such "abstractions" from life.[3] Like Carlyle, I see this shifting view as the major contribution of nineteenth-century historical fiction.

This historical sensibility produced an intuitive modern sense that we live in historical time, characterized by later historians like Mark Salber Phillips as the "historicization of everyday life."[4] In the analysis of later historiographers, this shift takes hold in the transition from the eighteenth to the nineteenth century, over the same period that new literary genres, particularly the historical novel, captured the imagination of Western Europe. In the wake of the French Revolution and in response to the explosive growth of natural history collections, a new generation of authors rejected static schemas, epic narratives, and the stability of earlier typologies of change in favor of comparing and analyzing local patterns *between* individuals, artifacts, epochs, and social systems. Though comparison was often used to support master narratives about the past (as testified by works like Adam Smith's *Wealth of Nations* [1776]), these new writers of history placed analogy above system building.[5] *The Age of Analogy* provides to literary history, scientific history, and historiography an account of how the literary past contributed this comparative turn to the nineteenth-century historical imagination.

If Benedict Anderson and Sue Zemka have convincingly argued that industrialization and mass media impacted the collective experience of historical time, this book explains how comparatism shaped that experience.[6] The chapters that follow recognize comparatism as a broadly distributed and widely influential approach to understanding the past, which, though now seeming (as Carlyle says) "like a truism," has gone unmarked as a nineteenth-century invention. I term this new habit of thought *comparative historicism* to distinguish it from "comparative history," a related but narrower movement in twentieth-century historical study. The latter drew its comparisons between national histories on the model of comparative literature and anthropology.[7] By contrast, comparative historicism names a broadly shared habit of thinking comparatively about previous ages and customs that, in turn, conditioned the formation of these later disciplines.

In this way, I provide a fresh look at interdisciplinarity, excavating a historical epistemology (as Lorraine Daston has put it) that was organized through analogies drawn between the past and present.[8] This sense of historical comparison, characterized by A. Dwight Culler as the "mirror of history," has had a long influence.[9] Hugh Grady suggests that such reflection remains "inescapable," insofar as "historicism itself necessarily produces an implicit allegory of the present."[10] But this insight is inescapable only to the degree that comparative historicism underwrites our understanding of the past, making similarities as well as differences between the past and present both meaningful and potentially allegorical.

Though comparative historicism took shape in the nineteenth century, it continues to condition how we humanists approach historical questions. Of the academic articles published between 1984 and 2014 and indexed under the "History" category in JSTOR's online database, sixty-four percent include at least one of the following terms (including cognates): *analogy, comparison*, or *homology*. I choose these three terms with care, because it was in the nineteenth century that all three took the mutual configuration that defines their modern use, a transformation central to this book. Forty-one percent of articles for the same period filed under "Language and Literature" contain these same terms, and that percentage is higher in journals that focus on literary criticism (sixty-six percent in *PMLA*, for instance). I encourage any humanist readers to pull out the last three essays they've written and look for these terms. Notice the work they do. They hold together important pieces of the argument, whether the critical connection between the objects under consideration or, more often, the relation between those objects and the larger historical, social, or formal canvas they are held up to. Often they map connections to the external conceptual model that helps explain those objects. These are only the explicit cases; multipart comparisons are foundational to our work.

This is a book about relationships in time, between writers and scientists, between books and artifacts, rather than about individual figures or works. To understand why comparative thinking became so important to the modern imagination, *The Age of Analogy* turns back to the nineteenth century and the writers who helped tease out its possibilities. The value of a full estimate of comparative historicism's influence on the nineteenth century can be gauged in the way it undermines confident assumptions about this period and its literature. Here I challenge three such assumptions: first, that Erasmus Darwin's reputation collapsed because he was a flake, a dilettante who confused Romantic-era science for literature; second, that there is little more to add to our understanding of the relation between Charles Darwin and literary history; and, finally, that the later Darwin developed a way of analyzing nature that naturalized social norms, with profound and disturbing influence on later thinking about humanity and its social relations. All three, I argue, are based on an insufficient grasp of how these writers and scientists worked to expand and develop comparative historicism as a way to draw connections between the scientific and the literary, and between the present and its past.

By turning back to the archive, I show that Erasmus Darwin, far from working at the fringes of the print market, labored to refashion contemporary literary models, particularly the epic, into a new mode of narrative analysis that could address the contingent world disclosed by contemporary natural science. Erasmus Darwin's epic *Botanic Garden* (1789–91) and his scientific masterwork *Zoonomia* (1793), written in advance of the comparative synthesis, relied on older theories of analogy and neoclassical poetic protocols. For these reasons, he was unable to marshal a compelling account of evolutionary change; put more accurately, the grand evolutionary patterns he described failed to persuade a subsequent generation of poets and naturalists.

When Charles Darwin reviewed his grandfather's efforts through the lens of the nineteenth-century comparative turn, he recognized an uncanny sense for where such thinking would soon lead. As I show, the intervening link that connected them—the new literary mode that made it possible to narrate the evolutionary past convincingly—was not literary language broadly conceived or even the novel *tout court*, but rather, the historical novel, particularly as authored by Walter Scott. The influential account of social history in Scott's novels—detailing both the tension between "history from below" and political history and the interrelationship of present and past societies—drew on Scott's substantial involvement with comparative naturalism. He was both an avid antiquarian and a philologist, and his investment in scientific history culminated in his appointment as president of the Royal Society of Edinburgh. Scott's historical fiction forged new strategies of narrative comparison. The literary techniques that Scott aligned, particularly the way his narrators wield sympathy and interpolate intent in order to suture contingent plots to historical perspective, gave Charles Darwin a way to imagine more flexible and persuasive stories about chance events in the past. Immersed from his youth in Scott's novels and the explosion of historical fictions that followed, Charles Darwin succeeded where his grandfather had failed, adapting the narrative idiom of comparative historicism to revise his grandfather's ideas. In taking up the new historical sensibility that Scott's fiction advanced, he engaged in comparatism and its habit of treasuring the possibility that things can always be otherwise. By these means, Charles Darwin advanced a mode of comparative analysis that critically undermines the authority of the given. In writing "Zoonomia" at the top of his transformational notebook, Charles Darwin cleared imaginative space for his desire to revive and reform his grandfather's efforts, translating the theory of evolution into the network of imaginative comparisons that sustain the *Origin of Species*. A careful reading of this larger shift in historical thought will expand our understanding of comparative historicism's radical potential, as it unsettles the past and makes space for fresh encounters in the present.

In this study, comparative historicism operates "between the Darwins" in two senses: as a fresh perspective on the relationship between Erasmus and Charles but more importantly as a recognition that their major works bracket a fundamental shift in how the past was understood and described. In this sense, the Darwins, both Erasmus and Charles, conveniently frame the seventy-year period over which comparative historicism emerged as a dominant procedure for thinking about and writing about the past. This book is after something larger than the Darwins: it seeks to recognize a historical sensibility that still conditions our understanding of human and natural history. For this reason, and despite his extraordinary influence and success in translating his grandfather's thinking, Charles Darwin is just one of the writers I study. As the following chapters show, his approach to comparative historicism was widely contested, by both contemporary naturalists and historical writers anxious about the Darwinian inheritance. That inheritance was more differentiated than we often assume. Placing Charles Darwin within the larger comparative turn shows that contemporaries like Alfred Tennyson, George Eliot, and Richard Owen found radically different implications within the new historical program. So, if Charles Darwin thought that comparative analysis disclosed a past filled with uneven and discontinuous histories, Tennyson and Owen argued that such comparisons proved the unity of the natural plan and its singular divine law. Hence, while the first half of the present book gives a genealogy for the emergence of comparative historicism in the nineteenth century, the latter half offers a comparative analysis of three distinct visions (some more successful than others) of where it might lead.

Science, Literature, and History

The *Origin of Species* was the culmination of extensive field research into natural and domestic species and even wider reading. Darwin brought these together by writing at his desk at Down House in rural Kent, fifteen miles from London. Written text is the central technology of the *Origin*, both as immediate instrument and as a modality of social engagement, because it coordinated Darwin's extensive correspondence, profuse journals, and the hundreds of scientific monographs and articles he drew from. Here I underline the important fact that Darwin sat down to write the *Origin* only when he became anxious to intervene in the print community, worried that he might be "forestalled" by Alfred Russel Wallace. With its various editions, the *Origin* was only one node within this larger network of individuals, events, and texts; but it was a central one, and it came together as a singular work with extraordinary imaginative power.

The central object for the subfield of humanistic study called "science and literature" is to explain the role of imaginative language in science and to explore the impact of literary form on scientific practice. Charles Darwin has long been its major figure. An early concern for the relationship between liberal and scientific education opened in dialogue with Darwin's work, particularly in exchanges between the naturalist Thomas Henry Huxley (the "bulldog" who famously mauled Darwin's critics in the Anglican Church) and Matthew Arnold.[11] Their argument was revived in the mid-twentieth century, as a generation of writers and theorists, including Aldous Huxley, argued for the poetics of scientific practice, emphasizing not merely the literary qualities of descriptive science but also the way writers like Darwin engage the imagination.[12] All emphasized the importance of working across, even as they underlined, the practical differences between humanist and scientific inquiry. The ongoing conjunction of this field as "science *and* literature" captures this division between "two cultures" (as described by C. P. Snow), a continuing allusion to Huxley and Arnold and their foundational accounts a century before: "Science and Culture" versus "Literature and Science."[13]

Many literary historians have focused on the flow of scientific thought across this divide and into contemporary literature, but a subsequent generation of literary scholars, including Gillian Beer, George Levine, and Sally Shuttleworth, has worked to flesh out the common ground between literature and science implied in the ubiquitous conjunction "and."[14] One endpoint to this critical direction is expressed in the argument, made by Levine and others, that there was in fact "one culture" (contra Snow) within which both science and literature operated in the nineteenth century. The one-culture thesis was closely tied to a modern critical understanding of social forms as embedded in systems of communication that were universally available to the strategies of textual analysis, a perspective often described as "textualism." Ambitiously comprehensive, the one-culture thesis tacitly aligns the local with the global, aligning the individual agent with both nationalism and the imperial project (the "cultural integrity of empire" posited by Edward Said) in ways that make it difficult to differentiate between what happens at either smaller scales (the social units of local and semipublic spaces within the nation) or larger (the cosmopolitan, transnational, and often translingual communities inhabited by scientists and authors).[15]

This study explores the importance of literary modes to the scientific imagination without insisting on a convergent model of nineteenth-century British culture. Instead of one culture, I explore how a network of writers contributed to comparative historicism as one differentiated mode of historical understanding among many.[16] This focus on a single episteme or paradigm makes it possible to discriminate among the different contributions made by the authors I study. So, for example, Walter Scott's Waverley novels were points of convergence that gave focus to the narrative strategies that came to sustain comparative history, whereas Alfred Tennyson's *In Memoriam* stands as an outlier, a limit case for the capacity of verse to formalize comparative inquiry while retreating from its dispiriting implications.

The Age of Analogy is, in the first instance, a literary history, and so it examines works of literature, science, and history under the conviction that changes in historical understanding are disclosed through close attention to the innovative language that collects this movement on the page. At the same time, the text is not everything, and I consider the works that follow as participants in a larger network of people, objects, and modalities. In part, I look to ongoing work inspired by the history of science and sociology that provides a more systematic approach. Literary scholars including Adelene Buckland, Gowan Dawson, Jonathan Smith, and Jon Klancher have focused our attention on the material practices of both nineteenth-century science and the literary marketplace.[17] Klancher characterizes nineteenth-century Britain in terms of the incomplete professionalization of science and academic humanism, as an era that therefore supported a dialogue between their constituent fields, particularly in popular "arts and sciences" institutions and more generally in the eclectic pages of contemporary periodicals like the *Edinburgh Review* and the *Quarterly Review*. Conceiving science and literature as filiated systems of worldly engagement, authorship, and economic life, these scholars remind us that the field of science and literature must attend to differences in how scientists and literary authors work.

The stakes for our understanding of comparative historicism are greater than the recognition of important analogies between "science and literature," or even the case for substantive exchange within the social and economic practices that supported them. As naturalists and literary authors turned toward each other in their efforts to shape a historical understanding suited to their different ends, collaboration became central to their work and left a profound imprint on the modern imagination.[18] For this reason, many of the chapters that follow recuperate comparative historicism as a collaborative project, sustained collectively by authors and scientists, subsisting between time periods, and transcending geographic, social, and even spiritual divides. So, in my third chapter, I explore Tennyson's *In Memoriam* as a collaborative poem, reviving and embracing the verse forms of his deceased friend, Arthur Henry Hallam, in order to constitute the poem itself as an active field of shared authorship and distributed life.

As an empiricism that collaborates through the text, the descriptive science of the nineteenth century can be understood as participating in what I term *collective authorship*, an interactive network of print production in which the subjects (writers, readers, publishers, and printers) and the objects of inquiry (whether natural or human) interact over distances that are simultaneously physical, temporal, and imaginative.[19] I emphasize the intimate relationship between descriptive science and the printing industry, both scientific and literary, but I also recognize the importance of the world of print to scientific networks of fact and observation. As I further explain in the prelude that follows, *The Age of Analogy* places the study of the material production of texts in dialogue with theories of collaborative inquiry, and especially the actor-network theory of Bruno Latour. If historians have shown us how books are produced through the collective labor of authors, printers, publishers, booksellers, and buyers (rather than by the author alone), actor-network theory aligns the network of print production with the distributed networks that produce scientific knowledge: communities of scientists, technicians, bureaucrats, instruments, and scientific objects.[20] Collective authorship locates scientists and authors within a dispersed community of agents that includes the industries and readers that engaged their work, and the actors—real and imagined—they wrote about.

Though it has been criticized for being descriptive (rather than critical), this kind of network analysis is useful because it serializes the constituents of historical accounts, placing them at the same level, and so it helps explain how even semifictional actors—like Scott's historical protagonists or Darwin's "inherited characters"—make substantive contributions to our knowledge of the past. As I explain in the prelude, I locate this counterintuitive insight in Latour's later analysis of actor-network theory's thorough realism, which makes strong claims for how the world shapes its own description.[21] Description of the past does not prescribe its features. Realism, whether scientific or literary, finds common ground with imaginative historicism because, although all fiction is constrained by convention, plausible fictions are further structured by their hypothetical status. As reconstructions of what might have happened, historical narratives provide an exemplary account of what perhaps did. By these means, historical fiction weighs in on the world, pushing back on implausible accounts, and giving tentative access to a real past. In the chapters that follow, I explore how comparative historicism, as developed through this collective authorship, honed analogy as a method to construe these plausible fictions, and recognize the importance of the historical novel to both literary realism and nineteenth-century thought.

Collaborations between scientists and authors underline the unusual features of descriptive science. A renewed focus on practice—particularly in histories and sociologies of science—has only reinforced our sense that the ubiquitous pen and common printing press were the chief instruments of scientific production in the early nineteenth century. Yet our most important histories of British scientific empiricism place greater emphasis on the development of specialized instrumentation, the importance of novelty, and new protocols of visual representation. Thus, Steven Shapin and Simon Schaffer argue that experimentation in the early modern period rested on instrumental, social, and textual technologies that secured scientific consensus, tools that supported a new mode of "virtual witnessing"—the willingness of other naturalists to accept a scientific observation as if they'd experienced it themselves. Barbara Benedict and Katherine Park have provided a complementary account of the importance of acquisition and the engagement of wonderful specimens to a culture of curiosity in the early modern world. For Lorraine Daston and Peter Galison, such practices elevated the importance of specular engagement with exemplary objects in constructing a "collective empiricism" that allowed naturalists to collaborate across geographic and historical divides. The atlas succeeds the wonder cabinet in this evolving vision, as it "trains the eye to pick out certain kinds of objects as exemplary" and secures a visual *lingua franca* for the larger scientific community.[22]

These theories need significant adjustment to address the kind of imaginative history that Darwin produces in the *Origin of Species*. Like the written word, the visual is a speculative rather than specular category within its pages. On occasion, Darwin asks his reader to accept the virtual witness of his own eye, but he more commonly asks the reader to imagine something with him—a moment of adaptation, an incident in history—that is effectively invisible.[23] Comparative historicism is a core element of his textual technology, producing these histories for the reader by carefully tracing their analogy to other events (in the domestic, contemporary world) that could be described from life. The sole illustration in the *Origin*, the famous branching tree of life, is an abstract of this imagined past; it is not an exemplary object but an ambiguous schema that traces a network of similarities and differences between examples that do not exist. Rather than a visual and experimental science, Darwin's science is narrative and comparative, relying heavily on formal experiment and those common technologies—the pencil, the handwritten note, the printed page—that can conjure the imagined past.

Analogies give voice to patterns that have no name. Hence, in the nineteenth century they were enormously important to scientific fields, including geology, comparative anatomy, and botany, which relied on description and imagination to elucidate natural pattern. Rather than experiment, these "descriptive" sciences (so characterized in contrast to "normative" or "predictive" sciences) rely on the technologies of narrative description and comparison in order to make new patterns visible.[24] It has long been recognized that evolutionary science, in Darwin's time and in our own, has peculiar features that distinguish it from the traditional physical sciences—it is particularly "historical," relying on a reconstruction of past events rather than being "hypothesis driven" or "experimental."[25] Biologists David A. Grimaldi and Michael S. Engel have recently argued that descriptive science, though often a term of derision, remains vital because it works in advance of predictive sciences, deploying a "comparative method [that] assesses variation among individual things, and organizes it into systems and classifications that are used for making predictions."[26] They see Darwin as an important example, and their account is especially relevant to the *Origin of Species*, which returns repeatedly to a central analogy between meticulous descriptions of domestication and imagined histories of natural selection. As such, this study contributes to our understanding of the "specialness" of evolutionary science itself and contributes to a growing body of scholarship addressing the fundamental disunity of the sciences.[27] One focus of Darwin's study is on the constitutive tension in deciding whether a group of organisms should be grouped together or assigned to distinct subgroups (a tension Darwin described as the disagreement between "lumpers" and "splitters"), and Darwin shows that this tension is a signature of the different strategies comparatists devised to address the patterns of a common evolutionary history.[28] He uses imaginative history to explore how stories about what might have happened explain what we find in nature and in social systems. In this way, Darwin's work tested comparative historicism's ability to disclose new patterns and break from conserved narratives, offering a new way to understand complex interactions in time, a new way to let the past speak.

The New Historicism

In order to gauge the complexity of this comparative turn, so crucial to the nineteenth-century historical imagination in Britain, it is necessary to begin in the century before. The shift I study coincides with the transition, offered in Michel Foucault's now classic study *The Order of Things*, from the "classical" episteme, marked by unitary systems of order, into the "modern," marked above all by the "profound sense of historicity" that shaped later schemes of organization.[29] My study has more modest aims, focusing instead on a handful of figures spanning a seventy-year period. While the synchronic series of Foucault's analysis explicitly precluded a study of the causes of this transition, I look between the Darwins to sketch out the problems that confronted eighteenth-century schemes of historical organization in Britain and motivated a shift toward comparative historicism. As I explain at greater length in chapter 1, at the close of the eighteenth century British historians relied on three major views of the past: a Christian tradition of biblical typology, including the eschatology of "church historicism," which sifted analogies between biblical prophecy and secular events; a Whig-progressive synthesis that understood the restoration as part of a continuous narrative of constitutional development; and the "stadial history" of the Scottish Enlightenment, which analyzed the universal stages of social and economic development that characterize modern and ancient society. But in the aftermath of the French Revolution and renewed anxieties of domestic unrest, these unitary historical narratives could no longer provide a persuasive account of the uneven past. In a convergent development, the accelerating aggregation of scientific specimens and artifacts, which was driven by imperial expansion and the collaborative efforts of naturalists and lay collectors, generated expansive individual and institutional collections that did not fit into rational or Linnaean taxonomies. Comparative historicism emerged as a way to address these growing congeries of objects, abandoning the conceit of a governing system of order in favor of local, comparative explorations of the patterns and differences between individual artifacts and incidents. From the beginning, comparative historicism sought to engage historical alterity, substituting studies of both shared and distributed encounters for unitary narratives. If, in Indian historian Dipesh Chakrabarty's influential analysis, we still characterize "historicism" in terms of wholeness and progressivism, this means we have not yet registered its comparatist countermovement: an impulse toward complexity and difference, resonance and counterpoint, that continues to drive our best work.[30]

Many historians afford imaginative literature an important role in this new sensibility. John Tosh, for one, sees the massively popular historical novels of Walter Scott as prime movers in this change, arguing that Scott provided a new sense of the "autonomy of the past," characterized by organic and autonomous social systems. Tosh likely has in mind how a novel like *Waverley* (1814) pits the tribalism of the eighteenth-century Scottish Highlands against the Whig ascendance of the English parliamentary system; or perhaps how *Ivanhoe* (1820) counterpoises feudal Norman society to the social networks of its subjugated Saxon vassals. Certainly, the number of historians and authors who responded to Scott's historical vision is immense and included Thomas Babington Macaulay, Harriet Martineau, and Edward Bulwer Lytton in England; Leopold von Ranke, Wilhelm Heinrich Riehl, and Willibald Alexis in the German states; and Jules Michelet, Hippolyte Taine, and Honoré de Balzac in France—to name only three national traditions. Yet Scott's most important influence, I argue, was not his sense for the autonomy of the past but a new investment in the comparison of different social customs and historical periods, a comparison that emphasized connection to the present and its alternatives.[31] Comparative historicism emphasized the interrelationship of societies and customs, the complex and gradated connections between the present and the past, over distinctions between epochs and ways of life. It mobilized a comparative analysis over time and within specific periods in order to expose similarity within difference. It did not so much sharpen the difference between "ages" as show the process by which societies came to feel aged at all.

Historical fiction was central to this new relation to the past. Recuperating comparative historicism as an important historical episteme, *The Age of Analogy* helps us to understand the deep engagement of nineteenth-century science with contemporary literature and the narrative strategies of comparison furnished by literary history. In this book, I engage comparative historicism as a network of interrelated textual modes and devices. These include the narrative features of contingent plot and retrospective description; affective modalities including sympathy, wonder, and the interpellation of intent; specifically "literary" devices like free indirect discourse and idiomatic translation; genres including the historical novel, natural-historical monograph, and elegy; and rhetorical figures including metaphor, allegory, and personification. Attention to these disparate forms demonstrates the influence of literary technologies on the modern historical imagination. Above all, I understand comparative historicism as the exploration of how different literary modes and social sensibilities intersect in time, its defining feature being the rapprochement of historical accounts through explicit instances of analogy and comparison.

The most important (if tacit) aim of these new comparative histories was to rehabilitate the discourse of analogy, developed within natural theology, biblical hermeneutics, and moral philosophy, as the insistently empirical and historical "comparative method." Until the later eighteenth century, the terms comparison and analogy do not appear in the same context. *Comparison* was understood as a rhetorical trope, a strategy of *contrast* significant, for instance, to the important subgenre of comparison tracts shaped by the tractarian debates of the seventeenth century and sustained until the 1800s.[32] *Analogy*, by contrast, was understood as an analysis of *similarity* that was part of a tradition of Christian and moral philosophy distinct from rhetoric.

This changed when a core community of philologists (including James Burnett, Sir William Jones, and Max Müller) and anatomists (including Georges Cuvier, Étienne Geoffroy Saint-Hilaire, and Owen), as well as geologists, antiquarians, and biblical scholars (especially the German "higher critics"), collectively and explicitly took up the relation between analogy and comparison in order to develop a new approach to the study of historical specimens. They founded the *comparative method*: an analytic mode that coordinated *similarity* and *contrast*. Literary historian Susan Manning has argued that this coordination between analogy and comparison produced an "aesthetics of correspondence" with deep implications for literary, historical, and moral understanding.[33] Historical fiction, particularly the massively popular novels of Walter Scott, along with associated modes of fictional retrospection, gave the new comparative method unprecedented historical depth, by furnishing a persuasive narrative form. So, as I have argued elsewhere, the famous opening of Charles Dickens's *A Tale of Two Cities* (1859) addresses the events of the French Revolution by rejecting the contrastual analysis of eighteenth-century historians, who previously insisted "It was the best of times, it was the worst of times, it was the age of wisdom, it was the age of foolishness, it was the epoch of belief, it was the epoch of Light." Dickens's novel proposes something besides the "superlative

degree of comparison only": a more moderate comparatism that embraces similarity as well as difference.[34] In this fashion, a new generation of historical writers developed startling new theories of history and differential process, and new
ways to talk about them, and so transformed the period's literature and its vision
of the past.

The argument of this book is located at the intersection of several important
critical narratives about the nineteenth century: the birth of historicism and a
"metahistory" that compared historicisms; the death of epic and the "rise of the
novel" to preeminence in the transition from the eighteenth to the nineteenth
century in Britain; the shift of "collective empiricism" toward the formation of
collections in series that supported comparison across scientific specimens;
the "demise of rhetoric," which afforded a temporary loosening of disciplinary
boundaries isolating comparison and analogy; the subsequent professionalization of literature, science, and history that gave the comparative method its
modern configuration; the secularization of belief, as conceived by Charles Taylor
in terms of multiplicity of view; and the growth of political pluralism.[35] These
movements helped to flatten previous distinctions, a great leveling made possible in part by the new modes of relational analysis studied in the following
chapters.

Comparative historicism is a strong critical procedure, but it is more fundamentally a way of writing about the past. Its central gesture is the movement
between histories, drawing connections between lives, experiences, and material objects that articulate history as a tense composite rather than an organic
whole. For this reason, I am less interested in particular kinds of plot—the marriage plot, the *Bildungsroman*, the history of progress—than in how comparative
historicism plays different plots against each other, pluralizing them. I see a deep
continuity between Leo Braudy's account of the thorough narrative impulse of
eighteenth-century histories and Hayden White's influential taxonomy of the
figural tropes that organize nineteenth-century historicism.[36] As I will demonstrate, the historical novel provided the nineteenth century with transformative
plots that weave comparisons drawn between incidents, persons, and things into
larger analogies between the stories they told. If, as Alex Woloch posits, "the
emplacement of a character within the narrative form is largely comprised *by*
his or her relative position vis-à-vis other characters," this holds equally for their
plots, and comparative historicism furnished a transformative way of reading
those relations over time.[37] So, even as many nineteenth-century novels formalize this complexity in lifting one plot above many—especially the marriage
plot—they do not foreclose the sense that things could be otherwise, that other

plots are strong and perhaps more faithful possibilities (a point recently argued by Hilary Schor).[38] *The Age of Analogy* puts heavy emphasis on literary authors, Scott as well as Alfred Tennyson and George Eliot, who advanced comparative historicism's embrace of narrative pluralism. Henry James's Pulcheria gives a particularly apposite verdict after she reads George Eliot's *Daniel Deronda* (an 1876 novel that critiques both national history and the marriage plot): "I never read a story with less current. It is not a river; it is a series of lakes."[39] Comparative historicism exchanges strong narratives for an interplay between stories, incidents, and actors that allows new patterns to emerge. The novel is key here because, as the capacious genre of genres, it was able to place the conventions of different literary and historical modes in dialogue, evaluating their capacity to disclose an authentic past. This critical engagement with the past allowed the novel to coordinate historical distance with imitation, even immersion. So, as James Buzard has argued, the nineteenth-century novel helped teach early anthropologists and sociologists how to balance observation with participation, modeling critical comparison as a way to interpret societies in time.[40]

Linda Hutcheon has observed that "the nineteenth century [gave] birth to both the realist novel and narrative history," and I see their conjunction in comparative historicism as foundational to the nineteenth-century imagination.[41] All of the authors I examine explored alternative historical perspectives and placed them in juxtaposition. I have already suggested that collective authorship helps to flatten the ontological hierarchy presupposed by the figure of the author (a point I elaborate in the prelude), and this complements comparative historicism's own serializing tendencies, which can be understood, in part, as a nineteenth-century response to a general crisis of faith in strong natural or social order. I take philosopher Charles Taylor's thesis of secularization as a model here.[42] The writers I examine differed widely in their relation to scientific knowledge, social change, political outlook, and religious conviction—but they were frank (with the possible exception of Erasmus Darwin) in acknowledging that the histories they detailed operated alongside alternatives. And they worked hard to find historical genres—the elegy, the historical novel, and (with less success) the epic—that could embrace this pluralist view.

We do not need to look far to find political implications in such thinking. The comparative sciences, particularly as applied to social groups and artifacts, developed a tense engagement with organicism that shaped the modern understanding of "culture" itself as an autonomous social formation.[43] Thus, Matthew Arnold, who once theorized culture as "the best that has been thought and said," came to see culture as the fruit of comparative analysis across times and societies.[44]

When modern anthropologists, sociologists, and historians describe the history of their disciplines, they broadly agree that the comparative subfields of each emerged after the turn I identify here, at the close of the nineteenth century, as outgrowths of organic theories of human society and evolutionary thinking after Darwin. It is now recognized that these efforts were often compromised by their naturalization of historical and cultural difference and the ease with which they served imperialist and nationalist systems of knowledge. Thus, as Edward Said has shown, the growth of comparative philology served to underline often gendered distinctions between "East" and "West" and authorized the imperial ambitions of Britain and France.[45] Similarly, in George Stocking's analysis, the late nineteenth-century rift between ethnography and anthropology that emerged across the Anglo-American academy was deeply conditioned by different readings of comparative anatomy and natural selection in the service of divergent attitudes toward the question of race. Throughout these debates, comparative analysis was used to adjudicate the relation between national difference and racial distinctions.[46] But comparative historicism, particularly as it was deployed by Darwin, cut against attempts to posit irreducible differences within nature or society. W. E. B. Du Bois, as both a central author of the Harlem Renaissance and a sociologist who studied African American communities, saw this clearly. After hearing Franz Boas's ground-breaking comparative study of American immigrant populations at the First Universal Races Congress, Du Bois put the point succinctly in an editorial for *The Crisis*, the official organ of the NAACP: "Anthropologists, sociologists, and scientific thinkers as a class could powerfully assist the movement for a juster appreciation of peoples by persistently pointing out in their lectures and in their works the fundamental fallacy involved in taking a static instead a dynamic, a momentary instead of a historic, a fixed instead of comparative, point of view of peoples."[47] Where Du Bois sees science, anthropology, and sociology as converging toward a more dynamic comparative historicism, *The Age of Analogy* understands this shared impulse as part of the longer evolution by which this historical modality was institutionalized across the distinct disciplines of the modern university.

Thinking through Analogy

Any history of comparatism runs into deeper questions about the continuity of the comparative method over time. History stands in dynamic relation to formal structure; as George Eliot has it, the "tendency to repetition . . . creates a form by the recurrence of its elements in adjustment with certain given conditions."[48] Despite the complex ways that "comparison" and "analogy" interact as terms of art in the period I examine, both describe multipart comparisons between two different sets of relations. In other words, in the modern period, both terms denote analogies in practice, statements that can be classically modeled by the schema "A is to B as C is to D." For this reason, a central object of this book is to explore the relation between how analogies are used and the descriptive vocabularies that form a conceptual context for their work.

At the sentence level, analogies are simply that, but their operation is framed by assumptions that have a profound impact on how they are understood, and even whether they are recognized *as* analogies. The first four chapters of the *Origin of Species* are saturated with analogies that coordinate degrees of similarity and difference, beginning with the first sentence: "When we look to the individuals of the same variety or sub-variety of our older cultivated plants and animals, one of the first points which strikes us, is, that they generally differ much more from each other, than do the individuals of any one species or variety in a state of nature" (7). Here, Darwin underlines the analogy between domestic varieties and wild species and emphasizes a divergence in their respective degrees of variation. Darwin reserves the word analogy itself, however, to describe only a handful of his most controversial speculations, particularly as they touch upon the common origin of life. I see this feature of his writing as evidence of the continued differentiation between the discourse of analogy and its successful and generally unmarked incorporation into comparative historicism. In the following chapters, I trace the application of analogies on the page to map the major features of this shift.

In the process, I uncover a significant gap in our understanding of how analogies actually work. In the prelude I elaborate this new understanding of analogy at greater length; here it is important to summarize my perspective. Whereas analogy is generally understood as a top-down, formal operation, an understanding I ascribe to *formal analogies* (which apply a previously understood pattern of relationships to a new context), I have found that many of the analogies most important to this study operate differently. In particular, the following chapters explore the importance of what I term *harmonic analogies*, analogies that work from the bottom up, exploring a pattern between two different sets of relationships, to see what common features the pattern picks out. These harmonic analogies allow significant shared features to emerge through contact between two different domains placed in serial relation—juxtaposed across time or space, in the world or in the imagination—rather than asserting the prearticulated features of what is already understood. Like George Eliot, I find the analogy to music useful. As she puts it in an epigraph within *Middlemarch* (1871–2):

> How do you know the pitch of that great bell
> Too large for you to stir? Let but a flute
> Play 'neath the fine-mixed metal: listen close
> Till the right note flows forth, a silvery rill:
> Then shall the huge bell tremble—then the mass
> With myriad waves concurrent shall respond
> In low soft unison.[49]

Harmony is not an inherent property of one string or another; it emerges only through interaction. And it is only after such harmonic patterns are established, in the instant that punctuates the expectant colons above, that a more general understanding of harmonic functions can develop. Eliot understood harmonic analogy (which she termed "careful analogical creation")[50] as a powerful way to understand *both* how we sound history (and so explore the common and different features of living in the past and present) *and* how we come to understand others in our own time. Similarly, the analogy between domestic and natural selection developed by Darwin in the *Origin* had reverberating implications for both natural and domestic change—most immediately, arguing that new species can be formed through either process.

For this reason, while it is only one kind of analogy, and perhaps less common, I put more weight upon harmonic analogy in the study that follows. This stands in part as a way to orient my argument toward current trends in scholarship and the ideological commitments (in the broadest sense) that we perceive as giving fiction the widest ground of possibility. *The Age of Analogy* engages harmonic analogy as another element of literature's capacity to escape the given, to provide contact with singularity and alterity. Analogy's engagement allowed the comparative method to search for new patterns in history, in social and natural systems, where older models failed. Only after the singularity of a new pattern is formalized as a definable attribute does it come to be seen as what early modern rhetoricians termed a *tertium comparationis*, a specific attribute with respect to which one considers the objects of comparison.

The most vital analogies operate in advance of such definition and so produce new forms and new kinds of contact. They dramatize the possibility of misunderstanding, and it is for this reason, I think, that Said celebrated novels like *Middlemarch* for their "contrapuntal" moments—instances of an understanding that is shared without being imposed.[51] As chapter 4 explains, the realization that one is in error—that, for instance, *Middlemarch*'s Dorothea has misjudged Rosamond's actions—motivates the search for a new shared understanding that both subjects might acknowledge. Harmonic analogy stages the ground for new contact between two systems, from engagements between individual perspectives to conflicting theories of natural order, and it is hard to say at the outset where this might lead. This dynamic can be recognized in various strategies of comparative engagement that disclose new commonalities and distinctions, from Shu-mei Shih's "relational comparison" to the analogies of "identity/difference" that, in Andrew Cole's view, laid the foundation for dialectical thought.[52] Analogy gave comparative historicism a method for disclosing new patterns and unsettling historical convention, for excavating perspectives, voices, and connections that were ignored by standard accounts. By these means, analogy furnishes history with (in Chakrabarty's words) "the capacity to hear that which one does not already understand."[53]

Implications for Comparative Historicism

One of this book's most significant findings is that there is a genealogy for comparatism and comparative literature that is not captured by an exclusively continental history. The ongoing debate about the status of "Comparative Literature in an Age of Globalization" (the title of a major 2004 report on the state of the field) can be seen as the climax of one story about the field's nineteenth-century foundations in German Romanticism and comparative linguistics. The standard account posits at the outset a tension between the national coordinates of individual "literatures" as the aesthetic expression of specific peoples (per Hugó Meltzl) and a "world literature" that was cosmopolitan and potentially global in scale (per Johann Wolfgang von Goethe). On this view, the problem of comparative literature can be seen as two intertwined questions. What are the political implications of the comparative method (particularly in connection to national literatures)? And what constitutes the (sometimes transnational) literature that is its object?

The Age of Analogy restores a third crucial term to this discussion, *history*, characterizing the modern comparative method as an analysis of the relation between agents, events, and objects in time but not necessarily in common time. The key innovation of comparative historicism was to turn away from unitary narratives that were progressive, stadial, or—to use sparingly two characterizations that are used far too casually—teleological or social Darwinist. As Natalie Melas has argued, comparatism was, at the outset, an attempt to engage alterity.[54] Moreover, this study discloses an alternative history for comparative literature, one that includes Sir William Jones, Walter Scott, and George Eliot, comparatists who worked against an understanding of the nation-state, or the national tradition, as a determinative category of study. So, when George Eliot reorganized the *Westminster Review* so that the Contemporary Literature section used subject-based divisions in place of national categories, she emphasized her intent to build a "comparative History of Contemporary Literature" outside the coordinates of national identity. Rather than taking "literature" as an object of comparative analysis, this study examines how literature, in such cases, shaped comparative historicism as it evolved into a dynamic, open procedure for encountering the unknown past.[55]

Many scholars have noted that analogies are a central feature of studies that examine the relation between literary text and historical context.[56] Foundational studies by Stephen Greenblatt and Catherine Gallagher gave us powerful new procedures for exploring those analogies without reducing literary works to their historical background.[57] Fredric Jameson's dictum to "always historicize" now seems sufficiently proximate to the modes of reading that characterize the new historicism that they can be taken as one differentiated movement, even though, thirty years ago, profound differences were seen. Those same arguments prove there is nothing novel in saying that the new historicism was not really "new" at all, but a longer understanding of comparative historicism shows this insight has a deeper and more important claim on our understanding of other societies and times than we may think. Comparative historicism was equally important to the *Annales* school of French history, which developed an approach to "quantitative" or "serial" history that emphasized the comparative analysis of time series, particularly population and economic data across long periods and between different regions.[58] This had a profound impact on Michel Foucault, who both acknowledged his debt to serial history and summarized his approach as a mode of "comparative analysis that is not intended to reduce the diversity of discourses, and to outline the unity that must totalize them, but is intended to divide up their diversity into different figures."[59] Organized through analogy and a new sense of the past, comparative historicism produced the "profound sense of historicity" that Foucault sees as one of the nineteenth century's most important legacies. We are still writing the "history of resemblance" (and the coordinate history of difference) in which analogy is centrally, often stealthily, engaged.

Summary of Chapters

The Age of Analogy explores the relationship between Erasmus and Charles Darwin both through their works and between their times; I am interested in the two both as they bookend the nineteenth-century turn toward comparative historicism and as that turn explains the relationship between their theories of historical change. Even as my study explores in detail works published between 1789 and 1874, the important shift comes in the contrast between modes of historical engagement common before the century's turn and after. Rather than juxtaposing texts and contexts, I provide a comparative literary history, which means that I look over time at the differences as well as similarities between specific works and between their characteristic modes of historical understanding. Because this study has strong cross-disciplinary interests, there is a vast archive of scientific, historical, and literary documents that could feature here. Within the confines of this book, I offer a handful of case studies as samples of what that archive contains. Any problem becomes visible only when examined at the appropriate scale, and I do not think comparative historicism becomes evident as the wide-ranging mode of understanding I recognize without placing these case studies at some distance from each other.

This introduction summarizes some of the central findings of this book, but in order to make this book legible for the widest audience, it minimizes theoretical speculations regarding analogy. My focus here is on how the book fits into ongoing discussions within specific academic fields—literary criticism, history (particularly the history of science), and comparative literature—while exploring what comparative historicism tells us more generally about modernity and its relation to the past. The prelude that follows is written for those interested in the theory of analogy itself. It sets my approach to analogy, particularly the distinction I draw between harmonic analogy and formal analogy, in the context of longer developments in the philosophy of being, logic and mathematics, biblical hermeneutics, and rhetoric. The prelude also provides an opportunity to entertain examples of analogy that range beyond the scope of my argument about comparative historicism in the nineteenth century. Drawing on developments in the history of physics, mathematics, and speculative philosophy, I explore analogy as a way to interpret natural order and the relation between different forms of life.

Each of the chapters that follow returns to the problem of comparative historicism, taking up the contributions made by specific authors and works as impulses within its larger development. The first half of the book reads the emergence of comparative historicism against the background of late Enlightenment theories of historical process and natural order. My first chapter foregrounds the speculative science and epic poetry of Erasmus Darwin as evidence for an eighteenth-century crisis in historical understanding and in the status of analogy. In works like his massive scientific study *Zoonomia* and his epic poem *The Botanic Garden*, Darwin argued that analogy was both central to scientific inquiry and a powerful tool for the poetic imagination. Darwin understood analogy as an engrained system of transcendent order, a pattern of design that secured the relation both between experience and reality and between natural processes at all levels of scale, from vegetable reproduction to human progress to cosmic evolution. He drew this particular brand of naturalism from both the vitalism of contemporary science and the scientific histories of the Scottish Enlightenment, which gave him confidence in positing universal patterns of development. More than his political sympathies, Darwin's faith that analogy secured an epic narrative in which social, technological, and natural progress cohered was overturned in the wake of the French Revolution and amid growing skepticism of the relation between literary imagination and scientific discovery. Yet his analogical vision inspired a subsequent generation of naturalists and poets to look for a more effective way to describe the patterns of order they found in the natural and social worlds.

The second chapter takes up Walter Scott's close collaboration with a network of antiquarians and collectors, particularly the linguist and early ethnologist John Leyden. His journeyman years of collection, collaborative authorship and publishing, and growing interest in translation and comparative philology, developed through the *Tales of Wonder* (1801) and *Minstrelsy of the Scottish Border* (1802–3), conditioned the comparative textual imagination of Scott's later historical fiction. I emphasize Scott's poorly-understood engagement with the comparative sciences, particularly antiquarianism, which I understand as a precursor to the science of anthropology. These engagements helped him shape a comparative understanding of historical experience that underwrites the famous "many-sidedness" of the author of *Waverley*. Scott's novels rely on linguistic and antiquarian vocabularies to juxtapose elite political history with a "history from below" and to counterpose national history with the alternative histories of the dispossessed. In doing so, he shaped the comparative method as an historical methodology that allows the writer to counterpose different narratives and so assemble thick descriptions of past societies. Scott prompted historians to consider the creative dimension of interpreting the past, from his contemporary Thomas Babington Macaulay to the early twentieth-century historiographer R. G. Collingwood.

The second half of the book explores the work of three different writers who worked to adapt comparative historicism to disparate questions of theological, social, and scientific history in the mid-nineteenth century. These retrospective fictions, following Scott's model, placed imaginative comparison at the center of historical understanding. But they also turned this historical methodology to new problems—the status of providential history and humanity's place within it, the grounds of individual understanding and the productive possibilities of error, the study of natural forms and their differentiation in time—that tested comparative historicism's reach. In my third chapter, I explore the verse form of Alfred Tennyson's innovative elegy, *In Memoriam* (1850), as an attempt to reconfigure historical comparison as a template for recuperative grief. A decades-long effort to evoke and work through the legacy of his deceased friend Arthur Henry Hallam, *In Memoriam* finds its major success in its insistently comparative and historiographic verse form. Tennyson's object was greater than restoring contact with a vanishing past, as he sought to bring the world of the living into a serial relationship with the dead, drawing on recent developments in geology, astronomy, and anatomy to fashion a version of natural theology that emphasized the continuity between our world and possible others. By these means, Tennyson refashioned elegy as a capacious retrospective genre, a "cooperant past" that engaged past experience as an alternative world, and (with important qualifications) restored Tennyson's faith in the benevolence of human experience and ultimate design. His final confidence that such engagement demonstrated "One God, One Law, One Element" closely aligned the poem with the homological theories of Richard Owen; both, I argue, rejected comparatism's differential possibilities in favor of a unitary and implicitly conservative vision.[60]

For both Tennyson and Scott, formal innovation shaped modes of retrospective writing that address loss through historical reassembly and recuperation. My final two chapters examine the moderation of this synthetic aim in favor of sharpened epistemological precision in the works of two of Victorian Britain's most influential writers, George Eliot and Charles Darwin. Eliot's initial work as translator and critic, both in her participation in the Rosehill Circle and as de facto editor of the *Westminster Review*, produced an extraordinarily broad perspective on comparatism as it functioned in linguistics, biblical criticism, and biology and put heavy emphasis on comparison's commitment to uncertainty and fallibility. Personal knowledge in Eliot's fiction, particularly as produced in *Middlemarch*, is found in failed dramas of reconciliation that emphasize the tenuous gains of sympathetic understanding. My fourth chapter revisits the obscure term *disanalogy* to understand why Eliot's sympathetic realism is rooted in productive error. Eliot exploits analogy's strong potential to be *disproven* in order to formulate a mode of representational realism that is falsifiable. In this

way, her fiction locates the real predominately in the experienced difference between self and other, reflexive experiences that mark the gap between perception and apprehension, representation and reality. In disclosing what is not
known, her critical mode of comparative historicism moves us toward a fresh
understanding of others and of the world we inhabit. For this reason, Eliot's fiction has a deep suspicion of all presupposed systems of understanding; indeed,
Catherine Gallagher describes her as "the nineteenth-century novelist who is
most skeptical about categorical thought."[61] As Eliot's later essays make clear, the
famous critique of Casaubon's "key to all mythologies" or Lydgate's "primitive
tissue" comprehends all comparative empiricisms that presuppose a single unifying principle, from Goethe's "primary plant" to Darwin's "origin of species."

If anything, Eliot failed to gauge the open-endedness of Darwinian science.
His studies inaugurate a new species of narrative naturalism rooted in "just so"
stories that emphasize the contingent patterns of history. The *Origin of Species*
solved the central organizational problem of contemporary natural history: the
contrast between marked similarities and unstable distinctions among contemporary and antecedent species. The comparative network of the *Origin*, woven between the warp and woof of analogies and disanalogies, shows how these
patterns of similarity and difference emerge as the legacy of a shared evolutionary history. Its central actor, natural selection, brings the epic sweep of geological history in the style of James Hutton or Charles Lyell to the individual scale of
the struggling individual, the family, the local world. Natural selection engrained
comparative historicism as the definitive narrative mode of evolution. Like the narrator of Charles Dickens's historical novel *Bleak House* (1852–3), it continually
asks, "What connexion can there be" between this variety and its contemporary?
Between what came before and what came after? What stories recount the history that links any two individuals? This crucial narrative technique, derived
from Darwin's extensive readings of Scott and others, provided an imaginative
vocabulary that refurbished the speculative theories of his grandfather Erasmus
into a functional, and massively influential, scientific model.

But the *Origin* was only an "abstract" of Darwin's evolutionary science; to
place the new program on a strong footing in the work that followed, Darwin
returned to Erasmus's work, and the uncanny sexuality of his botanic studies.
In *The Loves of the Plants*, Erasmus had invited readers to identify their own desires and motivations with those of the plants he describes, a sympathetic strategy that quietly drew the reader into the experience of a common evolutionary
history. Charles Darwin makes this strategy explicit in *On the Various Contrivances by Which British and Foreign Orchids Are Fertilised by Insects* (1862), which

demonstrates that the key to evolutionary inquiry is the study of natural intent and botanical "contrivance."[62] In doing so, he extended a strategy of sympathetic identification over historical distance—central to Scott's historical realism—into the natural world. To understand orchids and the insects that pollinated them, Darwin argued, we must tactically impute human motivations and desires to their intricate behaviors. Like his grandfather before him, Darwin saw a purpose for personification in the study of nature. Darwin's study of orchid pollination should revise our own understanding of the Darwinian stance. His key strategy was a new mode of "intentive reading": comparing specimens to examine the natural world *as if* designed but *without* necessary intervention or intent. By these means, the *Orchids* cultivates intent as an active agent in the network of collective empiricism. Derived from the effort of contemporary naturalists and theologians to find evidence of God's hand in nature, Darwin's agnostic stance reformulated the language of design as a "flat theology" that recognizes basic continuities between humanity and other forms of life. A short "Coda" that appends this final chapter suggests what this sense of interrelationship, and the model of analogy I develop in this study, might suggest about anthropogenic climate change and our attempt to understand and curb its effects.

Between the *Origin* and the *Orchids*, I rework the long-standing question of Erasmus Darwin's influence over his grandson's work and thus provide a family portrait of the intimate function of literary form within scientific inquiry. Nineteenth-century writers—poets, scientists, and especially novelists—showed us how to talk about the past as something various and yet interconnected. *The Age of Analogy* traces that story, tracking the evolution of these insistently textured descriptions of what came before. By the second half of the nineteenth century, it no longer mattered whether the writer was a political economist, a botanist, or an author of social novels; if they wanted to know why certain connections existed in the present, they turned to the past, and the intertwined narrative patterns that, to modern eyes, now constituted its basic warp and woof. This new appreciation for the interconnectedness of past and present meant a new sense of everyday life as being *lived in history*. We're still living there today.

Thinking through Analogy

What happens when I write to you, the reader? I know you are there. You know I am here. For an instant, we are together. In this moment, the medium of engagement—the book you are holding, the printed type on the page—borders the experience, framing a moment of contact. Historians, because they deal with the traces of real people, have long had a particularly strong sense that we contact others through the page, that we write histories to revive the past. Niccolò Machiavelli steps into his study and is "warmly welcomed" "to the ancient courts of rulers who have long since died"; for Jules Michelet, "History and historian merge" by means of the "resurrection of life in its integrity"; R. G. Collingwood explains that the past "so revived is not a mere echo of the old activity . . . ; it is the same activity taken up again and re-enacted."[1] In such moments of contact, our sense of distance from the past fades, loosening our confidence that the past is gone and that we cannot directly apprehend it. For a moment, we seem to bridge the profound differences in knowledge and existence that separate us from previous life, and especially from the dead. In philosophical terms, these moments suspend problems central to both ontology (the study of different kinds of existence or being) and epistemology (the theory of knowledge and what we can learn about the world).

The philosopher Georg Simmel, in describing this experience of reencounter, observed a waffling between presence and distance. Analytically, he explains, we know that distance is preserved by a sequence of operations: "First, I perform the [historical] mental act in question myself, and then I attribute this mental act to an historical person." First, we generate the experience in the present—it exists now and shares our ontological being, our presence in the current moment—and then we map it onto the past—as acquired knowledge, an epistemological formation. "But," Simmel adds, "this is only a retrospective analysis of historical knowledge into its elements; this distinction is not made within the process of historical knowledge itself." In the moment of contact, the present experience and the original event are the same thing. It is as if we share our being with the past (taking the same ontological status) and so receive knowledge about that past (overcoming the problem of epistemology). Ontology and epistemology are fused. We experience history as an "irreducible, integral act."[2]

Intuitively, it seems necessary that encounters with fictional worlds and imagined characters remove us further from this experience of contact. On the contrary, philosopher Paul Ricœur has argued that historians deploy narrative fictions to persuade readers of the coherence of history as a subject. In his view, it is "the plot of narrative . . . [that] 'grasps together' and integrates into one whole and complete story multiple and scattered events" (and here, I would add, scattered subjects).[3] In Ricœur's view, these narrative fictions are precisely what produce history as an "irreducible, integral act," and this effect is a feature of the remediation of narrative—the way a story reworks diverse aspects of the intelligible world into a more coherent whole, which subsequently, through the act of reading, we integrate into present experience. Yet the experience I am trying to identify is *immediate* in two basic senses. First, it marks a temporary collapse of historical distance, of the separation between then and now. And, second (and this is the corollary), the media of this experience—the language, the formal features, the book itself—cease to function as something lodged *between* present and past and serve instead to bring us into contact with history. They form our experience, producing fictional worlds as well as the world of the past. As Simmel puts it, while form has an epistemological status, making certain kinds of knowledge possible, "a form also has a metaphysical status. It is an ontological condition for the existence of certain kinds of things: aspects of the world . . . insofar as they are constituted by this form."[4] Even as the medium shapes the past, it liberates its messages.

Some forms are better at this than others, and the present study argues that analogy has an extraordinary ability to bring us into contact with other worlds. I understand analogy as the primary formal constituent of *comparative historicism*, a method of historical engagement, now about two hundred years old, that engages the past as a tense composite of societies, languages, and modes of thought. As the introduction explains, *The Age of Analogy* approaches comparative historicism as a network of interrelated textual modes and devices. Attending to those literary forms in the chapters that follow, I explore how these literary technologies continue to shape the modern historical sensibility. Central to this story is the interrelation of "analogy" and "comparison" as terms that designate interrelated methods of analysis. Throughout, I will be arguing that the various modes of comparative historicism rest upon articulations of analogy, multipart comparisons that establish a pattern of similarity between two different sets of relationships. If comparative historicism can be understood as the exploration of how different modes and accounts intersect in time, the study of its constitutive analogies merit special focus, particularly their capacity to formalize new understandings and gain fresh insight into the past.

This prelude offers a theory of analogy that explains its formal and historical character. At once the product of distinct discourses operating within rhetoric, hermeneutics, and theology, and a formal device that explicitly counterposes two different relationships, analogy is a complicated object of study. In order to address this complexity, I begin with a short overview (extended in the notes) of how both the word analogy and its sometime synonym comparison were understood within their respective conceptual frameworks. As Rita Felski and Susan Stanford Friedman observe, in their introduction to a recent collection of essays on comparatism, "Comparison is a mode of thinking, an analogical form of human cognition, that seems fundamental to human understanding and creativity."[5] Yet no theorist of either analogy or comparison has considered the history of their relationship over time. Here, I will consider these distinct histories in order to bring the nineteenth-century comparative method into sharper focus. Insofar as the comparative method often confused these two terms, later chapters give close attention to the interplay between both analogy and comparison, understood as alternative vocabularies for similar practices. In the second portion of this chapter, I explore the theory of analogy itself, using Niels Bohr's atomic theory to highlight a distinction between "mapping" theories of analogy and the more dynamic form of analogy that will be a central focus of this book. In order to further clarify the relation between these two forms of analogy, I track their distinction in Aristotle's theory of natural order. Aristotle recognized analogy as a crucial problem for the production of scientific knowledge, and the relationship between that knowledge and the natural order it captures. For that reason, Aristotle raises a problem for the relation between epistemology and ontology that I explore throughout this book. In each of the following chapters, a central question is whether analogy is simply a rubric for ordering the world, or whether it reflects the basic structure of that world. Is analogy simply a literary device? Or is it a natural pattern—a feature of how our world is structured? These questions get to the basic dilemma that constitutes the field of "science and literature" insofar as it hopes to show the value of literary forms in producing scientific knowledge, and to prove that imaginative language generates real knowledge about the world.

The remainder of the prelude explains why I think analogy requires us to reject this distinction between the "literary" and the "real" or, more precisely, why the so-called literary features of analogy are precisely what afford its ability

to capture natural patterns. For that reason, I spend some time exploring a collection of recent writings on the relation between ontology and literature—
sometimes described as "speculative realism" or "new materialism." I see these
efforts, drawn from the continental philosophy of Gilles Deleuze, Alain Badiou,
and Quentin Meillassoux, as part of a larger effort to circumvent the epistemological challenges presented by literary language in favor of a harder-nosed ontology, based on the model of formal languages and mathematics. Turning to
examples drawn from the history of logic, mathematics, optics, and semiotics, I
argue that this misunderstands the relation between literary language, formalization, and the sign. Ultimately, I argue, we can use Bruno Latour's study of
signification to understand analogy's peculiar entanglement within natural and
social patterns and, through that entanglement, explain analogy's ability to produce new understanding. In revaluing analogy's literary qualities, I hope to
show how they allow us to recognize patterns that we do not yet understand. In
this way, analogy brings us into contact with the world.

Analogy vs. Comparison

"Have I not instinct? Can I not divine by analogy? Moore never talked to me either
about Cowper, or Rousseau, or love. The voice we hear in solitude told me all I know
on these subjects."

Caroline Helstone in Charlotte Brontë's *Shirley* (1849)

To understand the close interplay between comparison and analogy in this period, it helps to recall their differentiated histories. As explained in the introduction, until the late eighteenth century comparison and analogy were understood as completely distinct terms of art. *Comparison* was a rhetorical strategy of
distinction that underlined difference.[6] As Jonathan Swift's Gulliver puts it,
"Undoubtedly the Philosophers are in the right when they tell us, that nothing
is great or little otherwise than by Comparison."[7] By contrast, *analogy* was understood as a distinct strategy for finding similarities, shaped by theories of
biblical exegesis and natural theology, and directed at patterns within natural
and textual accounts.[8] A. O. Lovejoy's *The Great Chain of Being* (1936) and Michel
Foucault's *The Order of Things* (1970) provide our most influential accounts of
this earlier history of analogy, as described by the *scala naturae* and Renaissance
semiology.[9] And yet analogy persisted after the early modern period. At the
close of the eighteenth century, it became deeply intertwined with comparison
through the formation of a new "comparative method" of scientific inquiry. A
range of thinkers, from philologist James Burnett to comparative anatomist

Geoffroy Saint-Hilaire, used analogy and comparison interchangeably to describe a new mode of relational analysis. As this comparative method matured, these terms of art again separated out. By the mid-nineteenth century, the term "comparison" recognized the study of both similarity and difference, while "analogy," when not operating under the aegis of the comparative method, marked an outmoded and speculative strategy of reasoning. So, in the preceding epigraph, Charlotte Brontë uses "analogy" as a historical marker in *Shirley* (1849), a novel set in the Napoleonic wars, to mark Caroline Helstone's antiquated philosophical vocabulary. To "divine by analogy" in the style of Joseph Butler's *Analogy of Religion* (1736) is to be triangulated within history by a dated set of intellectual coordinates.

Brontë drew this historicizing strategy from the historical novel, particularly the novels of Walter Scott, and as I will argue in subsequent chapters, this strategy is bound up (along with the critical procedures we use to read such moments) in the larger shift by which the comparative method was engaged to understand historical difference and pattern. I describe this larger shift as *comparative historicism*, which provides a fruitful way of examining how terms like analogy or comparison function in different contexts and at different moments in the past. Brontë helps us to recognize the nineteenth century as one of many ages of analogy, as well as to recognize the nineteenth century's special status as a period when analogy itself, when not operating under the sign of comparison, came to be seen as an outmoded style of reasoning. In other words, it was the century in which analogy aged.

At the same time, I am concerned here with the way that comparative analysis operates on the page, along with the syntactic and semantic features that allow comparative analysis to work. And when one looks closely at how it operates, it becomes clear that both comparative analysis, and the older mode of "analogical" reasoning it replaced, are constituted through multipart comparisons between two different sets of relations. In other words, both "analogy" and "comparison" are terms that index analogy as a practice, classically expressed in the schema "A is to B as C is to D," or A:B :: C:D. For this reason, a central object of this book is to explore the relation between how analogies are used and the descriptive vocabularies, whether described in terms of comparison or analogy, that frame their understanding. Throughout the following chapters, a key focus will be to identify the space between how analogies operate at the sentence level—the peculiar syntactic and semantic features of the words organized into multipart comparisons—and explicit, higher-order discussions that accompany them, specifying their meaning and significance.

This distance is clearest in writings that place analogy and comparison in immediate opposition. So, in the epigraph, Caroline Helstone uses "analogy" to cap off and mark a series of speculations in which she tells her companion, Shirley Keeldar, first, "You never would have loved Cowper . . . He was not made to be loved by woman"; second, "And what I say of Cowper, I should say of Rousseau . . . He loved passionately; but was his passion ever returned? I am certain, never"; and, finally, "And if there were any female Cowpers and Rousseaus, I should assert the same of them."[10] This is a multipart analogy, in which Cowper's "love" is mapped on to Rousseau's "passion" and then extended to all analogous "female" cases. To schematize, this relation can be described [Cowper : love :: Rousseau : Passion] and, by implicit extension, [. . . :: female Cowpers and Rousseaus : (their loves and passions)].

In George Eliot's novel *Middlemarch* (1871–2), by contrast, an almost identical scene of analogy is described as a "comparison." Dorothea and Celia Brooke are fighting over Dorothea's potential lover, Isaac Casaubon. While Celia finds him "ugly," Dorothea argues this makes him "remarkably like the portrait of Locke. He has the same deep eye sockets"; and when Celia complains he is "sallow," Dorothea retorts, "I suppose you admire a man with the complexion of a cochon de lait."[11] At stake are two analogies that describe, in Dorothea's view, alternative ideals: [Casaubon : Casaubon's "eye sockets" :: Locke : Locke's eye sockets] versus [Celia's "man" : his "complexion" :: "cochon de lait" : the suckling pig's complexion]. The former underlines the attraction of the mind and its supremacy to conventional beauty; the latter treats Celia's eventual lover (James Chetham) as a kind of chattel in the marriage market. Though Dorothea insists it is a "good comparison," Celia is shocked: "Dodo! . . . I never heard you make such a comparison before," and the narrator agrees: "Miss Brooke was clearly forgetting herself."

In both scenes two marriageable heroines use analogies, of the form [A:B :: C:D], to explore the possibilities of romance, even as they use contrasting vocabularies of analogy and comparison to specify that operation. The following chapters analyze many examples of such overlap and distinction in how both analogy and comparison are used to characterize acts of juxtaposition: chapter 2 explores the intersection between the "analogy of language" postulated by James Burnett and Franz Bopp's comparative philology; chapter 3 traces the careful distinctions between comparison, analogy, and homology drawn by anatomists including Geoffroy Saint-Hilaire and Richard Owen; chapter 4 traces the interplay of comparative analysis and analogy in the German "higher criticism"; and chapter 5 explores how Charles Darwin worked to differentiate several different classes of analogy within a host of specific comparative discourses. Chapter 1 is an outlier, insofar as Erasmus Darwin used the term analogy exclusively to describe the patterns of his wide-ranging studies of natural and social order. Throughout these chapters, I will be arguing that the entanglement of these vocabularies marks the larger nineteenth-century transformation by which different discourses of analogy and comparison were reformulated as the modern comparative method.

To return to the novels, it is tempting to speculate how the different vocabularies of *Shirley* and *Middlemarch* reflect a fine distinction in the two decades that distinguish their historical setting—between the intellectual vocabularies appropriate to the Napoleonic era and the first Reform Bill—or the equivalent distance between the periods in which the two novels were written decades later. To answer this question, we would need to weigh the value these terms had for both Brontë and Eliot as markers of historical distance and differentiate that historicizing impulse from the more immediate use of diction as a way to particularize character. But this question cannot be resolved, and the present study helps explain why. The fact that both scenes turn on an analogy drawn between the outlooks of two heroines, and so between their possible marriage plots *and* their respective historical circumstances, underscores the close interplay of comparative strategies within both historical fiction and contemporary historicism. To establish an analogy between Shirley and Caroline, or between Dorothea and Cecilia, is to frame a comparative study of both different fictional persons and the different historical impulses of the period. Such comparisons are central to the nineteenth-century novel and the historical sensibility of its public. For this reason, *The Age of Analogy* traces the play of analogies on the page, and the descriptive vocabularies assigned to them, to map the larger contours of this comparative turn.

Harmonic vs. Formal Analogy

By giving close attention to how analogies are used, I recognize an important gap in conventional accounts. Analogy is usually understood—on the model of I. A. Richards's description of metaphor as a relation between vehicle and tenor—as the "mapping" of relationships from "source" to "target."[12] As a major example, key theorists of how analogies are used in science make the case that the Rutherford-Bohr theory of the atom—otherwise known as the "solar system" model—took the known relationship between the sun and orbiting planets and *mapped* it on to the proposed relationship between the positively charged nucleus and the negatively charged electrons (figure 2a).

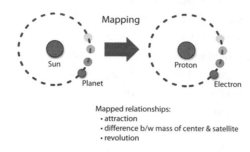

Mapped relationships:
• attraction
• difference b/w mass of center & satellite
• revolution

2(a): "Solar System Model" as Formal Analogy

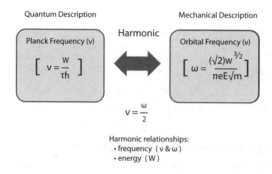

Harmonic relationships:
• frequency (v & ω)
• energy (W)

2(b): Bohr's Model as Harmonic Analogy

Figure 2. Two models of Niels Bohr's analogy.

But analogies do not always work this way, particularly, as I will argue, in the most interesting cases. Take Darwin's argument for natural selection. The first four chapters of *On the Origin of Species* are devoted to an extended analogy between what breeders do when "selecting" for specific characters and what happens in a state of nature. Darwin argues that continued selection over very long time scales will produce new species in either case; this is the important analogy between "domestic" and "natural" selection. On the traditional account of analogy, one might say Darwin "mapped" features of domestic selection onto a natural process, using domestic breeding as a model for what happens in nature. But, as I note in chapter 5, Darwin's contemporaries did *not* believe that domestic selection could produce new species, only new breeds. The consensus was that species had innate "limits of variation" and that breeders could not transgress those limits. To establish the productive analogy between domestic and natural selection, Darwin had to argue against contemporary wisdom regarding *both* natural *and* domestic species: insisting that domestic selection could produce new species *and* that natural selection does so in nature. This was an insight he gleaned by testing each half of the analogy against the other, teasing out further features of both. This is the labor of the first four chapters of the *Origin*, which constantly move between domestic and natural examples. Neither domain of the analogy should be understood as the "base" or "model" for the other; features of each side are adjusted and brought into better alignment to make Darwin's case.

In fact, even the standard account of the Rutherford-Bohr "solar system" theory of the atom substantially distorts a similar dynamic at work in the seminal papers that set out Niels Bohr's contribution to the theory. Bohr's revolutionary paper "On the Constitution of Atoms and Molecules," published in three parts in the *Philosophical Magazine* in 1913, never mentions the solar system, but it discusses a different analogy at length: the central "analogy" between Planck's equation for the quantum emission of spectra, and the "ordinary mechanics" of a classical Newtonian description of objects in orbit.[13] Through an equation that assumes the frequency of spectral emission is half the frequency of the electron's revolution, Bohr establishes the analogy between these systems: "Let us now assume that, during the binding of the electron, a homogeneous radiation is emitted of a frequency v, equal to half the frequency of revolution of the electron in its final orbit" (4–5). This analogy between one kind of frequency and the other launched the entire quantum mechanical revolution. It supports a network of further comparisons. Bohr notes, for instance, that ratios established between the frequencies of spectral radiation can be "compared" with the ratios between frequencies at which electrons orbit the atom (6).

It is moot to ask what kind of frequency relationship in Bohr's paper "maps" onto the other in the analogy; what happened in practice was that Bohr set them in a dynamic, harmonic relation and thus entangled the physical and quantum worlds in far-reaching ways (figure 2b). Despite his use of the term analogy, Bohr was not working within comparative history—he worked instead in a tradition of analogical philosophy advanced by John Tyndall in the nineteenth century and supported by the mathematical nature of physics. But his example is instructive. The point is not that "mapping" analogies don't exist. Standard accounts accurately describe how the solar system model of the atom (which developed in the years after Bohr's paper as a way to popularize his work along with Ernest Rutherford's efforts to theorize the atom) works in popular writing.[14] Instead, I see the formal operation of analogies like the "solar system model," which use a structure from one domain to characterize another, as standing in tension with the harmonic engagements provided by other analogies, like Bohr's actual equation, that place the two systems in dynamic and reciprocal relationship.

I characterize these two modes of analogy in terms of their formal versus harmonic features. *Formal analogy* works from the top down, applying a pattern of relationships that are already understood in one domain to a new context. *Harmonic analogy* works from the bottom up, exploring a pattern between two different sets of relationships, to see what common features the pattern picks out. Understood as processes, these modes of analogy stand in dialectical relation: a formal analogy can disclose exceptions that produce new serial encounters; a harmonic analogy, once firmly established, can indicate a new formal model for additional comparison—though neither makes a dialectic on its own.[15] Formal analogy is directional, harmonic analogy is reciprocal. The former can be programmatic—designed to explain specific features of a target domain— but the latter are dynamic and uncertain. It is precisely the unpredictable, emergent patterns of harmonic analogies that make them both powerful and difficult to use.[16]

As I note in the introduction, the "harmonic" allusion to music is instructive. Harmony is not the product of any single instrument. If we hear, for instance, a violin and a cello striking a harmonic note, this lasts only as long as each instrument sustains it. If either changes its tune, the harmony is destroyed. This is not just a feature of musical instruments; in principle, any two objects can be tested to see if they resonate in this way (this is the point of George Eliot's "great bell" epigraph in *Middlemarch*). The contingent ability of vibrating objects to harmonize preceded the vast efforts to formalize those harmonies through standard systems of tuning, from ancient Indian rāgas to the "well tempering" of seventeenth-century keyboard instruments. The chapters that follow explore how harmonic analogy functions when applied to a range of problems: from theorizing how translation provides a meeting ground between different social formations (in chapter 2) to exploring the possibilities of rhymed verse and elegiac poetry (chapter 3), and as a model for sympathetic understanding (chapter 4). In each case, the uncertain harmonic potential of analogy is set in dialogue with its power to formalize and stabilize new patterns.

There are many different ways one might characterize differences between analogies, for example, Paul Ricœur's distinction (developed from Saint Thomas Aquinas) between "vertical" and "horizontal" analogy.[17] I choose the distinction between *formal* and *harmonic analogy* to emphasize the contrast between, on the one hand, dynamic modes of formal generalization that do not merely constrain interpretation but also produce new models and consensus and, on the other, an open engagement between particulars that *can* (but does not *necessarily*) understand them as a series within or across time. This is only one implication of the way that analogies, organized by degrees of likeness and membership, violate the discrete categorical framework of classical reasoning. On the one hand, because they assert similarity rather than identity or distinction, analogies are antithetical to the most basic tenets of classical logic, from the law of the excluded middle (loosely, that any statement is either true or its negation true) to the principle of noncontradiction (that any statement and its negation cannot both be true). On the other hand, analogy's ability to move between traditional domains of knowledge affords its singularity, its ability to disclose new patterns. For this reason, analogy bears an uncertain critical relation to classical epistemology, given its capacity to tease out insights beyond the framework of deductive or inductive inquiry.[18] Though the "bottom-up" versus "top-down" movement of harmonic and formal analogy would seem to reflect the contrast between induction and deduction in traditional accounts, this is not necessarily so. As I observe in chapter 3, analogy's ambivalent relation to such logical procedures can be seen in the opposed views of Victorian philosophers Herbert Spencer and John Stuart Mill; the former argued that analogy "approximate[s] toward deduction," applying a model to a new case, while the latter insisted that analogy was a "not . . . complete induction" that generalizes from particular patterns.[19]

Between Spencer and Mill we see two different perspectives on analogy that emphasize its formal versus harmonic features. But their disagreement has its roots in the long history of logic's struggle to deal with analogy. Aristotle, the founder of classical logic, recognized the exceptional status of analogy by emphasizing its discomfiting relation to classical reasoning, and his response highlights analogy's peculiar relation to both natural patterns and our ability to learn about them. In Greek, analogy, or αναλογια—a similarity "according to a due relation or proportion"—meant a serial pattern of relation that held between distinct ratios.[20] Mathematically, it designated a four-part relation of proportions, in which two ratios were compared, for instance, ½ to ¾. Understood as a pattern between relationships, analogy was generalized by Aristotle and others to cover situations beyond mathematics. Aristotle's *Posterior Analytics*, for instance, explores harmonic analogy as a method to elucidate patterns that traverse the categorical boundaries of his highly organized system of knowledge. Usually, Aristotle emphasizes, inferences are applicable only to the specific domain they are drawn from, their "genus." You can't argue that a proposition from one genus applies to another, unrelated genus: "One cannot, there-fore, prove anything by crossing from another genus—e.g. something geomet-rical by arithmetic . . . except such as so are related to one another that the one is under the other."[21] Analogies present a problem for such thinking because, by their nature, they draw a connection between patterns in different domains. This latter point can be stated in terms of set theory. When analogies draw rela-tions between members of sets that are thought to be discrete, they put the nature of those sets at stake: whether the new patterns constitute a new set, whether the new set includes other examples, what the conditions of membership are, whether the set is real. In short, analogies make fuzzy sets. Harmonic analogy troubles set theory because it can challenge the central principle of discrete membership. And for this reason, Aristotle turns to analogy as a way to address violations of the hierarchical boundaries between different domains of knowl-edge: "Of the things they use in the demonstrative sciences some are proper to each science and others common—but common by analogy since things are *useful* so far as they bear on the genus under the science" (124).[22]

Analogy has a queer place in Aristotle's larger program: because it draws pat-terns between domains, it does not belong exclusively to one genus or another, and for this reason, it should be invalid. Elsewhere, Aristotle insists that knowl-edge is rigidly hierarchical. The natural world (as well as the rules of legitimate thought) is strictly determined by a nested series of descriptions that are orga-nized into groups and members of those groups—the categorical genera and the species that populate them. This is the ontological ground for deductive and inductive knowledge—methods of enquiry that work up or down the hierarchy of knowledge by moving from the general to the particular or from the particu-lar to the general. Aristotle's strict epistemology was grounded in his confidence that this hierarchy captures the ontological structure of the world. As far as Ar-istotle is concerned, this is the essential nature of science, of all knowledge. But in the *Posterior Analytics*, Aristotle gives examples where analogy finds patterns between discrete categories that are otherwise invisible (one example: "you can-not get one identical thing which pounce [talon or claw] and spine and bone should be called; but there will be things that follow them, too, as though they were some single nature").[23] At their most vital, analogies produce singular in-stances of pattern that break with the given framework of understanding. With respect to Aristotelian science, this means, in turn, either that (1) the order of nature, its ontology, is not strictly hierarchical, or (2) the epistemology of anal-ogy does not strictly cohere with the hierarchical order of nature, insofar as it finds patterns that violate that strict ontology.[24] For this reason, Aristotle's analysis recognizes analogy as a persistent problem for the relation between ontology and epistemology, between being and knowledge.

Analogy and the "Swerve around the Literary"

Recently, the being-knowledge problem has come to a head in a collection of writings on the nature of ontology and its significance for literary study. This work, alternatively described as "speculative realism" or "new materialism," rejects literary or linguistic approaches to the problem of epistemology in favor of a strict ontology inspired by formal language and mathematics. Though there are important differences between the "virtual ontology" of DeLanda and the "object-oriented ontology" of theorists like Tim Morton and Graham Harman, they share this focal concern for the problematic of language. If Alain Badiou is the most influential philosopher of this speculative turn, particularly through his influence on Quentin Meillassoux, Jeffrey Nealon notes that Badiou's philosophy is foundational to this larger "swerve around the literary."[25] By this, Nealon means that Badiou is one of several philosophers who have taken the "literary" as both a central concern of twentieth-century continental philosophy and as a crucial impediment to philosophy's continued advance. In Badiou's analysis, this emphasis on the literary, with its focus on poetic indeterminacies and its insistent "thinking 'vis a vis,' always in relation to the object or the world," has prevented philosophy from engaging the problem of "ahistorical truth."[26]

Recognized as an attempt to think *ahistorically* and *without* "relation," Badiou's philosophy presents an attempt to find an alternative to comparative historicism. And it does so by insisting on a basic distinction between the scientific and the literary, more precisely, by emphasizing the distinction between the two as "generic procedures" of philosophy ("the matheme" vs. "the poem") and the consequent difference between the truths they produce ("scientific" vs. "artistic").[27] This is at odds with the central thesis of this book, which holds that specific modes of literary description—in particular, the comparative strategies that constituted nineteenth-century historical fiction—changed how both historians and scientists access the past. Speculative realism holds that vernacular language, and particularly literary language, obscures the world. But I believe that imaginative language, and particularly analogy, helps us engage the world and its history. For this reason, a critical evaluation of speculative realism here secures a more general argument for the value of the literary object to scientific practice.

I am suggesting that analogies might help us to rethink the critical divide between description and understanding, between words and the world, between epistemology and ontology, that shaped the twentieth-century turn toward continental philosophy (schools that, until recently, were described as alternatively "postmodern" or "poststructuralist") and against which speculative realism has mounted its critique. At the close of this section, I'll explain why this critique of the "linguistic" turn is predicated on a misreading of structural linguistics and, in particular, a model of the "arbitrary" and "differential" sign that overlooks the articulation of structural linguistics with comparative philology and the study of the historical sign. Here, an appraisal of speculative realism and formal language explains why, in order to understand the scientific theories of a figure like Charles Darwin, we must reevaluate the ability of literary language to grasp the world.

In the analysis of Manuel DeLanda, natural selection effectively flattens the ontological distinctions upon which Aristotle placed total confidence (a point I take up and elaborate in chapter 5). In recognizing a harmonic similarity between genera and species where Aristotle saw a formal relationship, Darwin argued that genera and species share a common status as unstable categories that emerge within a process of continuous historical change. In his study of the ontological philosophy of Gilles Deleuze, DeLanda uses Darwin's theory as a chief example of what he terms "flat ontology": "While an ontology based on relations between general types and particular instances is *hierarchical*, each level representing a different ontological category (organism, species, genera), an approach in terms of interacting parts and emergent wholes leads to a *flat ontology*, one made exclusively of unique, singular individuals, differing in spatio-temporal scale but not in ontological status."[28] I think this accurately grasps a key feature of Darwin's claim "that all past and present organic beings constitute one grand natural system, with group subordinate to group, and with extinct groups often falling in between recent groups, is intelligible on the theory of natural selection with its contingencies of extinction and divergence of character."[29] Darwin effectively historicized Aristotle's natural hierarchy (at least in relation to organic life), showing that species and genera (as well as varieties, subvarieties, families, order, classes, etc.) marked more and less proximate relations in a long history of differentiation and subsequent evolutionary divergence. However, I think DeLanda's orientation toward language—or, more precisely, language's inconsequence—crucially misreads Darwin (and perhaps Deleuze), by insisting that such ontologies can be separated from the language that produces them. Here's how DeLanda explains his approach to reconstructing Deleuze's ontology: "I will not be concerned in this reconstruction with the textual source of Deleuze's ideas, nor with his style of argumentation or his use of language. In short, I will not be concerned with Deleuze's *words* only with Deleuze's *world*." This formulation assumes, at the outset, just what needs to be shown: that "ideas" are distinct from their "textual source" and "language"; that we can access the "world" without "words." I take this to mean not only that specific language is not necessary for the apprehension of a specific idea but more generally, that specific insights into the world have a status independent of any language and that this is what makes them *real* and Deleuze's philosophy *realist*.

I will take up this assertion in more detail, but first I want to emphasize its failure to gauge how a figure like Darwin, as a writer, contributes to our understanding of the natural world. Few scholars would question Darwin's extraordinary contribution to our understanding of nature. And yet Darwin realized that this access happens *through words*, not in spite of them. As I argue in chapter 5, Darwin, like his grandfather, emphasized the demonstrable fact that poetic or literary language, like metaphors, often captures something real about the world. He noted, for instance, that naturalists "frequently speak of the skull as formed of metamorphosed vertebrae . . . [and believe they] use such language only in a metaphorical sense: they are far from meaning that during the long course of descent, primordial organs of any kind . . . have actually been modified into skulls or jaws" (438). In light of Darwin's theory, "these terms [could now] be used literally," but the more consequential implication was that, in using such metaphors, naturalists disclosed a pattern of relation that moved toward a new understanding and a new way to describe the natural world, a more realistic account that science could not yet fully comprehend. Darwin recognized that such metaphors help scientists grapple with intuitions about the structure and nature of the world for which they do not yet have a formalized scientific vocabulary. It is for this precise reason, in fact, that Darwin himself relied upon the term "natural selection," which, he noted in the third edition of the *Origin*, is a "metaphorical expression" that "personif[ies] the word Nature."[30] In Darwin's view, such metaphors are *realist* in the sense that they give us purchase on real features of the world. This means that literary devices have an important place in the "flat ontology" that DeLanda ascribes to Darwin, and this in turn demands we think about the role that such language plays as part of the "interacting parts and emergent wholes" that constitute that ontology.

Darwin's writings show why a strong distinction between the "scientific" and "artistic" is untenable (indeed, this distinction would make "literature and science" impossible as a field of research), but I want to take up Badiou's claim that the "matheme" (in contrast to "the poem") is central to scientific discovery.[31] To summarize, mathematics (or, more accurately, the model of mathematics) operates as a primary "generic procedure" for speculative realism. And here, a caveat: because I am concerned with what we do with words, and how those words give us access to the world, I am not interested in what Graham Harman calls the "inscrutable reality" of things in themselves.[32] For this reason, I am intrigued by the importance of mathematics to speculative realism as a way to get at the world we encounter. DeLanda shares this evaluation of mathematics in his elevation of "Deleuze's world" over his "words." He is able to give Deleuze's ontology a free translation because, as he explains, it merely expands on the "mathematical resources" used by Deleuze, including projective and differential geometry, group theory and, more generally, a host of models derived from the study of dynamical systems. The central conceit is that these mathematical models describe a world that is independent of Deleuze's language. An obvious problem for this line of thought is that mathematics itself is a "formal" language (or quasi-formal language),[33] created by people to be self-consistent and to aid in our understanding of and engagement with the world. If the world is independent of language, it is independent of mathematics, too.

In chapters 4 and 5, I will revisit the problem of formal analogy in order to argue for a much looser sense of what formalism means, and to explore how literary and natural forms operate in time and respond to new conditions. Crucial to those discussions is the idea that formal analogy helps to produce and stabilize such forms, while harmonic analogy can pull them apart and rework them. Here, I want to focus on how analogy has helped to formalize mathematics in order to emphasize the continuing influence of analogical thinking on speculative realism. The question is whether mathematics, as a formalization, overcomes the problems of ambiguity and relativism generally recognized in literary language.

Part of the problem, I suspect, is a disagreement over what the formalization of mathematics entails. Meillassoux argues that the power of mathematics, even in the analysis of a poem's meter, is that "numbers are 'engendered' by thought in so far as it finds itself formulated by them, but in themselves they have no meaning—and in particular no meaning linked to the thought in question."[34] Accordingly, numbers have a special independence and a special purchase on reality, noticeable in their resistance to the author's semantic intentions. This, in turn, depends on a notion of the relation between number and vernacular language that Catarina Dutilh Novaes, a historian of mathematics, describes as "de-semantification": "On this view, to be purely formal amounts to manipulating symbols as blueprints with no meaning at all, as pure mathematical objects and thus no longer as signs properly speaking."[35] Elsewhere, Dutilh Novaes makes several key arguments about this process of "de-semantification": (1) it only emerges in the twentieth century, alongside a robust understanding of formal logics; (2) this came with a generalized acceptance that these formal logics provide "models" of the world, rather than capturing "something really *in* [its] phenomena"; and (3) their validity, with respect to the world, is "relative to the formal systems in question, and there could at least in theory be several equally 'good' formal systems" for a given problem.[36] Of course, as Dutilh Novaes notes, there are many different meanings for the word "formal," even in mathematics, but her work demonstrates that the evolution of both mathematics and logic show a steady formalization that is characterized by increasing schematization (through numbers and symbols) and abstraction (so that functions and operations could be applied to increasingly general "objects" rather than simply numbers or terms). Importantly, this comes at the cost of greater semantic *distance* from the world.

Here I am more interested in the important role analogy played in this formalization. One of the most startling coincidences in the history of both mathematics and logic is the apparently independent descriptions of logical algebra furnished by Gottfried Wilhelm Leibniz and George Boole in the seventeenth and nineteenth centuries. According to Dutilh Novaes, this suggests that the fusion of mathematics and logic was inevitable: "With the continuous advancement of algebra, the structural analogies between algebra and logic became increasingly manifest once more."[37] Like Bohr's model of the atom, these algebraic logics required the perception of a common pattern and adjustments that could draw that pattern out. Whether or not this was in fact inevitable, the recognition that there are competing formal systems suggests there are multiple possible paths. And while I am not qualified to evaluate whether the analogy that Leibniz and Boole drew between logic and algebra was formal or harmonic, it does appear that this analogy has subsequently influenced both logic and mathematics in far-reaching ways. In the long view, what this means is that analogy continues to shape the notion of formal language itself.

This book demonstrates why such claims should not be surprising and emphasizes the interplay between harmonic discovery and formal abstraction across a range of nineteenth-century sciences and literary forms. For this reason, I don't dispute the argument that mathematics gives us special access to the world (i.e., that it can be "realist"), but I also do not believe that this makes mathematics special. If math gives us access to the world, so does vernacular or "natural" language. In both, there is an interplay between efforts to abstract and formalize language, giving it wider reach, and the related need to adjust models and meanings to new applications. Moreover, literary figures like analogy and metaphor, as we have already seen, have a peculiar ability to extend our understanding to new phenomena.

History shows that analogies have played an important part in improving the sophistication of mathematics and its ability to make precise predictions and models. Even projective and differential geometries (which, in DeLanda's view, hold the key to Deleuzian ontology and its argument for the relation between structure and pattern) are based upon the study of how projections construct analogies between different geometric figures.[38] In the 1822 treatise that launched those fields, Jean Victor Poncelet begins by considering the "center of projection"—the spot within a spectator's eye where images from the external world are focused into a point, before projecting onto the retina. Poncelet explains the implications: "If one cuts a conic surface at any place with two parallel planes, the two sections will be completely similar. And further, if one takes a center of projection at any point in space, and extends vectors from that point to a system composed of lines or any surface, and then either further extends or shortens those vectors by the same proportion, one obtains a second system of points, lines or surfaces, similar to the first and similarly placed."[39] Poncelet's treatise explores at length how to treat this "center of similarity" (*centre de similitude*), which, in more general cases, he terms the "center of homology" (*centre d'homologie*).[40] But he underlines its basis in the study of the eye and a model of vision derived from the camera obscura. Historians of art, technology, and medicine have widely documented the use of the camera obscura to study vision during the early modern period.[41] In Kaja Silverman's view, the camera obscura demonstrated how analogy (originally the mathematical ratio of ratios) captured physical patterns of similarity between distinct systems.[42] Figure 3, from Johann Zahn's 1685 study of the eye, underlines this relationship; there is a shared proportion in the relation of each segment to the "center of similarity" (C), so that the proportion of segments AB and AC is proportional to LK:LC, IH:IC, and DE:DC [AB:AC :: LK:LC :: IH:IC . . . etc.]. The camera obscura underlines how Poncelet's "center of homology" generalizes the copula of analogy (::), formalizing its place in geometry in order to explore its possibilities. Though analogy classically meant a mathematical relation drawn between two proportions, the kinds of analogies we are talking about here (domain-transcending analogies, like those identified by Aristotle, which find a system of pattern

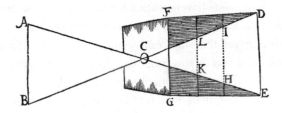

Figure 3. Camera obscura. Woodcut. From Johann Zahn, *Oculus artificialis teledioptricus sive Telescopium* (1685), 487000:1005, p. 95. Courtesy of The Huntington Library, San Marino, California

across distinct systems or areas of knowledge) mark a far more general and powerful mode of understanding. Through such analogies—drawn between electromagnetism and classical mechanics, or between the artificial eye and the human—real patterns converge in the world.

The Sign of Analogy

Despite my confidence that analogy provides access to new patterns and new understandings, I have not yet explained how analogies disclose new features of the world, even as they are articulated and explored through language. How do analogies bring ontology and epistemology together? How can a harmonic analogy find real features of the world that we did not previously recognize? I should stress that I do not believe analogies themselves are *real*, in the sense of having an observer-independent ontological status. Here, I break with Silverman's *Miracle of Analogy*, which argues that analogies show how "the world . . . is structured by analogy, and help us assume our place within it because it, too, is analogical."[43] Analogies are essentially epistemological: they help to recognize and understand patterns (whether in the world or imagined) for some (not necessarily human) observer. But I am a realist regarding *pattern itself*. Patterns exist in the world, and analogies (like a moiré pattern) have the ability to draw different patterns into harmonic interaction, revealing features of their structure.

A key sticking point is the continuing notion that language, conceived as a sign, is arbitrary and, hence, divorced from the world. If for some of us it remains counterintuitive to suggest that analogies cast through language (and so articulated as a series of signs) bring us into contact with features of the world, this is because the notion violates one of our long-held convictions—namely, that signs are arbitrary and differential. Speculative realism, like other efforts to "swerve around the literary," addresses a real problem with the understanding of language that humanists have wrestled for some time now. But I think the arbitrary notion of the sign rests on a basic misreading of structural linguistics, particularly as it engaged the nineteenth-century study of comparative philology. A consideration of that misreading can help build a new understanding of the sign that emphasizes its engagement with the world.

The first thing we find, if we return to Ferdinand de Saussure's influential *Course in General Linguistics* (the work that enshrined the notion of the arbitrary and differential sign), is that Saussure understood structural linguistics as a complement to comparative philology and grammar. Moreover, he credits later comparatists like William Dwight Whitney for "placing the results of comparative studies in their historical perspective" and making the study of *langue* possible.[44] His famous distinction between the study of *langue* (the static system of a given language) and *parole* (language as spoken and changing in time), as well as the consequent distinction between the synchronic and diachronic study of languages, captures this division of labor. Saussure meant to articulate structural linguistics and comparative philology together. The aim of structural linguistics, then, is to simplify the historical complexity, diversity, and associative nature of linguistic practice, in order to study language as if it were "a self-contained whole" (9). His entire theory of linguistics is predicated on this dialectical relation between system and history, including his analysis of the sign. This is why students of *langue* must "discard" everything they know about the historical nature of languages (81) and assume that signs are arbitrary, even if (as Saussure freely admits) signs are in fact conditioned by many historical factors: "Signs that are wholly arbitrary realize better than others the ideal of the semiological process" (66). For this reason, he also insists that his analysis applies only to linguistic signs, not, for instance, to graphic symbols that underline a connection to the world (68). His final articulation casts the relation between historical *parole* and ahistorical *langue* in terms of projective geometry: "To show both the autonomy and interdependence of synchrony we can compare the first to the projection of an object on a plane surface. Any projection depends directly on the nature of the object projected, yet differs from it—the object itself is a thing apart. . . . In linguistics there is the same relationship between historical facts and a language-state, which is like a projection of the facts at a particular moment" (87). In light of our previous discussion of Poncelet, we recognize that Saussure uses projective geometry to underline the complexity of the analogy between *langue* and *parole*. This means that the seeming arbitrariness of the sign is a formalization, whether understood as a "de-semantification" or a model, that suspends the meaningful and historical features of a language in order to examine its internal pressures—for instance, the way that shifts in pronunciation and agglutination force realignments in how words are parsed phonetically.[45]

In treating signs as discrete and independent parts of a system, Saussure makes no argument about the semantic and associative features of *strings* of signs: how signs, assembled together in sentences, relate to the world (or, in his analysis, the ideas these signs express). As Saussure admits, "Any creation must be preceded by an unconscious comparison of the materials deposited in the storehouse of language, where productive forms are arranged according to their syntagmatic and *associative* relations" (165, emphasis added). Saussure did not intend to divorce language from the world but to prove it has some degree of autonomy—that it was shaped by internal principles not wholly determined by its connection to semantic usage and the historical evolution of its forms.[46] The sign, in the ideal case, is arbitrary and differential, but words are certainly not.

The Age of Analogy explores how the "associative relations" of language are organized, through comparative historicism, in order to disclose patterns in the world. Analogies show that words *matter*, and by this I mean they not only shape the understanding they provide, but also provide one of our most powerful forms of access to matter itself—its behavior, and its interaction with human life and history. Recently, Bruno Latour has tried to rethink the nature of the sign precisely in order to explain its engagement with the world. Taking up the basic question of scientific representation in *Pandora's Hope*, Latour reinterprets the interaction between representation and reference as a nested series of mediations—a circuit of exchanges between features of the world and their representations. In the traditional account, a word ("A") signifies some feature of the world (w) (figure 4a). In Latour's account, words are just links in a longer chain of signification: some aspect in the world (w) is described by a representation ("A"), which can, in turn, be recognized as a feature (a) that is described by a higher-level representation ("B"), and so on (figure 4b). This is how formal analogies work: they describe a pattern in the world (w:x) in terms of a higher-level model ("A:B") that is already known. Latour's point is that these chains of "circulating references" extend through the world and close the "gap" between world and description by distributing it across the actual network of mediations that bridge this divide in practice, emphasizing exchange and contact between levels of representation. This is clearest in scientific practice, where observations in the field are recorded and organized into increasingly general representations that support scientific findings and theories. At each level, the smaller gaps between representation and object mark equally distributed sites

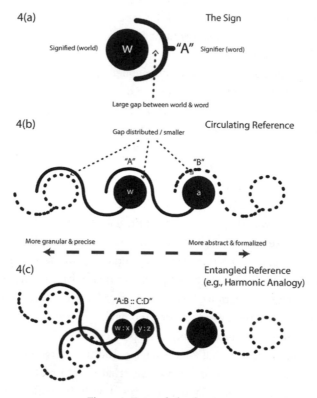

Figure 4. Entangled reference.

of uncertainty about both content and form—the possibility that other content fits better, the possibility that any content opens onto other forms, that either must be reworked to suit the act of adequation that language brings into play. The gap is not only a site of deferral and the play of difference but nascent contact with something new, the singularity of new encounter.[47]

Harmonic analogy amplifies the possibility opened by the gap between known form and unknown content by playing pairs of such relationships against each other. The symmetry of harmonic analogy (each half of the relationship models or represents the other equally) indicates a symmetry in the relationships themselves, an emergent pattern in the world. And this has to be so; there is no relationship between the expressions "A is to B" and "C is to D" unless the relationships they posit (w:x, y:z) support that symmetry through resonance of their patterns (figure 4c). The world has a primary claim on the copula of analogy. By bringing two different patterns in the world together, as interactive descriptions of each other, harmonic analogy opens the chain and interleaves distinct circuits of representation (bringing the pattern w:x into contact with y:z). Harmonic analogy both entangles the catenated references and is subject to the interaction of those patterns in the world; the signifying braid snaps taut as the world and its relationships tug representation into place. In this way, relationships in the world and their descriptions resonate and speak to each other as they speak to us.

While Latour does not focus on this interaction between distinct chains of reference, it is implicit throughout his analysis: in the way that collected samples brush up against each other, filed away for later comparative analysis, in the "pedocomparator" that standardizes the color of different soil samples and so formalizes the interrelation of their spectral patterns.[48] By distributing and so minimizing the "gap" between representation and represented, Latour's study compresses the epistemology of science. This complements the distribution of agency in actor-network theory, which imagines scientific research as a distributed network of human and nonhuman actors (or "actants"). As I argue in chapter 5, a flat ontology demands a flat epistemology.

A central thesis for *The Age of Analogy* is that if we closely consider the interrelation of literary forms, we may disclose the interrelation of actors (both human and nonhuman) in history. For this reason, a key strategy is to take examples of formal modeling—the interpretation of one system of relationships in terms of another—and flatten that dynamic, constituting (or reconstituting) a harmonic analogy that allows us to ask how both systems interact. So, in addressing Tennyson's *In Memoriam* in chapter 3, I ask how that elegy and its object, the poet Arthur Henry Hallam, might be placed in a more dynamic relation. Rather than exploring how Tennyson's poem reshapes the life and legacy of a dead poet, I ask how Hallam might have helped to write his own elegy, helping to constitute a space wherein Tennyson, poem, and Hallam meet and interact. In such instances, epistemological and ontological distinctions collapse, or at least compress, providing contact with other times, other people, and other forms of life.

In such moments, which move from thickness to flatness, from formal model to harmonic encounter (or, if you prefer, from depth to surface), the experience of contact is not reducible to the vast gulf that we usually find between self and other.[49] Ray Brassier has argued that such "irreductionism" is really just another version of correlationism, in which "the human and the non-human, society and nature, the mind and the world, can only be understood as reciprocally correlated."[50] But I see it the other way around. In Simmel's argument for the "irreducible, integral act" of historical experience, I see a limit case for experience in general: the way our experience of the world, and our way of being in it, are deeply entangled within our environment. After all, it is only after a long process of psychological, social, and intellectual development that these divisions—between human and nonhuman, society and nature, mind and world, knowledge and being—seem self-evident. In my view, these divisions, rather than "correlationism," reflect so many different ways of directing two perspectives on a singular moment of encounter—so many "dual-aspect monisms" (to borrow a phrase that has been applied to George Henry Lewes's study of the irreducible integration of brain and mind).[51] The literary object provides a quintessential case for the entangled dualism of knowledge and world. Either novels and their characters make a mark on our lives and show us something, or they don't. As a reader, you are either in their world, with them, or you are not. I believe this is what Derek Attridge means in underlining the "singularity" of literature: the ability of fictional encounters, as events in our lives, to remind us that we are immersed in a world that is beyond our complete control.[52] Analogy, in its movement between formal distance and harmonic encounter, can show us how.

CHAPTER ONE

Erasmus Darwin, Enlightenment History, and the Crisis of Analogy

[*The Botanic Garden*] seems, indeed, to have opened a new route through the trodden groves of Parnassus. The poet, who shall rashly venture to wrestle with such a rival, will be exposed to the derision of the *Arena*; for it will be long before a writer shall appear; who to all the prodigality of IMAGINATION shall unite all the minute accuracy of SCIENCE. It is a performance which makes a prominent figure in our national poetry.

Isaac D'Israeli, *Curiosities of Literature* (1794)

Was Erasmus Darwin, author of *The Botanic Garden*, a poet or a scientist? The most celebrated bard of the 1790s or the object of that period's most violent satires? An innovative and internationally renowned doctor or the surgeon who asked his wife to bleed him as he died? As Charles Darwin's grandfather, the most important British advocate of evolution before the publication of the *Origin of Species* (1859) or the crank who once harangued a crowd, while both drunk and sopping wet, on the virtues of circulating air? In sum: visionary or lunatic?

In each case, both. The complexity of Erasmus Darwin's life and the extraordinary range and mixed successes of his works, with their intermixture of distinct scientific, literary, and political traditions, make him an exceedingly difficult figure to work through: a unique product of a unique period. A doctor, inventor, scientist, poet, and founding member of the freethinking "Lunar Society" (hence, "lunatic"), Erasmus Darwin was notorious for his speculations about the origins of life *and* his energetic support for American and French republicanism. Critics and enthusiasts alike described him as a genius. Though shocked by Darwin's open deism, Samuel Taylor Coleridge was agog when they met, observing that "Dr. Darwin possesses, perhaps, a greater range of knowledge than any other man in Europe, and is the most inventive of philosophical men."[1] Darwin's floral epic, *The Botanic Garden* (1789–91) was equally celebrated when it first appeared, with at least nineteen printings in the first twenty years, and translations into French, German, Italian, and Portuguese. Its promise—to "inlist the Imagination under the banner of Science"—suggested a deep congruity between the literary imagination and scientific discovery and advanced a political thesis about the importance of science in guiding the social convulsions of his day.

This book asks how Charles Darwin came to terms with this complex legacy, and here I emphasize that this legacy was larger than Erasmus's scientific theories, with complex social and literary dimensions, especially for Charles. The Darwin family was scandalized by the posthumous *Memoirs of the Life of Doctor Darwin* (1804), written by Anna Seward about her fellow poet and sometime friend. Seward characterized Erasmus's various enthusiasms as the fruit of both genius and his pronounced, sometimes inhumane idiosyncrasy. When Charles, approaching the end of his own life, sat down to write a new biographical essay about his grandfather, he worked assiduously to adjust that picture and bring its features into proportion. But his most important contribution to the legacy of his grandfather was to refurbish his theory of evolution and put it on a sound conceptual and empirical footing. As I will argue in later chapters, a central element of that retooling was to forge comparative historicism into a method for exploring evolution as a mechanism in time.[2]

Erasmus Darwin and his eccentric writings are fascinating in their own right. More important for this chapter is how they mark a turning point in the use of analogy. Collectively, his works argue for a pattern of universal progress that is based on analogies drawn between society and nature, between the insights of poetry and the discoveries of science. Articulated through epic poetry and framed in terms of vitalist philosophy and the progressive history of the Scottish Enlightenment, Erasmus Darwin's belief in consistent and ubiquitous development was ill-suited to an age of uneven revolution. Ultimately, the thesis of universal advance proved untenable. A subsequent generation of comparatists, including Charles Darwin, would free analogy from the commitment to providential order and develop new strategies for analyzing the discontinuities of social and natural history. But in order to understand how this worked, and why it was necessary, I will spend this chapter explaining why Erasmus Darwin's own model for evolutionary development failed.

Erasmus theorized, by means of a cosmic *élan vital*, a universal impulse to improvement that moved across stellar, geological, and human scales.[3] But the ongoing political and economic crises of the Regency era refuted the thesis that humanity was swept up in a larger cosmic progression. By the same measure, Darwin's struggle to find a poetic language adequate to this vision—and, particularly, his use of elaborate allegories and ponderous heroic couplets—prevented his verses from adapting to the revolution in taste ushered in by Romantic-era experiments with alternative poetic forms. By the early 1800s, to

"Darwinize" became shorthand for scientific speculation without empirical support, while "Darwinian" served as a pejorative epithet for overwrought verse. This explains why, in later editions, Isaac D'Israeli amended his essay on Erasmus Darwin's *Botanic Garden* (given in this chapter's epigraph) to read: "The Botanic Garden *once appeared* to open a new route through the trodden groves of Parnassus" (emphasis added).[4]

If his poem failed to open up a "new route" for poetic form, the understanding it projects nevertheless opened up entirely new avenues of thought. The central problem of critical engagement with Erasmus Darwin has been the almost compulsive search for the political, literary, and social determinants of Darwin's "fall."[5] The discussion that follows takes the opposite tack, arguing that Erasmus carefully engineered the conditions of this reversal in fortune. A careful attention to the publication history of his major work exposes Erasmus Darwin as a calculating scientific Jacobin, a provocateur who wanted to effect a revolution in scientific, political, and prosodic understanding. If the revolution did not succeed within his lifetime, it gave focused impulse to the longer trajectory of Romantic verse, from William Wordsworth's attempt to write what Samuel Taylor Coleridge called "a true philosophical poem," to Percy Bysshe Shelley's effort to cast poets as "the unacknowledged legislators of the world."[6] In Tim Fulford's account, Darwin's verse experiments and natural philosophy helped to shape Coleridge's conversational poems and, ultimately, the greater Romantic lyric.[7] Like Fulford, I understand Darwin less as a precursor to British Romanticism than as an architect of its natural and political philosophy.

Analogy stands at the center of Erasmus Darwin's thinking about the relation between social systems and natural order: he repeatedly used analogy to articulate *both* the method through which philosophers explore pattern *and* the nature of that pattern itself. He was an important theorist of the "aesthetics of correspondence" that literary historian Susan Manning recognizes in eighteenth-century philosophy and literature.[8] One of Darwin's most important contributions to Romantic thought was the consistent effort to collapse poetic allegories into more immediate analogies that tie human experience to nature's organization. As argued in foundational studies by M. H. Abrams, Earl R. Wasserman, and Paul de Man, scholarship on British Romanticism has long identified the period with a transition from analogy, understood as a neoclassical framework for poetic understanding that uses allegories and personified abstractions, to the use of more immediate symbols.[9] This account overlooks Erasmus Darwin in two important respects: he both showed how to decode such allegories as vehicles for a more basic insight into the analogy between all life and provided a new understanding of analogy itself. Through Darwin's works, analogy forges connections that do not hold the objects of concern apart (as Abrams argued) or defer their relation in time (as de Man put it) but, rather, work to bring the objects of comparison into intimate and transformational contact.

In particular, as Martin Priestman has shown, Erasmus Darwin mined the allegorical mysteries of Greek, Roman, and especially Egyptian antiquity for insights into the natural order—and, especially, the nature of evolution.[10] In collapsing these allegories into a universal analogy drawn between animal, vegetable, and terrestrial development, he produced a new way of thinking about poetry's capacity to reveal natural pattern. By these means, he also championed a theory of evolution modeled on sexual reproduction. If this emphasis upon nature's sexuality now seems strange, it was firmly based in Darwin's studies of spontaneous biological development, in both animal embryos and plant reproduction, and supported by authorities ranging from Lucretius, the classical poet, to Carl Linnaeus, the famous Swedish botanist. Moreover, Erasmus Darwin— who studied in Edinburgh, sent his sons to train there, and retained close ties to its academic communities—recognized analogy as a common feature of the philosophy of the Scottish Enlightenment, a "style of reasoning" (in Ian Hacking's sense) with profound implications for the relation between literary production, scientific investigation, and the larger order of nature.[11] In addition, as a central figure of the Lunar Society and the "Midlands Enlightenment" described by Gavin Budge and others, Darwin served as an important conduit between two centers of Enlightenment thinking, helping to model a relation between analogy, science, and poetics that would shape the Romantic imagination.[12]

Darwin's emergent understanding of analogy, as it shaped both poetry and science, can best be seen in the unfolding narrative of his major poems—*The Loves of the Plants* (1789), *The Economy of Vegetation* (1791–2), and *The Temple of Nature* (1803). This chapter explores how, through the close of the century, Darwin continued to search for verse forms suited to his theory of evolution. The decisive turn in this development was his prose masterwork, *Zoonomia* (1794–6), which first explicitly set out his theory of evolution from a single "first filament," and advanced the theory that all life developed from a single original "embryon." Perhaps the most important impact of publishing this theory was to reevaluate the narrative of the earlier poems, collected as the two-volume *Botanic Garden*, as a coherent epic of nature. In so doing, *Zoonomia* recast the *Garden* as a serious evolutionary epic that only masquerades as a polite experiment with didactic verse. By unlocking scattered hints within his earlier poems and demonstrating a developmental analogy drawn between nature's order (presented in *The Loves of the Plants*) and a progressive social history (offered in *The Economy of Vegetation*), *Zoonomia* exposed Darwin's poems to renewed critical attention and conservative censure. This is less a criticism of his work than a recognition of his extraordinary forward-thinking attempt to create the cultural forms that would suit evolutionary models of natural and social life. James Secord, David Amigoni, and James Elwick have argued persuasively that the "evolutionary epic" remained a major genre for nineteenth-century science writing, as seen in such exceptional works as Robert Chambers's wildly popular *Vestiges of the Natural History of Creation* (1844), and here I emphasize Erasmus Darwin's importance to that tradition.[13] Yet, while Amigoni, Elwick, and Herbert Tucker suggest that the epic shaped nineteenth-century long fiction, particularly the historical novel, the chapters that follow underline the deep division between the complex and differentiated histories explored by Walter Scott's multiplot novels and the unitary narratives of the epic.[14] If the historical novel was a wild success in the nineteenth century, the epic generally struggled to survive; Erasmus Darwin helps us understand why.

Darwin's *The Temple of Nature*, a mature epic that continued to emphasize universal progress, could not resolve a more basic tension between the diversity of natural forms, the complexity of human history, and the thesis of consistent development.[15] The strategy of this chapter is to argue that Erasmus Darwin was forestalled by the absence of a less teleological and, so, more supple and successful historicism. But its absence did not mean that such a historicism had not been anticipated, and so in practice my approach is to follow Erasmus Darwin's effort to figure out what a new historicism might look like. If Darwin's work remained stranded halfway into the nineteenth-century transition in historical thought by which, as Herbert Tucker puts it his study of epic poetry, "structure was apprehended as the result and temporary instantiation of an ongoing process," this was in part because the theory of process on which Erasmus based his epic poetry was rooted in older historical models that did not admit diversification, uncertainty, and the contingency of change.[16] The most profound implication of this larger shift was *not* a more modern historical sensibility—the recursive fashion in which we have come to recognize our own experience as already historicized within the "homogeneous, empty time" described by the philosopher Walter Benjamin or the "abstract time" described by literary historian Sue Zemka.[17] More important was the transformed understanding of pattern itself. Rather than pattern testifying to the organization, the design, the metaphysical structure of the world, it reflected interconnected and individuating histories—evidence for *how* we came to be, rather than our being.

If eighteenth-century analogy sought its answers beyond history, comparative historicism found them in it.

While this book narrates the decline of totalizing approaches to history and the rise of more dynamic and comparative nineteenth-century modes, this chapter focuses on Erasmus Darwin because, in the complexity of his understanding of natural pattern and the diverse approach he took to describing historical change, he provides a lens for examining Enlightenment historicism, and highlights a crisis for those who saw analogy as the evidence of transcendent order. In the first portion of this chapter, I examine Darwin's reading of the Linnaean taxonomy, both in his early translations as part of the enthusiastic Lichfield Botanical Society and later in his *Loves of the Plants*. Despite Darwin's fascination with Linnaeus's speculation about the historical source of living taxonomies, the Linnaean system projected a static natural order. In the following section, I consider how Darwin reformulated Linnaean taxonomy as a process in time, imputing a history to Linnaeus's "sexual system." Next, I explore the historical vision of *The Economy of Vegetation* as a counterintuitive alliance between different ways of understanding history at the close of the eighteenth century. While offering different conclusions, distinct political visions, and contrary opinions about the nature of history itself, these different understandings of history, developed within Christian theology, Enlightenment philosophy, and political history, held in common a style of reasoning rooted in historical analogies. *The Economy* aligns these modes of understanding and so explores how patterns in natural, human, and religious history express the providence of a higher order.

The third part of this chapter examines Erasmus Darwin's naturalization of allegory through the lens of his collaboration on the illustrations to the *Botanic Garden* with William Blake and Henry Fuseli. Rather than engaging allegory as a vehicle for esoteric knowledge, Darwin understood allegory as an engine of scientific insight, arraying it with other modes of scientific and philosophical study to explore the basic patterns of nature. Finally, I study how Darwin brought this complex amalgam of stadial history and scientific theorizing together, exploring the publishing history of the *Botanic Garden* and *Zoonomia* (1794–96) as an interlinked effort to reformulate the static Linnaean system as evidence of an epic evolutionary process. Darwin's poetry can be understood as an attempt to reconcile disparate historical visions through consistent reference to a cosmic process of evolutionary development. In Darwin's view, the loves of the plants over time, and their resulting evolution, recapitulated a basic impulse to development, a seminal vitalism, that operates on stellar, planetary, and living systems. Notably, the term *comparison* is absent from this account, even though Darwin took Linnaean taxonomy as a proof of the analogies that structure life, demonstrating a network of physical similarity that would eventually come to be known as comparative morphology. For this reason, Darwin's works are important waypoints for strategies of analogical analysis that would later underwrite the comparative method. Ultimately, in treating Darwin's works as a concerted effort to shift the rails of historical understanding, I recognize him as less a victim of reactionary fate than a harbinger for the revolution to come.

The Loves of the Plants and Sexual Taxonomy

The general design of the following sheets is to inlist Imagination under the banner of Science; and to lead her votaries from the looser analogies, which dress out the imagery of poetry, to the stricter, ones which form the ratiocination of philosophy.

Erasmus Darwin, advertisement to *The Botanic Garden* (1789)

The magnitude of Darwin's achievement bears emphasis. Insatiate fans and fuming critics agreed that *The Botanic Garden* was the most important poem of the 1790s. It was an enormous critical success—its popular circulation held back only (as one reviewer complained) by the lack of a "cheaper edition" than the expensive illustrated quartos published in his lifetime.[18] Its synthesis of scientific knowledge and poetry was unequaled in both scale and ambition and relied on extraordinary generic hybridity. The result is a mix of mock-epic, pastoral, and neo-Lucretian philosophy, on the one hand, and of scientific account, travel narrative, and speculative philosophy, on the other. Gnomes and nymphs collide with James Watt and his steam engine. The poem revels in dilation. As Darwin happily admitted, his extensive commentary was "3 or 4 times the quantity" of the verses themselves.[19] The combinatorial sexuality offered in the second volume, *The Loves of the Plants* (which took Linnaean floral classification by number and type of sex organ and translated it into orgiastic combinations of nymphs and attendant swains), produced a critical riot (Horace Walpole: "the most delicious poem on earth"; Coleridge: "I absolutely nauseate Darwin's poem").[20]

This intellectual brew was the product of Darwin's extensive network of correspondents and connections—scientific, literary, and political. He corresponded familiarly with Benjamin Franklin and Joseph Banks, patronized Watt's inventions, was invited by George III to serve as his private physician, and applied, with some merit, for the position of poet laureate. If, as his grandson would later put it, thinkers tend to be either "lumpers" or "splitters" (i.e., generalists or particularists), Erasmus Darwin seemed the lumper of all lumpers.[21] For a time, it was captivating. The very Romantics credited with sealing the turn against "Darwinian" verse, Coleridge and Wordsworth, were early adherents of the doctor. And as late as 1802, as Coleridge meditated a treatise evaluating the larger study of literature and verse that would mature into his *Biographia Literaria*, he took Darwin's "Interludes" to *The Loves of the Plants*, along with Wordsworth's preface to the 1800 *Lyrical Ballads*, as the most important contemporary statements of the function of poetry.[22]

The most important alliance, for Darwin and for this chapter, was the supportive connection he drew between poetic composition and scientific research. Yet he was sensitive to the hardening of vocational distinctions in the late eighteenth century.[23] In particular, Darwin worried that his scientific and literary pursuits might affect his medical practice. This is a primary reason why, when Joseph Johnson published *The Loves of the Plants* in 1789, just months before the storming of the Bastille, he issued it anonymously. (Though it was announced as the "second" volume of a longer work, *The Botanic Garden*, the official first volume—*The Economy of Vegetation*—was not issued for two more years.)[24] Its immediate success was due, at least in part, to the way it opens with a sharply delineated formal model. The first edition included not only a preface but also an advertisement and proem, which carefully explain what the following verses would include and how they should be read.[25] First, the proem invites the reader to imagine the work as the "Camera Obscura" of an enchanted garden, wherein "lights and shades danc[ed] on a whited canvas, . . . magnified into apparent life."[26] Kaja Silverman, a historian of photography, has explored how the camera obscura shaped early modern thinking about the analogy between representation and the world: "The classical camera obscura . . . was a darkened chamber with a small aperture through which light entered, bearing a reversed and inverted stream of images that both originated in the external world and analogized it."[27] As Silverman notes, this provided a key model for how representations could document and preserve structural relations within the world, and for this reason Erasmus Darwin uses it to formalize an analogy between nature's patterns and the conceits that structure *The Loves of the Plants*.

The camera obscura captures the intimate relation between Darwin's poetry and the natural pattern it "analogized." Its model also initiates a series of boundary-setting moves. These gestures, unfolding through the prefatory materials, specify carefully the relationship between poetic form and scientific content and, conversely, the subordination of the literary imagination to scientific practice. Following this line, the proem claims kinship to Ovid (rather than Linnaeus), explaining how the "trivial amusement" presented in the work reverses the conceit of the *Metamorphosis*, translating plants into allegorized people (*CW* 2, viii). The deferential status of literature vis-à-vis science firmly established, the advertisement takes up the task of explaining how such "amusement" can yet promote scientific knowledge: its "general design" is to "inlist Imagination under the banner of Science."[28] This is possible because the "imagery of poetry" (the way that the natural world is "magnified" by representation through the camera obscura) can present "looser analogies," patterns that may "lead" us to the "stricter ones, which form the ratiocination of philosophy" and science (*CW* 1, v). What this means immediately, as the preface further explains, is that the poem will translate the "stricter" analogy of the Linnaean sexual system of botanic classification into the "looser" analogy of romantic liaisons between nymphs and swains (*CW* 2, v). The camera obscura frames this relationship, inviting the reader to think of representation as an art that regulates the relationship between the distinct domains of literary and scientific production.

In essence, the poet presents a formal contract to readers: they will get the substance of the Linnaean sexual botany by means of a familiar pastoral convention, the fruit of science under the sign of didactic verse. *The Loves of the Plants* participated in an explosion of domestic interest in botany in the late eighteenth century, a fad that delivered Linnaeus to the home, while providing intercourse between domestic gardens and far-traveling botanical specimens made available by empire.[29] Recent scholarship by Theresa Kelley, Monique Allewaert, and Greta LaFleur has shown the broad influence of botany in eighteenth-century discussions of classification, sex, and race in the Anglo-American world, at the same time that "botanizing," understood as a polite science practiced by both (human) sexes, furnished a controversial new vocabulary for sexuality.[30] The "trivial amusement" of *The Loves* is inseparable from this larger imperial context, even if, within its pages, Darwin railed against slavery and press-ganging, practices critical to British imperial power. The didactic model carefully limits the expansive ambition of the poem, insisting that imaginative engagement accrues value only as it invests the reader in Linnaean botany.

Contemporary reviewers were beguiled. The *Monthly Review* took up the invitation to visit the "Inchanted Garden": "We have accordingly walked in, and viewed the whole exhibition, and we have received from it so much pleasure and instruction, that we give our readers a warm invitation to follow us."[31] The *English Review* echoed others in its observation that its account "would have done honour to Ovid's Metamorphosis."[32] The *Critical Review* appreciated the poet's transmutation of "harsh and unpleasing" Linnaeus into "elegant and flowing language."[33] Erasmus Darwin was extremely sensitive to reception; in working on his translations of Linnaeus, he consulted with both the eminent naturalist Joseph Banks and the stringent critic Samuel Johnson—respectively the president of the Royal Society and the author of the definitive English dictionary—to secure their support. In the context of the early 1790s, when reviews were still dominated by the perspective of liberal dissenters, *The Loves* found the sweet spot between scientific improvement and polite entertainment.

At a more basic level, the central conceit of *The Loves of the Plants* (an imaginative translation that taught the reader the Linnaean system of organization) was supported by the didactic convention of the Linnaean taxonomy itself. Darwin, who spent years translating Linnaeus's *Genera Plantarum*, understood this well. Linnaeus freely admitted that the sexual classification of plants, based on the number and relationship of male and female organs in their flowers, was "artificial" rather than "natural."[34] The sexual taxonomy gave a convenient but artificial way to organize the high-level orders and classes of plants for easier memorization and study. This artifice is a surprising feature of a system that was enormously influential. But it reflected a problem central to contemporary schemes of classification, both for the natural world and for the rapidly expanding universe of textual production. The relation between organism and organization was a defining challenge for eighteenth- and nineteenth-century science. Most systematists assumed that there was, in fact, a true system underpinning natural order but agreed that the crude nature of scientific understanding made its accurate description challenging, if not impossible.

The dilemma was whether to develop "rational" systems of classification or to rely on arbitrary patterns. This crucial distinction, derived from bibliographic cataloging practices, addressed any large collections of objects.[35] Rational catalogs organized their contents according to a higher subject-based hierarchy of topics (what modern specialists in information science call an "ontology"). In contrast, many systematists preferred arbitrary systems of organization, like alphabetization, that did not have to fabricate an ontology; as Scottish botanist Colin Milne put it in his *Institutes of Botany* (1771), "every system must [in practice] be artificial."[36] Without a God's-eye view, a truly accurate and comprehensive ontology, whether of living organisms or social forms, was impossible.

Linnaeus understood this ideal classification as an impossible goal, as did Darwin. Though there are multiple levels of hierarchy in the Linnaean classification of plants, they operate at two basic levels of scale. On the one hand, there is the macro-scale division of plants into orders and classes, which follows Linnaeus's sexual taxonomy (the division by number and arrangement of sex organs). On the other, there are the micro-scale divisions of species and genera, which are tied to a host of more particular characters. So, in an earlier translation of the *Genera Plantarum* produced by Darwin and the Lichfield Botanical Society, large-scale organization by sexual morphology presents "artificial arrangements," whereas the minute description of individual species and closely related varieties rely upon "natural associations."[37] Linnaeus defends the sexual system by arguing for its simplicity, its memorability, and its capacity to organize the vast wealth of descriptions of the natural world (lxiv–lxv, lxxxvi).

Linnaean classification presents a theory of ordered information that understands the importance of efficient and arbitrary indexing. *The Loves of the Plants* capitalizes on this lesson, organizing its treatment around a distinction between the arbitrary use of pastoral convention and the scientific knowledge it organizes. This clarifies the formal contract presented to the reader in *The Loves of the Plants*: the poem both reflects and is subject to a more basic division between the conceit of artificial classification and the order that is assumed to structure the natural world. The contract insists that artificial representation is valuable because it presents (even as it distorts) true but irreducibly complex natural relationships. *The Loves of the Plants* effectively leverages its value on this sequence of mediations, which firmly subordinate artificial arrangement to natural truth. The extensive editorial footnotes and appendices to the poem reinforce this conceit, characterizing the verses, through elaborate references to contemporary science, as playful allegories appended to more substantial and far more extensive botanical knowledge. So, in his description of the "imperial DROSERA" with her "five sister-nymphs" and "*five* fair youths," the poet observes: "A zone of diamonds trembles round her brows; / Bright shines the silver halo, as she turns; / And, as she steps, the living lustre burns" (*CW* 2, 1.240–2). An extensive note explains that Drosera is commonly known as sundew and has both five stamens and five pistils, and it then elaborates: "The leaves of this marsh plant . . . have a fringe very unlike any other vegetable productions. And, which is curious, at the point of every thread of this erect fringe stands a pellucid drop of mucilage, resembling a ducal coronet. This mucus is a secretion from certain glands, and like the viscous material round the flower stalks of Silene (catchfly) prevents small insects from infesting the leaves" (*CW* II, 24–5). The note both explains the glittering effect that gives the plant its colloquial name and situates the mechanism and purpose of that "coronet" within a wider botanical understanding.

The strategy of subordinating art to science also justifies the seemingly arbitrary progression of the poem itself. Scientific explanation in the *Loves* emphasizes the superficial quality of the allegorical vignettes. Occasionally the vignettes contain a short tale, or a note diverts widely to take in observations about geology or husbandry, but they retain their accidental, happenstance quality. The proem modestly describes the verses as "connected only by a slight festoon of ribbons," and for some readers, this was not enough. In his *Curiosities of Literature*, D'Israeli complained that the poem lacks "a little plot or fable . . . to connect, and to animate the whole."[38] As initially presented, Darwin's "ribbons" correspond to the incisive lines that divide the botanical plates at the front of poem, delineating a system that organizes the sequence of floral families and orders within a segmented grid, rather than a coherent story. The grid, based on the sexual taxonomy, reminds the reader that the "festoon" of incidental narratives in the poem is immaterial against the synchronic order of the sexual system. The narrative vignettes provide an illusion of temporality in a poem organized outside of history.

Stadial History and *The Botanic Garden*

Orbs wheel in orbs, round centres centres roll,
And form, self-balanced, one revolving Whole.

The Economy of Vegetation (1791)

But history intervened. When *The Economy of Vegetation*, officially the "first" volume of *The Botanic Garden* (and actually printed in 1792), its precedence promised a theory that would explain the static system of the subsequent *Loves of the Plants*. It projected a principle of cosmic development, exercised through ministering agents of fire, water, earth, and air, that could explain the longer history of terrestrial development—both social and natural. Darwin's explicit debt to Lucretius has prevented us from recognizing how the sequenced publication of *The Loves of the Plants*, *The Economy of Vegetation*, and *Zoonomia* worked to shift the genre of *The Botanic Garden*. His developmental theory was not fully articulated until the publication of *Zoonomia* in 1794. Together these works invited readers to read Darwin's verses as evolving from pastoral vignettes to epic narrative. Darwin gave his final statement on this generic shift in his posthumous *Temple of Nature*, which aimed "to amuse by bringing distinctly to the imagination the beautiful and sublime images of the operations of Nature in the order, as the Author believes, in which the progressive course of time presented them" (*CW* 9, preface). It is an accurate summary of Darwin's larger ambition—an attempt to translate didactic verse into an epic narrative structure that could support his theory of evolution. Put simply, evolution allowed Darwin to plot natural order in time. It also erased the Linnaean distinction between the "natural" order of species and genera and their "artificial" arrangement in families: all life became one natural family. And yet the fusion of didactic and epic verse could not provide a convincing solution for that plot. The problem was that the sweeping cosmic history he described, in which a universal evolution tells a profoundly coherent story, came at the expense of a more variegated and sensitive account of the different trajectories of that story's actors—the plants, the stars, and humanity itself.

In weaving a comprehensive historical understanding for *The Economy of Vegetation*, Darwin drew threads from various contemporary theories of the past. These included the progressive narratives most closely associated with "Whig history," the patterns of stadial change emphasized by the historians of the Scottish Enlightenment, and a background Christian understanding of providential creation. Darwin intertwined these disparate understandings with different degrees of success and commitment, but his effort presents a remarkable abstract of the eighteenth-century historical imagination.

When contemporaries first confronted the *Economy*'s description of nebular creation, they read it as an authoritative statement of the congruence of Christian belief and scientific understanding. Walpole called it "the most sublime passages in any author," alluding to its close connection with Milton's cosmogony in *Paradise Lost*.[39] The passage begins with the strong departure of an initial inversion ("*Let* there be *light*") that periodically returns in the following lines, as the opening dactyl establishes God's missive as a strong, hermetic phrase:

> Let there be light!" proclaim'd the Almighty Lord,
> Astonished Chaos heard the potent word;
> Through all his realms the kindling Ether runs,
> And the mass starts into a million suns;
> Earths round each sun with quick explosions burst,
> And second planets issue from the first;
> Bend, as they journey with projectile force,
> In bright ellipses their reluctant course;
> Orbs wheel in orbs, round centres centres roll,
> And form, self-balanced, one revolving Whole.   (*CW* 1, 1.103–12)

Here, the divine command is a seeding impulse that initiates a cascading series of stellar and planetary formations. God's missive, derived from Genesis, is carefully incorporated into a model of universal evolution that recognizes repeated patterns of explosive reproduction, a mass spewing suns, suns ejecting earths, smaller planets flung from larger. All cohere in a broad trajectory of development and (from the planetary perspective) progress to the current state of things.

I will shortly explore in more detail this fusion of Christian, stadial, and progressive historicisms, but here I wish to mark the passage as an important example for how self-organizing nature and its analogical patterns can model poetic form. Strictly speaking, the initial irregularity disturbs the meter. In *Zoonomia's* discussion of the association of sense that underpins poetic verse, Darwin stresses that verse's "little circles of musical time" owe their power to the pattern they establish and emphasizes their regularity: "Whether these times or bars are distinguished by a pause, or by an emphasis, or accent, certain it is, that this distinction is perpetually repeated; otherwise the ear could not determine instantly, whether the successions of sound were in common or in triple time" (*CW* 5, 374). Hence initial inversions, by destabilizing the iambic pattern, should confuse the ear and frustrate the poem's ability to produce pleasure.[40]

Anna Seward, however, once argued that Darwin's frequent use of inversions and the imperative mood calls attention to larger patterns.[41] The comparison that is mapped between the lines containing initial inversions in lines 103, 107, 109, and 111—each of which forms the first complete phrase that founds a new independent clause—marks the ramification of the initial command throughout various corresponding cosmic valences, from the cosmos as a whole to the particular organization of solar systems, and finally to an archetypal description of all these motions, where "Orbs wheel in orbs, round centres centres roll." The lines form a poetic cosmos of their own, the irregular pattern of the first line echoing as the passage unfolds. As we will see, the passage inaugurates a form of metric and phrasal involution that is picked up in key moments of the poem—as in later descriptions of seed formation and terrestrial evolution. The effect is a resonance of meaning, in which sound and sense, substantive and predicate, reverberate in inverted pairs that cohere in "one revolving Whole" of self-organizing correspondences and extensions.

The environment within which Darwin's verse imagines itself operating—whether conceived as rhythmic chaos or fixed meter—is itself an emergent phenomenon, construed after the fact, as the contextual counterpoint to the poem's action. To describe Darwin's poem as written in "heroic couplets" or "iambic pentameter" is necessarily to make a generalization beyond the specific approximations and departures of individual lines, a generalization the poem invites. Meter itself is an evolving system, an environment of rhythmic noise that offers fertile ground for seeded patterns of reorganization and rhythmic departure.

On this view, Darwin's vision of nature offers a subtle understanding of form, as it reads natural order *and* poetry as emergent phenomena—broad patterns that structure networks of meaning—and so (potentially) beyond the control of the singular author or divinity.

Walpole's reaction suggests that Darwin's first readers were not anxious about these implications. Instead, they read the *Economy* more comfortably as a synthesis of contemporary religious conviction and a rapturous perception of the divine imprimatur registered in the patterns of the natural world. In the eighteenth century, figures like Isaac Newton argued that the study of natural science, the study of natural theology, and the study of scripture went hand in hand. As Newton put it, in his *Observations upon the Prophecies of Daniel, and the Apocalypse of St. John* (1733): "For understanding the Prophecies, we are, in the first place, to acquaint ourselves with the figurative language of the Prophets. This language is taken from the analogy between the world natural, and an empire or kingdom considered as a world politic."[42] Kenneth Newport notes that Newton was part of a "scholarly elite" that maintained strong ties with a larger tradition of "Church historicism," one that mapped biblical prophecies onto contemporary historical events in an attempt to set a date for the return of Christ.[43] Whereas Newport sees a basic contradiction between the "spirit of rationalism" reflected in the *Principia* and the thesis of a bible that was of "supernatural origin, faultless and timeless in its message," Newton argues that this seeming difference is bridged by a productive analogy between the "world natural" and the "world politic."

Joseph Priestley—a famous scientist, dissenting preacher, good friend of Darwin's, *and* a fellow member of the Lunar Society—also saw a deep congruence between science, history, and the power of millennial prophecy. He advocated a "rational Christianity" that combined faith in the historicism of the scripture—that Jesus was a supremely virtuous but historical (rather than divine) figure—with confidence in the coming of the Christian Millennium and the accuracy of prophetic biblical accounts.[44] In *An History of the Corruption of Christianity* (1782), Priestley ascribes to a "historical method" that has the capacity "to trace every such corruption [of Christian doctrine] to its proper source, and to shew what the circumstances in the state of things, and especially of other prevailing opinions and prejudices, made the alteration, in doctrine or practice, sufficiently natural, and the introduction and establishment of it easy."[45] Though Priestley sees secular history as a corrupting influence, in looking to contemporary "opinions and prejudices" as a "natural" context for changes in Christian doctrine, Priestley's historical method looks forward to nineteenth-century historicism, with its emphasis on the explanatory power of social context. In this way, Priestley's historical understanding registers a sense of the increasing pressure that contingency and social custom place on unitary narratives of continuous historical progress, at the same time that he works to resist this influence.

Despite his anxiety regarding the "corruption" of doctrine, Priestley expressed confidence that "Christianity has begun to recover itself from this corrupted state, and that the reformation advances apace."[46] Writing in 1782, almost a decade before he would be burned out of house and homeland during the Birmingham riots, Priestley yet retains contact with a general progressive optimism that history reflects a general pattern of advance. As such, his belief in spiritual progress aligns with the "Whig interpretation of history," Herbert Butterfield's term for contemporary political historians who "studie[d] the past with reference to the present" and "demonstrated throughout all ages the workings of an obvious principle of progress."[47] This view, which saw the ascendancy of Whig governance during the "glorious" revolution of 1688 as the culmination of an ancient constitutional tradition, was presented in the eighteenth century by the writings of Paul Rapin de Thoyras and William Robertson (especially the latter's *History of Scotland* [1759]). Colin Kidd has argued that Robertson's study was "one of the central 'achievements' of the Scottish Enlightenment" in the way it incorporated the Scottish history into "a compelling modern Whig synthesis."[48] Insofar as such progressivism seemed to provide a political complement to the Enlightenment's self-conception as the advancement of scientific knowledge, it is easy to understand why Whig history provided such an important background for contemporary thought, a new, more persuasive analogy between a "world politic" and a "world natural."

It was not the only analogy available. Rather than seeing Whig progressivism as the central narrative of Scottish Enlightenment thinking, it is more accurate to understand it as one historical pattern among several. David Hume's classic *History of England* (1754–61), for instance, was a polemic critique of the Whig political settlement, while Adam Smith's *Wealth of Nations* (1776) wrestled with the precarious nature of economic advance against the headwinds of sovereign and parliamentary intervention in the marketplace. Such histories sifted the different circumstances of historical societies for conserved patterns of progress, growth, and natural law. This historical understanding, characteristic of the Enlightenment, is variously identified as a "stadial," "conjectural," or "universal" history. Karen O'Brien describes such "stadial history" as "a natural process of development in which societies undergo change through successive stages based on different modes of subsistence."[49] The historians of the Scottish Enlightenment shared Erasmus Darwin's confidence that it was patterns, cycles and epicycles, that form history into "one revolving whole." Examples include Adam Ferguson's *Essay on the History of Civil Society* (1767) and Alexander Tytler's *Plan and Outline of a Course of Lectures on Universal History* (1782), in which the histories of different epochs and societies were reduced to common plans of maturation, passing through stages defined by universal features of agricultural, political, and individual development. So, for Adam Smith, all societies are ultimately tied to a central movement from agricultural to mercantile society, from slavery to freedom, and from undifferentiated to specialized labor.[50] O'Brien identifies this historical approach with a mode of comparative analysis that differentiates societies in service of general patterns: "Comparative treatment of different societies revealed the underlying principles and perversions of natural laws operative in history."[51]

Like the comparative historicism of the nineteenth century, stadial history was organized through comparisons between, for instance, classical Rome and contemporary Europe, Asian society and British culture, the New World and the Old. But it differed from nineteenth-century comparatism in its focus on strategies of contrast, turning on the question of what makes one society different from the other. This strategy of distinction allows the stadial historian to assign societies to different places within an overarching taxonomy, locating them within a universal pattern driven by models of economic, philosophical, and political evolution. As Adam Ferguson summarizes the case in his *Essay on the History of Civil Society*, "The latest efforts of human invention are but a continuation of certain devices which were practised in the earliest ages of the world, and in the rudest state of mankind. What the savage projects, or observes, in the forest, are the steps which led nations, more advanced, from the architecture of the cottage to that of the palace, and conducted the human mind from the perceptions of sense, to the general conclusions of science."[52] For stadial history, as for Priestley, attention to the particularity of history is secondary to capturing a higher-order pattern, the providential "steps" that are registered erratically in the historical record. Such analogical approaches to history can be distinguished from comparative historicism by the way they take historical particularity as a means rather than an end. Insofar as universal historians engage comparisons harmonically—as a comparison between peer societies—they establish the contrasts between different periods. Insofar as they institute formal comparisons—with one subject serving as a model for the other—they illustrate the analogy between the particular society or epoch and the larger model that describes a universal historical pattern. Serial distinction drives universal formal pattern.

While the universal histories of later eighteenth-century philosophers were not explicitly concerned with describing the analogy between the "City of God" and the "City of the World," the universality of their historical understanding was rooted in the conceit of an orderly cosmos articulated by predecessors like Thomas Hearne, whose *Ductor Historicus; or, a Short System of Universal History* (1705) studies the congruence between the divine and secular past.[53] When Priestley observes that "analogy is our best guide in all philosophical investigations," he marks this larger influence and an alignment between different ways of reading history in its analogy to a higher order.[54] The larger point is that, even as Erasmus Darwin suggests how order, or "self-balance," emerges naturally (whether in verses or in natural systems), he insists that this order is coherent—that there is *"one* revolving Whole," not several.

Flattening Allegory

Colossal, Anubis stands astride the river Nile (figure 5). Gathered over his head, his hands seem both to beg Sirius, shining above, and to gather the storm clouds that boil at either edge of the frame. Beyond, the arms, garments, and whiskers of Jupiter Pluvius dissolve into a curtain of warm rain that feeds the headwaters of the swollen river. On the right bank, the pyramids of Giza testify to ancient engineering and an empire founded upon the rich alluvial deposits of the fertile Nile Valley. Symmetries, echoes, and reflections organize the engraving. Bank mirrors bank, and the standing figure's gesture is tracked in the muscle of his shoulders, echoed above in the line that connects his knotted brows to his winglike canine ears, and below in the broad incisive sweep of Jupiter's wings. Strong bilateral symmetry anchors a firm vertical organization, which in turn asserts the secure relationship between the guiding influence of the heavens, intermediate agents, and the terrestrial sphere below. At the corners of the plate, the most striking reflection of all: "H Fuseli RA inv" looks across to "W Blake sc."

"The Fertilization of Egypt," one of several engravings on which William Blake and Henry Fuseli collaborated for *The Botanic Garden* (1791), illustrates the differences between Blake's mysticism and Darwin's ecstatic science. Blake's monumental engraving is arguably the most stunning of the nonbotanical plates commissioned for that work. Comparison to Fuseli's original sketch, held by the British Museum, shows how much Blake himself invents. The indecisive background behind Fuseli's strong foreground composition is fleshed out; hazy intimations of mountains to the left become the sharp doubled lines of the pyramids. This iconographic filigree multiplies the vectors of allegorical inter-

Figure 5. "The Fertilization of Egypt." Engraving. By William Blake, from *The Botanic Garden* (1802), 602000, vol. 1, p. 126. Courtesy of The Huntington Library, San Marino, California

pretation and poses new challenges. Does Anubis gather the clouds or simply invoke the power of Sirius? Does the Egyptian God intercede with the Greco-Roman or operate in parallel, or in some Trinitarian relation with the star? To what degree does Blake's shared billing claim artistic status alongside Fuseli, the plate's "inventor," and with Darwin, the poet?

Blake's engraving suggests the interconnection of disparate mythological accounts (classical, Christian, Egyptian) by their common access, through human experience, to metaphysical order. For Blake, this interconnection testified to what he called the "divine analogy," by which, as Ann Mellor puts it, "man sees himself as God."[55] Blake's engraving, which transforms the half-erased graphite traces of Fuseli's earlier drafts of Anubis into Jupiter Pluvius and the tracking lines of his outstretched wings, marks an alignment between Greco-Roman and Egyptian gods and projects a mystical sensibility that sees a deeper connection between distinct mythologies. In this way, Blake's engraving for the "Fertilization of Egypt" embraces the contemporary effort of comparative mythologists, especially Sir William Jones, to uncover the historical relation between different religious traditions.[56]

Whereas, for Blake these common mythological patterns (pictured here in the alignment of Roman and Egyptian gods) indicated the larger coherence of mystical Christianity, for Darwin they demonstrated the ultimate concordance of mythical belief as analogous evidence for the epic of nature. Though lesser-known today than either Blake or Fuseli, Erasmus Darwin was a colossal figure in his own time, and the verses depicted in the "Fertilization of Egypt" illustrate the extraordinary reach of his thinking about analogy. Closing canto 1 of *The Economy of Vegetation*, Darwin explains how Sirius (the dog star) and, by connection, Anubis (the dog-headed god) were associated with the annual monsoon:

> Sailing in air, when dark MONSOON inshrouds,
> His tropic mountains in a night of clouds;
> Or drawn by whirlwinds from the Line returns,
> And showers o'er Afric all his thousand urns;
> High o'er his head the beams of SIRIUS glow,
> And, Dog of the Nile, ANUBIS barks below.   (CW 2, 3.129–34)

In one traditional account of Romantic poetry, Romanticism is defined by the rejection of such neoclassical allegories (personifications marked here by their elevation into small caps and by Blake's stylized illustration) in favor of more immediate and organic symbols, generally drawn from a naturalistic vocabulary. On this view, neoclassical allegory fails because, even as it suggests an analogy between the human form and the object of description (whether god or weather system), these allegories inevitably distort and, so, hold apart the poem and the world it describes. Clouds have only so much in common with people, after all. Even so, as Paul de Man observed, the rejection of such "associative" or "formal" analogies did not mean the abandonment of analogy altogether: instead, Romantic poets, and later critics like Abrams and Wasserman, sought "a more vital form of analogy," rooted in the language of symbol, that insisted on the "fusion" of poetic description and world rather than their difference.[57]

De Man, in a familiar turn to semiotic critique, rejects the possibility of such contact between description and the world, arguing that symbols attempt (nostalgically) to close the perceived distance between "sign and meaning"—a distance seen clearly in allegory. This is a partial reading, one that recognizes only the formal capacity of literary tropes to separate out description and object.

By contrast, the "vital" analogies described by Wasserman indicate how poetic figures, including allegory, can serve as a conduit through which different aspects of the world come into contact.[58] In particular, as Erasmus Darwin recognized, allegories can incorporate knowledge about the natural world and so assemble analogous fragments of the larger pattern of nature.

Assembled as mythical accounts within *The Botanic Garden*, the analogies *between* these figures establish a harmonic relation and so bring the objects of description into contact. This point is made clear in long, digressive footnotes that unpack the dense syntax of Darwin's descriptions. The note to the preceding passage explains how, in early summer, just as Sirius begins to rise above the horizon, the sun's heat creates a low-pressure system over Africa, and moist air drawn from both the Red Sea and the Mediterranean converges above. The coincident rise of Sirius at the same time of year gave a convenient astronomic marker for the flooding of the Nile, and Anubis a coordinate graphic symbol for its approach. The association of these elements in ancient allegories, then, had long served as a way to disclose their association in the world.

In later works, the flooding of the Nile takes yet greater significance; in his *Temple of Nature*, Darwin gives extended attention to the relation between his own theory of evolution and the myths of chthonic evolution, prompted by the Nile's fertility, in which "Bird, beast, and reptile, spring from sudden birth, / Raise their new forms, half-animal, half earth" (*CW* 9, 1.409–10). As Darwin glosses it, "This story from Ovid of the production of animals from the mud of the Nile seems to be of Egyptian origin . . . showing that the simplest animations were spontaneously produced like chemical combinations . . . but distinguished from the latter by their perpetual improvement" (*CW* 9, 37–8). So mythical allegory is excavated for scientific knowledge and then, in the longer view of Darwin's vision, refigured as evidence for the universal process of development. This underscores Susan Manning's argument that "analogy and allegory [often] fight for dominance," insofar as allegory "does not readily accommodate the contingencies of historical circumstance."[59] Much like Blake's mysticism, this narrative of development provides what George Eliot later termed a "key to all mythologies" (a mistaken belief in universal pattern that she criticizes in *Middlemarch*).

In effect, the notes to *The Botanic Garden* ask what we might make of such allegories if we flatten their relationship, taking them as coordinated descriptions of the world rather than attempts to create distance between sign and meaning. In flattening these allegories, Erasmus Darwin discloses the analogies between cosmic, chemical, and biological processes that support his theory of evolution. The belief that such patterns are continuously realized in the world draws on a line of naturalist thinking distinct from Blake's interest in Christian metaphysics. Initiated by Aristotle's discussion of natural pattern in the *Posterior Analytics*, Bacon gave this view added emphasis in the *New Organon*, arguing that it uncovers the larger patterns of nature:

> Finally, we much absolutely insist and often recall that men's attention in the research and compilation of natural history has to be completely different from now on, and transformed to the opposite of the current practice. Up to now men have put a great deal of hard and careful work into noting the variety of things and minutely explaining the distinctive features of animals, herbs, and fossils . . . [which] contribute little or nothing to an intimate view of nature. And so we must turn all our attention to seeking and noting the resemblances and analogies of things, both in wholes and in parts. For those are the things which unite nature, and begin to constitute sciences.[60]

Bacon is generally seen as the anti-Scholastic crusader who substituted the collection of empirical description for the abstract theorization of Aristotelian naturalists. But as the preceding passage makes clear, the ultimate purpose of such description is to expose the accurate grounds for a careful analysis of the "analogies of things" that "constitute sciences." Analogy, in fact, constituted a crucial feature of Baconian induction.[61] Whereas in the divine analogy the pattern of terrestrial organization reflects the divine system, in Baconian analogy phenomenal patterns reflect the "unity of nature." For Bacon (and for Darwin), analogy is an attribute of the world, not ascribed to it; it is not applied to nature by the scientists but is "of" the "things" themselves.

Though seemingly far different in their understanding of how natural and divine order relate, and in their respective theories for how representation explores that relation, both Blake and Darwin shared a commitment to governing orders of analogy that places them on the other side of a radical nineteenth-century shift toward narrative models of comparative historicism—the subject of the following chapters. Where Blake sees an ineffable, divine connection, Erasmus Darwin held that our basic kinship with nature itself allowed us to recognize common pattern. Yet there is a basic continuity in their metaphysics, that is, their theories of natural order, insofar as they turn on the analogy between what we see (or read) and a higher order. For both Darwin and Blake, history serves as a convenient ground for positing the totality of that higher analogy. For both writers, that formalizing turn to higher order effaces historical particularity, sweeping a more granular history aside in favor of the patterns that transcend historical incident.

Jerome McGann has observed that a desire to escape historical particularity hardly divests a text of its historicity.[62] Together, Blake and Darwin stake out the terrain of analogy as a late eighteenth-century discourse that provided an important context for later thinking about relational history. The Romantic period saw a differentiation in ways of thinking about analogy that continues to shape our understanding today. The key change came as the genres of analogy were separated out into structural versus historical strategies of comparison. The structural functions of analogy, which had elucidated the imminent features of order in the divine and natural world, were relegated to the philosophy of speculative analogy and held at arm's length by nineteenth-century theorists of the patterns found in social systems and natural history. So Blake's "divine analogy" and the traditions of biblical hermeneutics and natural theology came to be seen as outmoded styles of speculation, whereas for many other Romantics (particularly in the account of de Man and, more recently, Colin Jager), analogy retained importance as a way to interpret the human significance of natural patterns.[63] The comparative sciences, by contrast, reformulated the procedures of analogy into the empirical "comparative method" and took a renewed interest in the relation between structural pattern and historical change. Comparative anatomy, philology, mythology, and historiography, for example, interpreted comparisons as emerging through historical perspective and so imagined alternative trajectories and future orders. Erasmus Darwin's complex relation to analogy—seeing it as both a method and a pattern in the world, both a system of mythical knowledge and a window into natural history—provides a microcosm of that larger movement. In his attempt to flatten mythical allegories into a more consistent understanding of nature's history, Erasmus Darwin marks an initiating movement from structure into history. This turn to history remained incomplete as long as it remained rooted in an epic of universal development.

Zoonomia and Darwin's Insurrection

> The GREAT SEED evolves, disclosing ALL;
> LIFE *buds* or *breathes*, from Indus to the Poles,
> And the vast surface kindles, as it rolls!

Erasmus Darwin, *The Economy of Vegetation* (1791)

Analogy provided a central strategy by which the "world natural" initially presented in the *Loves of the Plants* could be placed in dynamic relation to the "world politic" explored in the *Economy of Vegetation*. The two-volume *Botanic Garden* reformulates the static Linnaean system of order presented in *The Loves* as part of a continuous thesis of progress. Criticism of Erasmus Darwin's poetry has followed Anna Seward in attributing the out-of-sequence printing of *The Loves of the Plants* and *The Economy of Vegetation* to an accident of publication.[64] But the reversed publication of the two books had far-reaching effects. It is only through a more careful attention to this publication history—not only the sequence of printing but also the rearrangement of the works internally—that we can recapture a sense of the surprise Darwin's contemporaries felt in realizing he was after evolution all along.

After the first edition, all subsequent printings of *The Loves of the Plants* background the proem, so important to establishing the representational model, by moving it to the close of the prefatory material. And from the third London edition (1791), which coincided with the first publication of *The Economy of Vegetation*, this arrangement is further reorganized in order to accommodate both volumes in a single edition as *The Botanic Garden*. In so doing, Darwin adds an apology that defends the "subsequent conjectures," which feature prominently in the expanded work, and condenses the advertisement, which (in earlier editions) subordinated literature to the pursuit of science. This apology changes the terms of the original contract by asserting the continuity of poetic and scientific practice, arguing that, because "natural objects are allied to each other by many affinities, every kind of theoretic distribution" of knowledge—even the poetic— "adds to our knowledge by developing some of their analogies." Unlike the first edition's claim that there is a substantive difference between the "looser" analogies of poetry and the "stricter" analogies of science, the poet now argues that poetic analogy "adds" directly to scientific knowledge.[65]

This reformulation explained *The Botanic Garden* as far more than the "trivial amusement" Darwin advertised in 1789. Walpole, for one, was bewildered. Though he admired sections of *The Economy* as "the most sublime passages in any author," he observed that the poem as a whole, "crowded with most poetic imagery, gorgeous epithets and style," did not "please [him] equally with the 'Loves of the Plants.' "[66] *The Loves of the Plants* is a didactic poem that explicitly takes Ovid for its model. *The Economy of Vegetation* bends the framework of *The Botanic Garden* toward cosmological epic. The tension shows.

Darwin pursues two distinct but parallel strategies in *The Economy of Vegetation* in this larger effort to expand the scope of the work and add narrative extension. The first, following Pope's *Rape of the Lock*, was to adopt the Rosicrucian system of elements as a structural model for the separate sections of the poem. The four cantos are divided among the nymphs of fire, nymphs of water, sylphs of air, and gnomes of earth to explore the technological developments most appropriate to each medium. This framework, while it expands the overall scope of the poem, nevertheless suffers the same narrative deficit as the Linnaean taxonomy. Though it makes room for individual inventors and historical figures that punctuate a thesis of technological advance, it is still essentially synchronic and structural. The elemental conceit does not provide a narrative scaffolding adequate to *The Economy of Vegetation*'s epic ambitions. The epic had long made room for the catalog of persons and events—yet this catalog had to be subordinated to the contours of plot.

The need for epic plot gives urgency to Darwin's search for a unitary principle of development, a story that could support the web of analogies he draws between cosmic, terrestrial, and social patterns. Darwin drew the seeds of this progressive natural order from his reading of Linnaeus himself. Though Darwin had been flirting with evolutionary theory at least since 1777 (he briefly changed the family motto to *E conchis omnia*, "everything from mollusks"), his work translating Linnaeus in the following decade had a profound influence on his thinking. Linnaeus's speculations on the origin of plant species emerge in the same section of his *Genera Plantarum* in which he defends his sexual classification. Supposing that "the GREAT FATHER created at the beginning but one species of each genus," Linnaeus imagines "these first species of plants . . . to have been fecundated by species of other genera; [and] it would then follow, that more species would arise" (lxiv). So the great variety of plant species found today could be derived from hybrids of a few original lines. While Linnaean sexual classification defines the artificial relationship between families, orders, and classes, sexual reproduction helped theorize how natural categories of species and genera emerge.

Erasmus Darwin called attention to Linnaeus's theory of speciation in the first printing of *The Loves*, both summarizing the Linnaean account in the preface and marveling at the "ingenious" imagination that produced it (*CW* 2, v). And it forms the germ of Darwin's coordinated accounts of terrestrial and cosmic development, as they are both modeled upon the analysis of vegetable development. *The Economy of Vegetation* describes how, in plant seeds, "Maze within maze the lucid webs are roll'd / And, as they burst, the living flame unfold" (*CW* 1, 3.383–4). An endnote explains the connection to animal life: "The analogy between seeds and eggs has long been observed, and is confirmed by the mode of their production. . . . the uterus of the plant producing or secreting it into a reservoir or amnios, in which the embryon is lodged, and that the young embryon is furnished with vessels to absorb a part of it, as in the very early embryon in the animal uterus" (*CW* 1, 106 n. 26).

Extending this conflation of the vocabularies of animal and vegetal reproduction, Darwin is interested in how this model extends beyond plant and animal life. So, in describing the terrestrial formation, Darwin describes how the
"great seed" effectively shaped the modern earth:

> So, fold on fold, Earth's wavy plains extend,
> And, sphere in sphere, its hidden strata bend;—
> Incumbent Spring her beamy plumes expands
> O'er restless oceans, and impatient lands,
> With genial lustres warms the mighty ball,
> And the GREAT SEED evolves, disclosing ALL;
> LIFE *buds* or *breathes*, from Indus to the Poles,
> And the vast surface kindles, as it rolls!   *CW* 1, 4.401–8

It is an intriguing description of terrestrial development because of its recursive
ambiguity. Does the "great seed" refer here to the planet itself or to some
general first organizing element from which all life derived? What is the relation between ecology, organism, and evolution? Erasmus Darwin's theory
of fertilization—the way that seeds invade and transform the world around
them—envisions the natural world as self-organizing in an intricate and distributed fashion, reducing the singularity of the discrete organism and placing
it within a larger course of fertile development. This is true on the historical
view—all putatively discrete creatures derive from a common event—and on a
causal view, as demonstrated in the slippage between the self-sufficient way that
"LIFE *buds*" even as it collectively "kindles" the "vast surface" of the earth in cor-
porate action. Ambiguous agency is lodged in the uncertain referent, "it," which
refers most immediately to the turning surface of the earth but also to the collective impact of the "LIFE" that rolls across and transforms it.[67]

Such passages, ranging from Darwin's description of the formation of the
cosmos, glossed earlier, to his more explicit exploration of evolution in *The
Temple of Nature*, provide consistent evidence for his unique species of vitalism.
As Peter Hanns Reill and Catherine Packham explain, "vitalism" (understood
as the belief that life is organized by a "vital" force distinct from the physical
forces of nature) emerged in the latter eighteenth century as an important counterpoint to mechanistic theories of living systems.[68] The Scottish Enlightenment was a hotbed of vitalist thinking in eighteenth-century Britain, particularly,
the faculty and students of the Edinburgh Medical School (where generations
of Darwins studied, from Erasmus to Charles).[69] Erasmus Darwin's vitalism is
peculiar because, while vitalism usually explores the self-organizing powers of

living bodies—plant or animal—his poetry *also* finds vitalism operating on *in-animate* bodies of cosmic and terrestrial scale (as we have already seen). In other words, Darwin's poetry projects a "material vitalism," Jane Bennett's term for vitalisms that extend to nonliving matter.[70] In this, Darwin was not unique; in 1789, William Herschel, in his revolutionary study of nebular formation (cited in a footnote to Darwin's "Orbs wheel in orbs" passage), argued that his new catalog of stellar nebula "resemble[s] a luxuriant garden, which contains the greatest variety of productions, in different flourishing beds." If, Herschel suggested, such botanical gardens allow us to infer development from comparing plants of different ages, then why not solar systems and nebula as well?[71] Similarly, Darwin's friend, the geologist James Hutton, used vital principles to argue for the coherence of the geological (and meteorological) systems of the earth as self-sustaining cycles in his "Theory of the Earth." "Is this world to be considered thus merely as a machine, to last no longer than its parts retain their present position, their proper forms and qualities?" Hutton asked, "Or may it not be also considered as an organized body? Such as has a constitution in which the necessary decay of the machine is naturally repaired, in the exertion of those reproductive powers by which it has been formed."[72] As contemporaries, Darwin, Herschel, and Hutton all practiced a visionary science that used the vitality of life to model the self-organizing development of their objects of study. Generally, they were more concerned with vitalism as a powerful model for self-organizing systems than the problem of its empirical investigation. In Robert Mitchell's view, this would make them "theoretical" vitalists, but Erasmus Darwin outstripped his peers in the reach of that theory, synthesizing the speculations of both Herschel and Hutton into a singular thesis of fertile cosmic evolution.[73]

The peculiarly narrative drive of Darwin's vitalism helps to differentiate his thinking from contemporary German theorists of vitalism and organic form, particularly Immanuel Kant and Johann Wolfgang von Goethe. Whereas the German *Naturphilosophie* used idealism to frame out formal assumptions about the relation between part and whole, between type and *urtyp*, Darwin's theory emphasized dynamic content and material development. Robert Richards has contrasted German organicism with mechanistic models for life circulating in the Scottish Enlightenment. Darwin's material vitalism (which influenced both Goethe and Schiller) indicates an intermediate position, focused on material processes of development that are not characterized by common patterns of physical form but, rather, by common narratives of progressive development.[74] The thesis of a universally progressive nature, continuous across matter and living systems, emphasized transformation over the self-sustained organic whole—process over purposiveness. It also indicates an alternative genealogy for British Romanticism, with its complex relation to form and organicism, outside of German Idealism. Erasmus was not an idealist—he was a systems thinker.[75]

Darwin's most focused statement of this theory of development appears in *Zoonomia* (1794–6), though its focus on the biology of living systems leaves scant room for the discussion of planetary or stellar development. *Zoonomia* also first formally recognized its author, Erasmus Darwin, M.D.F.R.S., as also "Author of the Botanic Garden," giving critics of his previous works an authorized target. In expanding Linnaeus's speculations for *Zoonomia*, Darwin considered the analogy between animal spermatozoa, first described by Antonin von Leeuwenhoek in the *Philosophical Transactions of the Royal Society* in 1679, and the seeds of plants.[76] Darwin held to Leeuwenhoek's belief that the analogy between the "filaments" observed in sperm and in the germ of plant seeds proved that the embryo is composed exclusively of the sperm, which organizes the surrounding tissue into a growing fetus after conception. On this view, the egg was simply nutrient for the embryonic sperm, while the sperm was the direct analog of the vegetable seed. As Darwin explained, observation of asexual reproduction—budding in plants and in polyps—already showed that the bud transforms a responsive environment by absorbing and reordering its elements as it matures into a growing organism (*CW* 6, sec. 39). Whether polyps, seeds, or plants, all held to a common pattern.

This focus on the transformative power of initiating "filaments"—modeled on sperm and presumed male—gave a deeply androcentric bent to Darwin's theory and led to all kinds of hairsplitting about the nature and degree of hereditary contribution by both parents. But even more important is the way his theory generalized reproduction through fertilization as a basic evolutionary process in all self-organizing natural systems. If specific kinds of matter had the capacity, in interaction with their environment, to develop and grow in complexity, then there was no limit to the spontaneous capacity of nature to produce complex forms. Familiar with the implications of "deep" time from his friendship with James Hutton, Darwin further hypothesizes that it would not "be too bold to imagine, that in the great length of time, since the earth began to exist, perhaps millions of ages before the commencement of the history of mankind, . . . that all warm-blooded animals have arisen from one living filament . . . with the power of acquiring new parts, attended with new propensities, directed by irritations, sensations, volitions, and associations" (*CW* 6, 240).

The crucial leap to cosmology comes as Darwin imagines that, as David Hume speculated, "the world itself might have been generated, rather than created; that is, it might have been gradually produced from very small beginnings, increasing by the activity of its inherent principles, rather than by a sudden evolution of the whole by the Almighty fiat" (*CW* 6, 247). It turns out there may be no need for God's "potent word."

Erasmus Darwin's evolutionary theory had far-reaching implications. It naturalized artificial taxonomies by suggesting that they helped describe a history of hybridization and descent. Perhaps more radically, it fertilized the cosmos. Weaving reproduction into the fabric of vital development, Darwin suggested that the analogy between humanity and plant life—treated as an allegory in *The Loves of the Plants*—was really a substantive connection that opened onto the basic physical processes of cosmic progress. By expanding beyond the premise of *The Loves of the Plants*, *The Economy of Vegetation* violated the contract by which the pastoral conceit would be firmly subordinated to the Linnaean system it was designed to communicate; now *Zoonomia* wrapped them up in a larger thesis about evolutionary development, an extended argument for a world in continuous and progressive flux.

Conservative reviewers howled. Darwin, one observed, "see[s] no difference between man and monkey"; by attributing sensitive interaction to the "organic frame" rather than the human soul, "literally and seriously he animates all the plants, and makes them capable of that love, that he had before, by an agreeable fiction attributed to them."[77] The problem, as they now recognized, was that Darwin's later works sprang the trap, collapsing the "agreeable fiction" into a substantive theory of the progressive order of nature and human society. In response, Darwin's critics doggedly confused sexual dimorphism with sexual prurience, and so swept his cosmology into a generalized complaint against sexual license. Thomas James Mathias, in the expanded third edition of his popular satire, *The Pursuits of Literature* (1797), complained that Darwin "Raise[d] lust in pinks; and with unhallow'd fire / Bid the soft virgin violet expire."[78] And perhaps most famously, Canning and Frere bewailed, in their own satire, "The Loves of the Triangles," that "partial Science should approve / The sly RECTANGLE's too licentious love!"[79] If epic is designed to both explain the history of a collective social body and project its triumphal future, then Canning and Frere proved that Darwin had confused nationalist Britain with a more cosmopolitan state. By these means, they succeeded in tying Darwin's verses to his liberal politics.

I find the identification of Darwin's progressive thesis with revolution less interesting than the general recognition that Darwin meant for his body of work to be read as a unified system and that this system resolves into a coherent argument for progress. His critics had a point. The utopian thesis of collective political, organic, and natural improvement powerfully confused biological development with social progress. This circumscribes the stories available to *The Botanic Garden* as well as *The Temple of Nature*. Even the sometimes tragic vignettes of the *Loves* prove to be eddies within the larger narrative current (as at the sorrowful closing of both cantos 1 and 3, with the stories of Tremella, who freezes while ineffectually praying, and Cassia, subject of a long jeremiad against slavery). As Darwin puts it, at the end of the last canto: "The Loves laugh at all but Nature's laws" (*CW* 2, 4.490). In its implicit addresses to an audience that comprises not merely all of humanity but all of nature, the *Botanic Garden* gestures to a unitary history without granularity or uncertainty. It is a fatal problem, insofar as it leaves the poem unable to address the most important event of its day. Despite Darwin's noted advocacy, and the most suspicious readings of Canning, Frere, and others, the French Revolution finds no place in the poem, even as the revolution reflects the poem's guiding problem: whether order emerges from freedom.

Conclusion: "Philosophical Arguments of the Last Generation"

It is tempting to imagine that, as originally conceived, *The Loves of the Plants* was precisely what it purported to be: a witty and entertaining tour of Linnaean taxonomy that respected formal boundaries. And yet it is evident that, as Darwin cultivated the larger *Botanic Garden*, his ambition and his insight grew. Even as the success of the initial volume fanned the epoetic fire in his belly, his imaginative engagement with the vegetable loves, and his concentrated effort to explain their histories, generated a table-turning insight into the history of life itself. His imaginative, formal, and quintessentially literary engagement did not merely reflect science—it produced it. And it did so by developing new strategies to describe pattern and differentiation.

This concern for the formal texture of evolutionary narrative would be adapted and further developed by later theorists of evolution, particularly his grandson. Historians of science have long marveled at the remarkable coincidence between Erasmus Darwin's speculations and the epoch-making theory presented by Charles Darwin decades later. It was certainly not lost on Charles's contemporaries: Bishop Wilberforce, in his scolding review of the *Origin of Species* in 1860, alluded to how Canning and Frere dismantled "these philosophical arguments of the last generation."[80] But perhaps the most important impact of Erasmus Darwin's work was to test the narrative of *The Botanic Garden* as an epic of nature. In so doing, it explored new formal strategies for describing how natural patterns were shaped by time and helped generate new theories and new modes of description. As both a poet and naturalist, Erasmus insisted on the deep continuity between scientific inquiry and the literary imagination.

The more radical thesis of this book is that literature provides science with new kinds of stories and, hence, new ways to explore and understand natural history. In confronting the question of differential history through the medium of their chosen genres, the writers I examine in the following chapters were forced to work out what historicism could do and how it would function on the page. It was at the level of the word that they were forced to confront and modify the formal structures within which this new historical understanding would operate. This produced important changes, not only in the forms of description available to naturalists but in the way they saw nature itself. While this sort of argument is commonplace in literary studies, it remains radical in scientific history, where discovery has been traditionally understood as the product of empirical observation and inspiration and, more recently, as the collective labor of networks of allied actors. *The Botanic Garden* makes clear that, in understanding how changes in scientific understanding come about, we must look to the formal features of scientific writing and its print artifacts.

Yet, to borrow a distinction from David Duff, the bewildering thing about *The Botanic Garden* is that it advertises a smooth mixture between poetic and scientific genres, between verse epic and didactic treatise, that in practice is extraordinarily rough.[81] Herbert F. Tucker observes that epic, which incorporated genres like pastoral, georgic, and epistle, had to "eat in order to live," but in Darwin's hands, the didactic poem proved indigestible.[82] One possible trajectory for this study would be to pursue comparative historicism's epic encounters forward, tracking its expansive development in experiments like Elizabeth Barrett Browning's *Aurora Leigh* and Eliot's *Spanish Gypsy*. Alternatively, it might explore the legacy of Erasmus Darwin's analogical thinking in Romantic theories of nature, from Wordsworth's "analogy betwixt / The mind of man and nature" to Shelley's "permanent analogy of things."[83] But in attempting to sample the complex range of discourses that expanded the scope of comparative historicism in the nineteenth century, it is necessary to select, if not comprehensively survey, the cultural forms that had the strong purchase on the imagination of the past. In the chapters that follow, the novel, the elegy, and the scientific monograph take chief importance as genres of comparative historicism.

Erasmus Darwin's works make it clear why comparative historicism demanded formal innovation. Analogy, as a method heavily determined by its formal traditions, was not sensitive enough to tell the story he imagined. In principle, analogies are limitless—they draw upon an infinite range of tempting similarities. Analogy requires some dimension of constraint—a way to isolate the useful patterns from the expansive field of possible connections. The historicisms studied in this chapter, Christian, Whiggish, stadial, could be characterized as styles of reasoning that choose specific kinds of analogies as their basic principle. For Erasmus Darwin, a fertile theory of vital evolution provided this organizing principle, and yet sex couldn't solve everything. To use sexual fertilization to model self-organization across natural systems was to mask precisely the differential paths that would support his larger theory of change. Though his theory of evolution was designed to turn taxonomy into history, its recourse to the same conserved model for the past frustrated that purpose by reducing history to a monotone narrative of development.

For later writers, the primary dimension of constraint was history itself. Restricted to the field of history and focused on the comparison of specific narratives of convergence and divergence, rupture and reunion, analogies laid the foundation for an insistently comparative approach to studying the past. In the following chapter, I explore this comparative historicism in the novels of Walter Scott. By focusing analogy on the relation between disparate plots and alternative visions for the future, Scott uncovered a way of talking about the past that emphasized uncertainty and the unsettling sense that things can always be different.[84] For his part, by flattening strict hierarchies of knowledge—between science and poetry, between word and thing—into a more continuous plane of historical play, Darwin verged upon the kind of leveled understanding that is central to comparative historicism, without quite being able to take that field. Only half a century separated Erasmus Darwin's theories from his grandson's major works, but that crucial period witnessed the wide-ranging revolution in the strategies of analogical address that generated comparative historicism.

CHAPTER TWO

Crossing the Border with Walter Scott

Sir Walter Scott, in the same manner, has used those fragments of truth which historians have scornfully thrown behind them in a manner which may well excite their envy. He has constructed out of their gleanings works which, even considered as histories, are scarcely less valuable than their's.

Thomas Babington Macaulay, "On History" (1828)

How does Walter Scott make history? Over the past two decades, studies of the influence of Scott's fiction have made us increasingly aware of his massive impact on the nineteenth-century Anglo-European culture of letters. The thesis that Scott changed the shape of the novel, the publishing industry, and the social understanding of history is almost unquestioned. And yet, precisely because of his wide influence and the extraordinary range of fields he operated in (as novelist, poet, editor, reviewer, publisher, historian, essayist, biographer, architect, etc., etc.), it remains difficult to give a clear description of Scott's key innovation. Studies of Enlightenment history, and of historical and national fiction before Scott's first novel *Waverley* (1814), continue to demonstrate that many of the features we find central to the historical novel—from the dense cultural interconnection of historical societies, to the importance of mediatory figures in translating the past for a modern audience—were already in circulation.[1] In trying to figure out how Scott's fictions produce a sense of history for his readers, it's useful to turn back to the reaction of his contemporaries.

For most nineteenth-century historians, Scott marked a clear break from eighteenth-century traditions by filling in the complex relationship between historical periods, especially the complicated relationship between the past and the present. The Waverley novels (as his historical novels, including *Waverley*, are often designated) negotiate the relation between history and modernity through moments of transition and contact and, in this way, destabilize confidence in a history composed of isochronous periods. As Thomas Carlyle described this change, distinguishing Scott from both Enlightenment historicism and what R. G. Collingwood termed the "scissors and paste" critical history that preceded it, "these Historical Novels have taught all men this truth, which looks like a truism, and yet was as good as unknown to writers of history and others, till so taught: that the bygone ages of the world were actually filled by living men, not by protocols, state-papers, controversies and abstractions of men. Not abstractions were they, not diagrams and theorems; but men, in buff or other coats and breeches."[2] Carlyle—by no means a devoted fan of Scott's works—still ratifies the sense that Scott taught a new way to approach history, a "history from below" that emphasized the relation between common life and national events, drawing attention to both the particularity and the humanity of the past.[3] In his view, Scott marked a sea change in the relationship between the subject of history and the individual lives that constituted its movement.

This chapter connects a study of Scott's impact on historical understanding with John Stuart Mill's claim that the "dominant idea" of the early nineteenth century was "the idea of comparing one's own age with former ages."[4] Mill well knew that comparisons were also central to "philosophical" histories written in the previous century; his point is that by the early nineteenth century, comparison constituted a more general condition for how contemporaries understood their relation to the past. Mill's essay is concerned with the troubled immediacy of that past, in both the shape of contemporary politics and the sense of modernity's immanent rupture with tradition. This sense of increasing discontinuity stands in opposition to the progressive, providential, and stadial histories described in my previous chapter. In that account, Erasmus Darwin coordinates disparate eighteenth-century historical perspectives in his effort to "inlist the Imagination under the banner of [progressive] Science" and advance a case for the analogy between biological development and universal evolution. Where Erasmus Darwin saw analogy as evidence for a universal pattern, the new comparative histories used analogies—between "living" and dead men, between "one's own age" and those "former"—to tease out complex distinctions in historical perspective. As an important architect of this new comparative history, Scott shaped the historical imagination of the nineteenth century, exchanging Enlightenment models of history for complexly graduated relations within and over time.[5]

Scott, a lawyer, became famous and wealthy through his writings. Though best-known today for his novels about immediate Scottish history, beginning with *Waverley*, *Guy Mannering* (1815), and *The Antiquary* (1816), as well as for later experiments with deeper medieval history, especially *Ivanhoe* (1820), he published these novels anonymously. At that time, Scott was already famous, both as an editor of Scottish folk ballads, in his multivolume *Minstrelsy of the Scottish Border* (1802–3), and even more as a poet of fabulously successful metrical romances, particularly *The Lay of the Last Minstrel* (1805) and *Marmion*

(1806). Throughout his work, Scott showed a close attention to historical practices—economic, social, and literary—in their relation to modern society.

The dense material continuity of the past, as well as its unsettling impact on the present, is a key feature of Scott's *Guy Mannering*. Like Maria Edgeworth's earlier *Castle Rackrent* (1800), the estate usurped by the professional class serves as a critical microcosm for the Whig historical narrative in Scott's novel. When the unwitting heir to the Ellangowan estate first returns, he is overwhelmed by a collision of physical spaces, half-forgotten memories, and snatches of song that recall his family. As an unsuspecting heir, Harry Bertram muses upon the unsettling familiarity of the ruins of his family's old castle: "Is it the visions of our sleep that float confusedly in our memory, and are recalled by the appearance of such real objects as in any respect correspond to the phantoms they presented to our imagination?"[6] In the succeeding interview with the usurper, Gilbert Glossin, Bertram considers the possibility that he may recognize the eerily familiar grounds. In his *own* struggle to distinguish past and present, Glossin assiduously generalizes history. In order to avoid further stimulating Bertram's memory, Glossin describes the estate's structures as "The New Place" and "The Old Place" and substitutes, for the true family motto "Our Right makes our Might," the janglingly modern "He who takes it, makes it." Glossin also pretends to have forgotten all "legendary antiquities" within earshot of an incorrigible resident ballad singer who can't help piping in with a legendary song about Bertram's ancestors (246–8). Glossin seems the embodiment of Whig progress, a representative of the professional class who has effectively overthrown the old aristocratic order and sees history as a narrative of inevitable progress. On this view, Glossin's claim to possession is based on the force majeure of the Whig status quo, which justified political ascendancy on the basis of the inevitable superiority of a combination of mixed government, civil protections, and economic franchise that weakens the feudal order.[7] In this way, Glossin's reading of history is closely aligned with the narratives of progress considered in my previous chapter. Glossin's confidence that history is behind him seems confirmed by the civil armature of the Scottish legal system, which arrests Bertram for assault and (temporarily) saves Glossin's modern settlement.

For Glossin (as for Jason Quirk in *Rackrent*) history is merely a present possession. Yet throughout the scene of Scott's novel, the threat (ultimately fulfilled) is that the past will possess the present. Glossin himself initially mistakes Bertram for the ghost of Bertram's father; and though he quickly guesses his mistake, he is unable to shake an overwhelming sense of guilt and dread. Bertram is ultimately acquitted and resumes his estate because the modern fabric is always in tension. Like the ballad of Bertram's ancestors, the past cannot be put to rest. The point is that the movement of history is unstable, the forces of modernization skirmishing with the unsettling and powerful survivals of the past. Behind any sharp comparison between past and present is a more unstable network of filiations. If might tends to make right in the long view of history, the modern settlement yet continues to be reworked through the continued register of historical violence.

A central contention for this book is that we must look to nineteenth-century fiction to understand the historicist turn, because those fictions had a powerful influence on the historians who followed, both the authors of traditional histories and the wider community of scientists and authors who tried to make sense of the past. Scott's influence on contemporary historical thinking was massive and immediate. In emphasizing the importance of Scott for later historical understanding, I join a range of other historians and critics.[8] The historian John Tosh has argued that it was Scott, rather than Leopold von Ranke or Johann Joachim Winckelmann, who did most to launch the new historical awareness that characterized nineteenth-century thought. Tosh describes the premise of nineteenth-century historicism, credited to Scott, as the "autonomy of the past" characterized by discrete and organically interrelated social systems.[9] But to read history in the *Waverley* mode was to read with a more filiated and relational sense of social history, both for the particularities of the social moment and for its points of contact and departure from modern norms and cultural forms: rather than autonomy, his novels demonstrated interconnection; rather than integral wholes, social assemblages. This complexity formulates the challenge of coming to terms with the historicism of Scott's novels, which fashioned history as convergence of disparate views about the past and its meaning.[10]

In an article that established the program of historical research he would pursue for the remainder of his life, Thomas Babington Macaulay—the most influential British historian of the nineteenth century—argued that Scott managed to register such events by assembling new sources, comparing him to the artisan who crafts a superior work of art from the leavings of his master:

> At Lincoln Cathedral there is a beautiful painted window, which was made by an apprentice out of the pieces of glass which had been rejected by his master. . . . Sir Walter Scott, in the same manner, has used those fragments of truth which historians have scornfully thrown behind them in a manner which may well excite their envy. He has constructed out of their gleanings works which, even considered as histories, are scarcely less valuable than their's. But a truly great historian

would reclaim those materials which the novelist has appropriated. The history of government, and the history of the people, would be exhibited in that mode which alone they can be exhibited justly, in inseparable conjunction and intermixture. We should not then have to look for the wars and votes of the Puritans in Clarendon, and for their phraseology in Old Mortality; for one half of King James in Hume, and for the other half in the Fortunes of Nigel.[11]

Macaulay offers recognition in place of advocacy. The implication of Macaulay's praise for Scott is not only that Scott's historicism should be incorporated into the practice of history but that it *already* has been in the fourteen years since the publication of *Waverley*. Scott's impact on the historical imagination is already so widespread that he sits alongside Hume and Clarendon on the shelves lining the historian's workshop. Ultimately, Macaulay would "reclaim those materials" and forge a more unified Whig history, and so establish himself as master over Scott's apprenticeship. As Catherine Hall observes, Macaulay's powerful *History of England* (1848–55) "would stand as a universal history, for England was in his mind *the* modern nation, a nation that had progressed, was civilized, and could act as a beacon providing the model for other nations to follow."[12] It is a vision that presumes his successful integration of Scott's "fragments of truth" into the master narratives of Enlightenment historicism (explored in my previous chapter). Yet here Macaulay recognizes the historical novel, particularly as it explores the border between Scottish and English cultural identities, as assembling a sense of history that was heterogeneous and fragmentary, defined by conjunction and intermixture rather than unity, organized as mosaic rather than tapestry.

This chapter explores how Scott's "gleaning" worked, examining the specific procedures he used to sift the "fragments of truth" and shape the modern historical imagination. A closer consideration of Scott's investment in the past, as it evolved not only in the novels but in his work as translator, literary collector, editor, poet, and essayist, finds him immersed in virtually all of the historicizing disciplines of his day. These include the literary history of his ballad collections, the comparative philology of his studies in translation and vernacular languages, the material history of his antiquarian researches, the economic history explored in his essays on monetary policy, and the proto-anthropology of his many studies of traditional beliefs. Scott had a finger in a range of disciplines that were providing a more complex, nuanced understanding of social change than previously available. These different engagements shared a commitment to comparative methodology as the key to unlocking, on the one hand, the changing relation between different social forms like languages and, on the other, the concrete differences that allowed accurate placement of historical artifacts and texts relative to the present. The culmination of these different engagements, Scott's novels brought together several key features of contemporary social and historical thought: (1) that social worlds are historically conditioned conglomerates, characterized by complex relations between their material culture, social traditions, and, especially, idiomatic languages (as documented by physical objects, literary artifacts, and dialects); (2) that these worlds can be teased apart by antiquarian, bibliographic, and translational strategies; (3) and that the distinctions between overlapping epochs, especially between the past and present, can be bridged by acts of imaginative and sympathetic investment: handling the objects, imitating the literature, and speaking the patois of the past.

The remainder of this chapter is devoted to Scott's efforts as an antiquarian, as an imitator of antique literature, and as a translator, in order to elucidate the complex historical model articulated at the intersection of these disparate efforts. These three modes—antiquarianism, re-creation, translation—are understood to contribute to the editorial, writerly, and readerly positions that characterize the Waverly novels. The "self-reflexive" dimension George Levine has identified in Scott's novels can be examined more precisely as the interplay of these modes.[13] In exploring these filiated dispositions I describe a new mode of historical comparison that coalesced at the center of Scott's history, one that, unlike the contrasting strategies of the Enlightenment historians before him, would emphasize commonality as well as difference, interpenetration as well as distinction. By the same measure, and in place of Erasmus Darwin's universal analogy of progress, Scott articulated an uncertain network of analogies—a history of difference and differentiation as well as resonance—that exposed modernity as a tense and uneven composite.[14]

The first portion of this chapter sharpens the contrast between Scott's historical understanding and the Enlightenment histories he drew from. Ian Duncan's extensive study of the novelist's immersion in the philosophy of Enlightenment Edinburgh, especially after David Hume, is an important interlocutor here, as it builds upon earlier studies by Peter Garside and Kathryn Sutherland that examined the contending schools of "philosophical history" that conditioned Scott's thinking.[15] In Duncan's account, Hume's description of the skeptical mind, and its peculiar "double consciousness," furnished a " 'novelistic' model of the imagination" for Scott's fiction.[16] Along these lines, we might read the scene described earlier, in which Bertram senses a tension between "visions" of the past and "real objects . . . presented to our imagination," as marking his skeptical intuition of the troubled relation between present experience and all narratives of the past and, in this way, an allegory of historical fiction's relation to history. Yet I read Scott's break with Enlightenment historiography as far more consequential than his engagement with eighteenth-century philosophies of mind, insofar as it fashioned new narrative models for comparative historicism. Bertram's discussion with Glossin hinges on comparisons—of places, of idioms, and even of mottos—that provide an archaeology of the past in its relation to the present. Whereas skeptical philosophy privileges present experience over testimony, exploring the formal relation between them, comparative historicism attempts to set present and past artifacts alongside each other, flattening their relation and upsetting notions of relative value and authority. I begin by revisiting the Enlightenment models of unitary history, discussed in the previous chapter, in order to add an account of how Scott's novels exchange an eighteenth-century concern for the autonomous and rational human subject for a grounding interest in the idiosyncratic features of human experience and physical artifacts in time.

As agents of history, objects and people are peculiar in Scott's fictions, bound up within eccentric social codes that require translation. Comparison mediates that difference in his novels; analogies both establish the alterity of the historical subject and help translate that difference for the reader. In the second portion of the chapter, I explore antiquarianism—understood as the science of interpreting material history—as a key source of Scott's comparative methodology. And in the following movement, I suggest that antiquarianism, particularly as developed through the editing the *Minstrelsy of the Scottish Border*, generated techniques for forging historical narratives from comparative study. Finally, I explain how this mode of fabrication solved a problem inherent in Scott's efforts at translation and comparative philology: how to produce and reproduce the cultural idiom of the past. Whereas translation had generally been understood as governed by a law of diminishing returns, the fabrication of the past through imaginative translation offered a way of understanding historical fiction as something more than a debased currency: an analogue of the past that, in its continuities and differences, mediates between history and the reader. Simultaneously forgeries and histories, zones of contact that operate at the border between past and present, and between distinct social formations within that past, Scott's novels provided a way to marshal narrative fiction under the banner of comparative historicism. In place of unitary narratives, Scott's fiction turned to the complex patterns between narratives and to plotting their relations. By these means, Scott fabricated a new way of reading history.

The Subject of Enlightenment History

The people, tho' under the rod of lawless, unlimited power, could not forbear, with the most ardent prayers, to pour forth their wishes for his preservation; and, in his present distress, they avowed *him,* by their generous tears, for their monarch, whom, in their misguided fury, they had before so violently rejected.

David Hume, *The History of England* (1756)

Scott's impact on the historical imagination—the way, as Ann Rigney has put it, he "opened up the past as an imaginative resource"—remains a central problem for our understanding of nineteenth-century fiction, historiography, and descriptive natural history.[17] On the one hand, it was immediately evident that a new chapter had opened in literary history and in the understanding of the past. On the other, no one was able to understand precisely how this worked. As the "Scotch novels" continued to stream from the presses, reviewers recognized a profoundly disruptive publishing phenomenon, one that transformed the status of the novel, the author, and book publishing in ways we are still coming to terms with. The success of the Waverley novels, as everyone quickly realized, was history in the making. And yet reviewers struggled to make sense of them.

John Wilson Croker—the most influential critic of his day—gave an almost uniformly positive review of *Waverley* for the *Quarterly Review,* but he later criticized *Guy Mannering* as inferior, both on its own merits and as compared to *Waverley,* and argued that *The Antiquary* united the "merits" of the former with the "faults" of the latter. Worse, he implied that the "author of Waverley" *lied* in claiming that *The Antiquary* "completes a series of fictitious narratives intended to illustrate the manners of Scotland at three different periods": it was a "system," Croker argued, "he never thought of, and in which, if he had designed it, . . . [he] failed."[18] Croker's observation about the preface to *The Antiquary* is acute. Though the preface attempts to recast the first three novels as stadial history, in the style of *Castle Rackrent,* this doesn't explain their varied settings and disparate subjects. Croker's own analysis argued that the Waverley novels were riding the tide of literary history, and observed that novels in general had made a "natural" shift from generality to particularity of character. Where the narrator of *The Antiquary* proposed the novels be read collectively as a universal history, Croker suggested they be placed in the longer generic evolution of literature toward the ordinary and particular, an argument that anticipates Northrop Frye's "Historical Criticism," but completely overlooks the significance of history in these *historical* novels.[19]

Initially, neither Scott nor Croker could explain the central feature of those novels: the innovative way they articulate disparate historical perspectives through literary form. As I will argue later in this chapter, it was not until 1820, after Scott had written half a dozen, that he finally wrote a persuasive description of how the historical novel works. The struggle to define those novels indicates their ingenuity. Here, I should admit that the new historicizing sensibility of the nineteenth century, its sense both that elements of any historical period were defined by complex organic interrelationship *and* that the nineteenth century was one such period, is not derived from the writings of any single individual. As James Chandler puts it (adopting Hans Blumenthal), the nineteenth century is the epoch of the "epoch," and Scott's fiction, while epoch making for the history of the novel, was only one contribution to this larger shift in historical perspective.[20] Yet recent studies by Duncan, Lauren Goodlad, Rigney, Katie Trumpener, and Chandler himself have made Scott a focal concern for understanding the nineteenth century's perspective on history, taking up his influence on Romantic historiography, his placement of national history within the imperial imagination, the Romantic understanding of modernity he helped shape, his impact on Victorian realism, and the continuities between progressive and Enlightenment history explored in his fiction.[21] As I note in the introduction, one need only look at the generation of historical writers who followed Scott and cited his formative influence to get a sense of the diffusion of his commitment to the social density of history.

Moreover, the influence of Scott's historicism was global in scale.[22] Robert Crawford has argued that Scott was "the first English-language novelist aware that he addressed a global audience," and this may help to explain the diverse ways in which his novels were read.[23] The reasons are necessarily complex but must include some combination of the formal innovations and content that set the novels apart and the various perspectives on social and historical belonging that Scott's novels offered. The mutability of Scott's fiction—the way it could be adapted to widely different social contexts—cautions against the assumption that Scott's novels succeed insofar as they capture universal features of human experience—Carlyle's "living men [whether] in buff or other coats and breeches." In fact, Scott's novels insist that the choice of buff or "other" dress encloses deep distinctions, registering divergence in social position, custom, and often perspective. I say this even though, in the preface to *Waverley*, Scott's narrator insists that the homogeneous ground of history is located in "those passions common to men in all stages of society, and which have alike agitated the human heart."[24] This conventional statement regarding the universal features of the Enlightenment subject comes near the end of a sophisticated introductory chapter that, in detailing the implications of all the genres he will not write in, uses the language of contemporary reviewers to elaborate a sociology of the literary market. As such, the author's statement about human nature stands in the place of an unarticulated theory of social change and heterogeneity—a capacious sociology of the past for which Scott did not yet have a vocabulary—but which later writers would recognize in his fiction.

In chapter 1, I discuss at length the distinction between stadial, Whig/progressive, and Christian historicism in the eighteenth century. As I noted, these widely different modes of understanding conform in one key respect—their commitment to the idea that unifying narratives govern the complexity of historical events, a commitment that underwrites Erasmus Darwin's poetry, and his consistent argument for the analogy between biological development and cosmic evolution. This recourse to a governing narrative is evident even in the least totalizing of Enlightenment historians, David Hume. Here, I am interested less in the relation between experience and fiction that Hume provided for Scott than in Scott's departure from a Humean model of historical description. Hume's skeptical analysis of experience is as important to the Gothic novels of Anne Radcliffe as the historical novels of Scott; but where Radcliffe encounters a history organized along lines of national and stadial difference (as Duncan has shown), Scott finds complicated filiations, the zones of contact, which such differences obscure.[25]

The contrast with Hume is instructive. Hume's engagement with the past in *The History of England* (1754–61)—a work far more successful in his own time than his philosophical treatises—operates on a sympathetic model akin to Adam Smith's *Theory of Moral Sentiments*. For this reason, the particularizing potential of Hume's history, its ability to grasp the idiosyncrasy of the historical subject, was limited by an Enlightenment concern for the universal features of humanity, characterized by a fusion between the potential for rational thought and subjection to universal passions.[26] In his own account of that work, Hume admits that he (like the "people" in the preceding epigraph) can "shed a generous tear" for James I, but only because he "was the only historian, that had at once neglected present power, interest, and authority, and the cry of popular prejudice"—in other words, only insofar as he was able to encounter James I outside his character as a historical actor, disclosed as an expression of the unities of human nature.[27] As Hume explains in his *Enquiry Concerning the Principle of Morals* (1777), "no passion, when well represented, can be entirely indifferent to us; because there is none, of which every man has not within him, at least, the seeds and first principles."[28] Hume's distaste for the atypical draws the sharpest contrast with Scott's celebration of the idiosyncrasy of everyday life. Hume's spectacular model of sympathy, drawn from neoclassical aesthetics, produces a "normative response" to the drama of history—the motivation, in Adam Potkay's analysis, for Hume's disinterested spectator—and is founded on the universal franchise of human experience.[29] In Leo Braudy's analysis, this movement toward an aesthetics of disinterest, verging on objectivity, can be seen in the succession of volumes in Hume's *History*, which proceed backward from Stuart, to Tudor, to Norman eras.[30]

By contrast, sympathetic investment in Scott's fictions turns on the capacity to identify with customs and experiences marked by their historical alienation from our own. Scott's attention to the denaturalized specificity of the quotidian invests historical understanding with critical discretion. The importance of the particularized sentiment is demonstrated by Jonathan Oldbuck's intervention in the funeral service of Steenie Mucklebackit in *The Antiquary*, which had a long afterlife in the imagination of charity visits in nineteenth-century literature. Oldbuck insists on carrying the head of Steenie's coffin in order to forestall friends of the family who, as "an act of duty on the part of the living, and of decency towards the deceased" would have forced the father, hobbled by his grief, to perform that traditional function.[31] The narrator's gloss on this event offers an instructive contrast that would have suited Hume: this "compliance with their customs, and respect for their persons, [gave Oldbuck] more popularity than by all the sums which he had yearly distributed in the parish for purposes of private or general charity," a lesson in domestic government that aligns the novel with the traditional pedagogy of political histories. Yet the narrator's sententious gloss is strikingly out of step with Oldbuck's motivations. Oldbuck is, in fact, compelled to set aside his classed status as spectator. In this way, the scene powerfully dramatizes the incompatibility of social custom with the qualia of individual anguish. Sympathetic engagement forces Oldbuck to insert himself into the scene as a vested participant. The effect is to mark a "respect" that flattens his position and apprehends the dignity of both the father's grief and his conscientious friends. By means of (and not in spite of) these customary rites, Oldbuck helps us to apprehend the moment in its historical particularity.

In Scott's fiction it is not enough to "shed a generous tear" or give a coin, because such gestures, organized around a model of the universal subject, cannot demonstrate a sympathetic understanding of the uneven exchange between social customs in history.[32] Scott's protagonists must participate, as well as observe. In this way *The Antiquary* makes an important contribution to the ethnographic turn explored by James Buzard in the nineteenth-century novel.[33] Such scenes produce a "dialectic between absorption and reflection," a "double consciousness" that, in Duncan's analysis, characterizes Scott's fiction.[34] The scene also demonstrates how spectacular models of sympathetic engagement, drawn from the Enlightenment, evolve into an engaged and less mediated mode of understanding. Within the Mucklebackit cottage, patrician fantasy shades into what Rae Greiner terms "sympathetic realism," as the particularity of historical encounter produces a sympathy of mind, heart, and hand.[35]

The Antiquary illustrates how Scott's novels address Enlightenment models of the historical subject even as they recast these models into new modes of historical investment and individual experience. Because of Scott's incomparable impact on historical thinking, any study that looks at nineteenth-century history writing without looking to Scott is forced to orbit around his absent center.[36] His novels take up the methods of comparative inquiry to explore moments of contact and cultural translation, and they produce history as a composite of people and events, a network of conflicting narratives. Hence, while Scott's novels produce history through techniques of "narrative accrual" (Jerome Bruner's term), they yet resist the sense that these narratives are "centered around a Self acting more or less purposefully in a social world."[37] By emplotting and aligning distinct narratives with different trajectories, Scott suggests the impossibility of their reference to a single governing narrative, a master plot of history. Jonathan Sachs has characterized this contrast as the difference between an earlier understanding of historical incidents and characters as "exemplary" and universal, and a more robust historicism that "explain[s] events and human characteristics with reference to particular times and places."[38] Scott's fictions demonstrate the importance of comparatism as a way to embed this diversity into the novel as the interplay of historically differentiated characters, events, and narratives. If Scott's protagonists have been variously characterized as "passive" (Alexander Welsh), "middle of the road" (Lukács), "negative," "feeble," and "blank" (Hazlitt), and "insipid" (Scott himself), their flexibility serves this larger purpose.[39] In order to shift the attention of the novel from narrative to narratives, from a unitary to a comparative history, the story of any one protagonist—and his formation as a more modern subject—must take a back seat.

The Forensic *Antiquary*

Organized fossils are to the naturalist as coins to the antiquary; they are the antiquities of the earth; and very distinctly show its gradual regular formation, with the
various changes of inhabitants in the watery element.

William Smith, *Stratigraphical System of Organized Fossils* (1817)

As made clear in the line of comic antiquarians who feature in Scott's fiction,
from Jonas Dryasdust, to Jedediah Cleishbottom, to Jonathan Oldbuck (the "Antiquary" himself), an obsession with antiquity produces endless humor. Conceived
as a comic figure, the antiquarian constantly leads the reader down byways and
detours and away from the common road of civic history. Certainly, this is how
Scott's critical description of antiquarianism works in his comic account of *The
Antiquary*. An important contributor to the trope of "squire and parson" antiquarianism R. G. Collingwood knocked in *The Idea of History* (1946), Oldbuck is often
driven to obscure true history by his curiosity and a concupiscent desire for antique artifacts. So, among many examples, Oldbuck buys a worn Scottish "bodle"
or copper piece from a peddler in the belief it is an ancient coin.

The elaborate editorial apparatus that characterizes Scott's writings—the
reams of footnotes, appendixes, introductions, prefaces, and postscripts, even
the famous "postscript which should have been a preface" in *Waverly*—are the-
matized comically in those novels as the products of antiquarian investment. In
Jonas Dryasdust, Jedediah Cleishbottom, and even the bluff Captain Cuthbert
Clutterbuck, Scott's editorial personas internalize the critical standards of antiquarianism at the same time that they demonstrate their inadequacy. The irony
of their language and their scrupulous attention to a fictive narrative, our sense
that they are inside history but can't get inside the sensibility of their novel,
communicate that the work is larger than what they make of it.

In such scenes, humor seems to police the boundary between antiquarianism and a more serious history. As such, humor performs a kind of "boundary-
work," a phrase sociologist Thomas Gieryn has used to characterize the effort
of scientists to "create a public image for science by contrasting it favorably to
non-scientific intellectual or technical activities."[40] Recent scholarship has
shown that such boundary-work is a more general feature of how academic
disciplines construct authority. So, Susan Manning has linked the lampooning
of antiquaries to Enlightenment arguments on behalf of "philosophical" history. In Manning's memorable formulation, "Antiquarians were the misers of

historiography. As the antiquarian's delight in the detail of his collection impeded narrative progress in history, so the miser's accumulated hoard of coins was anathema to the circulating economic and sympathetic currencies of Civil Society."[41] On this view, the critique of popular antiquarianism helped sharpen the disciplinary boundary that would secure history as a distinct academic field. In the same fashion, antiquarianism rubbed against popular anxieties regarding bibliomania (which antiquarianism subsumes for both Scott and Oldbuck) and contrasting efforts to formalize English literature as a prestigious object of study. As Ina Ferris argues, "In the age of bibliomania the status of bibliography as a scholarly genre . . . was further compromised, and it was regularly represented in the periodical press under tropes of inflation and presumption."[42] However, as Deidre Lynch has argued, the bibliomaniac who "remakes the literary heritage as his cabinet library" rendered the literary heritage visible and ultimately available as an object of collective investment.[43] Similarly, to position antiquarianism as a counter-Enlightenment is to recognize an important contribution to the formation of Romantic historicism—a point Michael Gamer has made in rich detail with respect to antiquarianism's gothic sibling.[44] Boundary-work obscures traffic between specialist and nonspecialist research, the productive exchanges between scientist and nonscientist, academic and layman, that often supports important work (a point I will elaborate in my final chapter).

In the preceding epigraph, William Smith adapts these productive possibilities to natural history, conceiving his geology as a kind of antiquarianism.[45] His case for the use of fossil formations to date and then map the history of geological deposits of Britain, in the "map that changed the world," relies on the visibility of antiquarian research as a historical science organized around the interpretation of material objects.[46] Only a year after the publication of Scott's *Antiquary*, the approving analogy Smith draws between disciplines is sure. Smith's description casts antiquarians and geologists as speculative realists, arguing that the objects of the material world give a firmer guide to past realities than the testimony of historical authorities.

Scott did not generally concern himself with natural science.[47] There is slim evidence that, after his election as president of Royal Society of Edinburgh in 1820, he took the role seriously. Scott's discussions of his new role, from his opening speech before the society (which young Charles Darwin attended) to his private letters, emphasize the irony of his position. As he described it to his friend William Laidlaw: "I am made president of the Royal Society, so I would have you in future respect my opinion in the matter of *chuckie-stanes*, caterpillars, fulminating powder, and all such wonderful works of nature. I feel the spirit coming on me, and never pass an old quarry without the desire to rake it like a cinder-sifter."[48] Amid this catalog of scientific objects, the "chuckie-stanes" (or quartz pebbles) fix Scott's attention, as he zeroes in on how they are gleaned from rock quarries. The origin of crystallized rock, including quartz, was an important point in the geological debate between Wernerians and the Huttonians, who argued whether such rocks could be formed by aqueous sedimentation alone (Werner) or required subterranean heat and pressure (Hutton). A key episode in Hutton's science was the discovery that quartz had fused after infiltrating the granite at Glen Tilt.

As Adelene Buckland has noted, the kind of fieldwork Scott describes here became an increasingly prominent practice for geologists in the decades following Hutton's discoveries.[49] This sense of geology in motion catches wind in Scott's imagination, as it echoes the staging of antiquarian fieldwork, and he foregrounds this connection in his letter to Laidlaw in order to cast his new position in a comic vein. Scott knew that Laidlaw would get the joke: Laidlaw's father was Scott's regular companion during the roving "border raids" that took him into the Scottish lowlands, "gathering [as son-in-law and biographer John Gibson Lockhart puts it] wherever they went, songs and tunes, and occasionally more tangible relics of antiquity."[50] Those "raids" are at the heart of Scott's many puns on "reiving," a trope drawn from the centuries-long tradition of raids across the Scottish-English border for goods, cattle, and hostages. Scott's own expeditions—gathering stories and ballads and purchasing rare manuscripts and other artifacts—are generally recognized as the genesis of his antiquarian researches, and especially the *Minstrelsy of the Scottish Border* (1801–3), the three-volume collection of traditional and modern ballads, for which Laidlaw's father helped gather additional materials. Underneath Scott's joke about the connection between geology and antiquarianism rested a sense of shared scientific ambition. He may have been wry about the figure he cut among geologists, but he was serious about their common belief that the key to reconstructing the past was to gather material in the present.

This is because antiquarianism *was* a science. Noah Heringman has recently given a detailed account of the relation between eighteenth-century antiquarianism and natural history, "two allied 'sciences' often practised within the context of a single knowledge project."[51] The *Transactions of the Royal Society of Edinburgh* included a section for "Papers of the Literary Class" through 1798, such as "On the Origin and Principles of Gothic Architecture" by chemist Sir James Hall.[52] In later volumes such articles were folded in to the general content; examples in the *Transactions* include "On the Origin of Cremation, or the Burning of the Dead" (1818), and "Notice of an Undescribed Vitrified Fort, in the Burnt Isles, in the Kyles of Bute" (1826). Insofar as antiquarianism was opposed to the "philosophical" history of the Enlightenment, its materialist bent suited it to the nineteenth century's increasingly empirical study of the past. An investment in the objects of the proximate past gave localized purchase on national history. For this reason, historiographers have argued that antiquarianism ushered in a new investment in material history and particularly its appreciation for artifacts and documents excluded from the traditional source criticism of critical history.[53] This is the "gleaning" Macaulay ascribed to Scott: his historical novels proved these untapped materials could be used to reconstruct past societies.

Scientific antiquarianism drew its historical methodology from two main sources: critical humanism and the contemporary antiquities market.[54] First, antiquarianism gathered from textual criticism the comparison of historical works to scrutinize them for evidence of accuracy, alteration, and relative date of composition.[55] Desiderius Erasmus's *Novum Instrumentum omne* (1516) was a landmark text in historical criticism, because it drew together extant Greek and Latin versions of the New Testament, correcting and printing them alongside each other, so that readers could compare the two for accuracy. One important result of Erasmus's comparative approach was to exclude additions to the Latin vulgate that were not present in Greek originals, for instance, the "comma Johannem" that explicitly stated the Trinitarian doctrine. Though later scholars have railed against many of Erasmus's decisions (e.g., Erasmus's reliance on more recent over older Greek sources), he is seen as the transitional figure between "precritical" and "critical" Bible studies.[56] As I note in my fourth chapter, this careful comparison of sources would reach new levels of sophistication in the Tübingen school of "higher criticism" practiced by Friedrich Schleiermacher, David Friedrich Strauss, and Ludwig Feuerbach, who compared biblical sources to contemporary historical accounts in order to discern their historical accuracy.

Scott was not a student of this brand of historical criticism, but he was more than a bystander in its British analogue: the raging debates over the historical accuracy of oral and ballad poetry as a depiction of ancient manners. The main focus was James Macpherson's *Ossian* "translations." Purportedly from Gaelic manuscripts, these poems were the subject of a decades-long debate over their authenticity and fueled interest in the search for ancient folk literatures through-out Europe. The lack of contemporary sources for comparison, and Macpherson's refusal to produce his own, made the debate notoriously difficult to settle.

Scott was happy to argue the *Ossian* question with friends and in print, and like virtually all Romantics, intrigued by the Thomas Chatterton forgeries (as we will shortly see), but he was more immediately invested in the slightly older battle between Thomas Percy and Joseph Ritson over the historical re-creation of bardic culture from oral and historical sources. Percy, in his "Essay on the Ancient English Minstrels," which he added to the second edition of his *Reliques of Ancient English Poetry* (1767), argued that an English minstrel tradition once existed in the medieval period under the patronage of the royal court. Ritson, with devastating accuracy, reexamined Percy's sources and demonstrated a series of ambiguous and misleading descriptions. In one of his footnotes, Ritson edits Percy for accuracy with some violence (the bracketed comment is Ritson's): " 'These instances,' therefore, 'are [NOT] sufficient.' "[57] Under Ritson's analysis, Percy's case for a flourishing English-language poetic tradition within the French-speaking Norman court fell apart (Percy's fourth edition tacitly admits the point; the title changes from "An Essay on the Ancient *English* Minstrels" to "An Essay on the Ancient Minstrels *in England*" [emphasis added]). Ritson expended such energy on the analysis of Percy's sources because it gave insight into the editorial practices Percy used to edit the poetry. "The learned collector," Ritson often complained, "has preferred his ingenuity to his fidelity, without the least intimation to the reader."[58] These highly public arguments vested Ritson as a walking figure of the irascible antiquarian. At the same time, this scrupulous comparison of sources, and strict fidelity to original materials, established Ritson as the foremost British expert on literary antiquities and set the terms for careful antiquarian scholarship. (Scott was careful to credit Ritson's inspiration for the care he took in his edition of the *Minstrelsy*.)

The emphasis that Ritson placed on forensic inquiry—careful analysis for accuracy and imposture—places his critical practice in dialogue with the substantial forensic investment of the contemporary antiquities market. Forensic technologies thrived in the antiquities trade, where they helped evaluate the authenticity of artifacts and discriminate forgeries. Heringman has documented the importance of eighteenth-century antiquarianism as a "science of preservation" that organized a new "prehistoric" investment, and so laid the foundations of nineteenth-century historicism.[59] The most important tool of the collector was a sophisticated and culturally broad understanding of the past—one that included not only the major events and typical forms of expression that might properly be registered by an artifact from that period but, even more important, the historical technologies and social customs that explained how they were created and used.

Scott's novels, especially *The Antiquary*, demonstrate how the increasing sophistication of antiquarianism, understood now as a comparative and forensic science, could frame out comparative historicism for subsequent generations.[60] These forensics were particularly sophisticated in coin and medal collecting. Such minted artifacts were important objects for material history because, as both instruments of the state and portraits of the political imaginary, they straddled the divide between political history and the material "fragments of truth" that Macaulay describes as the ejecta of traditional history. This "science of medals" was revolutionized by the systematic study of Abbé Jean-Jacques Barthélemy in his function as keeper of the cabinet for Louis XV.[61] Contemporary coin collecting manuals demonstrate how advanced this science had become by the eighteenth century. Louis Jobert's *Science des medailles,* translated as *The Knowledge of Medals* in 1715, gives an influential summary of rules for analyzing both the age and authenticity of ancient medals and coins in order to estimate their value. He does so by emphasizing the intersection between material and traditional history, arguing that a strong knowledge of history is the first requirement for coin collecting.[62] Extensive experience, combined with a strong historical background, allowed the collector to develop general rules regarding, for instance, in what periods coins were stamped versus cast, what kinds of metals were common at different times, what kinds of markings were more typical for coins from a specific town, and even whether a coin that is stamped in one metal before being coated in another is original (as with the "Posthumi"), a recent but acceptable "enrichment" (as when a collector is completing a series of coins), or an illegitimate forgery.[63]

Scott was himself an avid collector of coins, and though the only collecting manual remaining in his Abbotsford library today is a translation of Guillaume Beauvais's *Essay on the Means of Distinguishing Antique from Counterfeit Coins and Medals* (1819), the several prospecting scenes in the *Antiquary* demonstrate Scott's fluency in the science. The arguments about antiquity between the chief antiquarians of that novel, Arthur Wardour and Oldbuck, mark a shift in antiquarianism's operative mode from an older regime of curiosity to a more precise historicism rooted in forensic science. When Wardour, victim to a long con by a German swindler, finds his first horde of silver and gold coins in an old copper horn, Oldbuck notes that they "are so mixed and mingled in country and date, that I cannot suppose they could be any genuine hoard," and further, that their vessel, an old copper horn, seems to be a common Scottish snuff mill (185). Contributing an early chapter in the history of archaeology, the antiquarians of the day began to realize that the analysis of provenance and disposition gave a window into both the value and history of a find. William Stukeley's *Palaeographia Britannica* (1743–) carefully documented various archaeological discoveries in England, scrupulously tying them to extant historical accounts in an attempt to re-create the significance of their composition in situ. Similarly, Abbé Barthélemy looked carefully into the source of new coins, asking for descriptions of the original owner and, if the coins came from a horde, of the condition and composition of the site.[64] The type and number of coins in a collection could provide useful information about its source, helping verify whether they were put together by a more recent collector, by an imposter, or whether they represented a real cache from antiquity. Forensic antiquarianism shifted from assaying the object to weighing the relation between objects—with a consequent shift toward relational notions of value.

In order to establish the value of Wardour's find, Oldbuck proposes they consult John Pinkerton's *Essay on Medals* (1789), a resource that carefully connected a critical history of minting to the material practice of coin collecting. For reasons of national pride, Wardour refuses, inveighing against Pinkerton and his criticisms of the accuracy of early Scottish historians. This is Wardour's significant loss; Pinkerton would have informed them that the "unicorn of James II" included in the cache was a certain forgery—the first gold unicorn minted under James III.[65] A more careful forensics would have exposed the fraud before any damage was done. Instead, Wardour is hoodwinked, and the second time round, when he finds an additional horde of silver ingots, Oldbuck is careful to examine each of them for their relative position within the horde and any identifying markings. The fact that only one is stamped with a "Spanish" mark is an unrecognized signal that the collection is melted silver plate from the Neville family rather than a buried New World horde (which would instead have consisted of the ubiquitous "pieces of eight," or struck silver coins). Derived from the Spanish *plata* and designating both silver in general and silver dishware, "plate" is particularly mutable in *The Antiquary*, serving a variety of functions: at times it is currency (one of Wardour's creditors asks for the family plate to discharge a debt), at other times, a magic totem (as with the censer Wardour provides for Dousterswivel's incantations). Matthew Rowlinson sees such examples as staging antiquarianism's problematic status as a field that translated curiosity "as an object whose social identity is constituted in the process of exchange."[66] The perceived problem of the "indeterminacy" of a complex market, however, can better be understood as the sufficient condition of antiquarianism's forensic method. It was precisely in order to regularize these uncertainties in the new market for old curiosities—what Chandler terms the "paradox of 'novel antiquities'"—that antiquarians worked to clarify the material complexity of historical practices.[67]

As an area of concern, and in opposition to philosophical history, antiquarianism was defined by casting the significant past as an intersection of practical fields instead of a singular narrative—enriching and making more complex the network of things and people that could be said to figure in the past. This broadened area of inquiry, in turn, required antiquarianism to develop new disciplinary methods for figuring out both how the elements of that past fit together and which elements were foreign. For this reason, forensic antiquarianism was thoroughly rooted in comparative practice, both in the comparison of textual accounts to material evidence and in the analysis of variation and distinction between physical artifacts. Jobert emphasizes this point, arguing that the "sight of different Cabinets causes the same thing in the knowledge of Medals, as Experience does in Arts."[68] Similarly, Pinkerton observes that "a real and practical knowledge of coins . . . is only to be acquired by feeling a great number, and comparing the forged with the genuine."[69] Antiquarian comparison generally emphasized distinction over commonality. In part, this was because rarity determined price. "Single," that is, unique coins, presented the quintessential case. On the one hand, their rarity virtually ensured their strictly priceless status, because, in Jobert's words, as "they exceed such a certain Price, they have no Other to be put upon them than what the Desire and Ability of the Purchaser gives them."[70] At the same time, distinction could provide evidence of falsehood. Coins "which Antiquaries have never seen in the Cabinets, even of Princes," were candidate forgeries.

The peculiar thing about forensic antiquarianism is that, as it became increasingly effective at figuring out how fake artifacts were made (and virtually all coin-collecting manuals give extensive information regarding not only what characteristics to look for but also the forging techniques they identify), it became the most important public venue for learning how to execute forgeries. Antiquarians provided both a theory of the attributes that make something interior or exterior to a specific period—hence, historical—and, more consequentially, the techniques that could be used to imitate those attributes, in essence, a gradated theory of both historicization and historical re-creation.

Faking the *Minstrelsy*

A man who will forge a poem, a line or even a word will not hesitate, when the temptation is greater and the impunity equal, to forge a note or steal a guinea.

Joseph Ritson (1790)

The theory of forgery provided by forensic antiquarianism, which understood forgery as a (questionable) mode of historical re-creation, gave Scott an important model for historical fiction. As a kind of forger, Scott reproduces a past based on an understanding of what actually happened. Scott emphasizes this connection in his 1830 essay on ballad imitation, where he argued that Chatterton's counterfeit ballad "Sir Baudwin" was marked by forgery, much as "the newly forged medals of modern days stand convicted of imposture from the very touches of the file, by which there is an attempt to imitate the cracks and fissures produced by the hammer upon the original."[71] Scott's criticism here is not an indictment of forgery, per se, but Chatterton's inept handling of his tools, as "the art used to disguise and misspell the words only overdid what was intended."

Scott could hardly indict literary forgery outright, because he himself relied extensively on the "touches of the file," especially in his practice as contributor to and editor of the *Minstrelsy of the Scottish Border* (1801–3). The collection and publication of the *Minstrelsy*, a work that kept him tramping across the Scottish midlands and combing through the bookstalls and archives of Edinburgh, was a turning point in Scott's career and his first literary success. It also cemented a friendship with his close collaborator, John Leyden. It was Leyden who suggested that the *Minstrelsy* include an entire section of "Imitations of the Ancient Ballad," which they filled with poems written by Scott, Leyden, and various contributors, including Anna Seward and Matthew "Monk" Lewis. The primary effort, an attempt to collect and combine all extant versions of printed and oral Scottish ballads, required extensive editorial intervention. Forensic antiquarianism furnished the protocols of this folk ethnography. Under the shadow of Ritson, Scott and Leyden insisted that they were scrupulous in their effort to transcribe and edit these old ballads. In the introduction to the second edition of the *Minstrelsy*, Scott argues that he was forced to select the best passages from across various versions of recovered ballads because discrepancies exist between those extant versions. In addition, he apologizes that "some arrangement was also necessary, to recover the rhyme, which was often, by the ignorance of the reciters, transposed, or thrown into the middle of the line,"[72] but insists he has been careful "never to reject a word or phrase, used by a reciter, however uncouth or antiquated" (172).

These statements, where they are not flatly contradicted by the text (and Scott in fact regularly replaced "uncouth or antiquated" words), serve to obscure the extensive liberties Scott took with ballads, inserting new lines and new events wherever he felt appropriate. Scott's version of "Sir Patrick Spens" provides the most intriguing example of these editorial interventions. The version popularized by Percy was known for the hopeless but unknown mission of the sailor— relayed in the contents of a "braid letter" that is never explained. Scott's new version was more than twice as long and included extraordinary new features: the contents of the king's letter; the mission Spens was given (to bring the king's daughter from Norway); a drunken spat between Spens and his Norwegian hosts; Spens's return from Norway in a huff (and apparently without the king's daughter); and the heroic efforts of Spens and his crew to save the sinking ship. A fragmentary story was now made largely whole, and in place of Percy's admission that it was a mystery "in what age the hero of this ballad lived," Scott now boasted that he had discovered "the cause of sir Patrick Spens' voyage . . . and it shews, that the song has claim to high antiquity, as referring to a very remote period in Scottish history." He confidently identifies it as a literary account of incidents that occurred shortly before or after the thirteenth-century death of Alexander III.[73]

It was too good to be true, and there was immediate controversy. Malcolm Laing wrote to Scott to ask for clarification of his statement, in the introduction to the poem, that it was "taken from two MS. copies, collated with several verses, recited by the editor's friend." Could he see these manuscripts, Laing asked? Scott demurred, forwarding only the manuscript he had received from William Laidlaw and retaining a second variant. The latter, as Scott describes it, was "picked up by Leyden with some other little things from a woman in Kelso."[74] Despite his fears that in Leyden's version "one verse contain[ed] lines . . . which I still think are an interpolation," he felt authorized by the fact that both Leyden's and Laidlaw's versions mention Norway, and so he used them, "assisted by the printed copies"—that is, the version published by Percy and later James Herd.[75] Scott further insisted that, as Laidlaw and Leyden were not well acquainted, it was unlikely they'd forge two confirming documents.

Scott's defense of the poem, and his reticence to produce all of his sources, failed to settle the issue. The ambiguous relationship between the version printed in the *Minstrelsy*, the variant related by Laidlaw and later published by

Robert Jamieson in 1809, and even later versions stoked continuing argument. The dispute culminated in the "Lady Wardlaw Heresy," in which Robert Chambers, the publisher and secret author of *Vestiges of the History of Creation*, argued that virtually all the historical Scottish ballads collected by Percy, Scott, and others were "romantic" rather than "historical," literary ballads written in imitation of traditional folk poetry in the early eighteenth century by Elizabeth Wardlaw, the author of the spurious ballad "Hardyknute."[76] Chambers argued, on the basis of a careful sifting of "Spens" for both anachronisms and echoes of Wardlaw's known ballad imitation, that all the other ballads that showed the influence of "Spens" must be forgeries, too. Strangely, he never entertained the possibility that anachronistic language—for instance, the use of "strand" instead of "sand" in Scott's "Spens"—could mark Scott's editorial intervention, rather than the art of an eighteenth-century forger. Chambers essentially put the historicity of Scottish poetry on trial, prosecuting it as a class. Later combatants would largely refute Chambers's account, but the result was to so muddy the provenance of "Spens" that Francis James Child, the definitive nineteenth-century historian of English-language ballads, would later throw up his hands, declaring, "This ballad may be historical, or it may not," and observing that "it is only editors who feel bound to look closely into such matters."[77]

Despite Child's frustration, such inquiries remain central to the study of literature and history today. Child's deeper insight is that the Romantic period elevated historicity, as opposed to critical accuracy or aesthetic beauty, as a major component of editorial practice. By analyzing historical records and internal descriptions of the past, editors could attempt to resolve questions of priority and disparity that could not otherwise be answered. One of the reasons Scott was evidently so pleased with his version of "Spens"—the third edition of the *Minstrelsy* moved the ballad from the third volume to front the entire collection—is that it serves as a founding object for a national Scottish literature, a literature about Scotland written in Scottish.[78] It was one of several old Scottish ballads recovered from Percy's "English" collection. By publishing this improved "Spens" in his collection of insistently Scottish verse, Scott repatriated the poem. The poem's historicity allowed Scott to do more than cast the relation between the *Minstrelsy* and the *Reliques* as a literary extension of the history of border reiving discussed at length in his introduction and featured in the ballads. Ratified as an historical account, "Spens" bears witness to Scottish as a literary language firmly rooted in a past that predates Chaucer by a century. Whereas Percy had been unable to prove to Ritson that English bards featured in the medieval court, Scott could argue that modern Scottish predates English as a mixed language adopted by both minstrels and, at times, the Scottish court.

Given the driving importance of "Spens" as a historical poem, Scott's editorial decisions, and his difficulty justifying them, become easier to understand. The linchpin of the controversy over the historicity of "Spens" is the reference to the "King's daughter of Noroway," a line that first appears in the *Minstrelsy*. This line had to be drawn from Leyden's copy, as it is absent in Laidlaw's version and, according to Scott, was not gathered from Hamilton's recitation. Child was unable to locate Leyden's manuscript, but today it can be found at the National Library of Scotland, bound in the second of two folio collections of "Ballads and Songs" from the Abbotsford library.[79] In Leyden's hand, the fragment is titled "The King's Daughter of Norway," but it is clearly a variant of "Spens." More songful than other versions, the short nine-stanza ballad includes a more extended statement of Spens's mission, delivered as the first of an evolving refrain (repeated in stanzas four and six: "Blow it wind or blow it weet / Our ship maun sail the morn o, / Our gude ship sails to Norroway / For the Kings daughter the morn"). No reference is made to Spens's letter, and the last half of the poem focuses on the battle to prevent the ship's foundering on its egress from Scotland. A comparison of this variant with the Laidlaw version, published in Jamieson, and stanzas 16–20 of the *Minstrelsy* version, which Scott identified as the Hamilton fragment (substantially the same as the version published by Peter Buchan in 1828), reveals three versions widely different in both style and incident.[80]

Scott faced several challenges in pulling them together. In order to regularize the Leyden variant's songlike qualities, he condensed the repetitive elements of the refrain into discrete lines (e.g., "Be it wind, be it weet, be it hail, be it sleet" in line 25) and removed most of the interjections, adding in their place a note that "in singing, the interjection O is added to the second and fourth lines" (3:64). Elements of the description of the boat's foundering on its way to Norway are moved to the return journey described in both the Laidlaw and Hamilton versions. But the most consequential change is the movement of the lyrics drawn from the first and second refrains of Leyden's variant, which contains the king's directive regarding his daughter, into the text of the letter sent to Spens, now rendered as "To Noroway, to Noroway, / To Noroway o'er the faem; / The king's daughter of Noroway, / 'Tis thou maun bring her hame."

Chambers later complained that "faem" was an anachronistic description, and he was right; the first, third, and fourth lines of the stanza are from Leyden, but the second is Scott's own. By inserting his own line into Leyden's refrain, and inserting that into Spens's letter, Scott interpolates the lyrical refrain into epistolary communication. Leyden's voice, or at least Leyden's stenography, is incorporated. By these means, Scott and Leyden reformulate a collective voice that exists in choral space as a private statement, effectively opening its own literate dimension within the oral lyric of the ballad. In Percy's version, the poem existed in parallel to the culture of letters but outside of it; even though the letter is mentioned, the poem is not privy to its contents. That earlier version is energized by the gap between folk and literate culture, organized through the subjection of Spens and his sailors to a sovereign power vested in the letter we cannot read. Through Scott's interpolation, the "historical ballad" paradoxically transforms into something more like the "romantic ballad" that Chambers accused it of being, even as this secures the poem's ability to testify to a real historical event. This thickens the even narrative mode of the folk ballad into the intermedial awareness of modern literature; the poem effectively turns a corner and runs into *Pamela* and its ministering editor. By means of such creative interpretation, the ballad editor uses contemporary tools not to reconstruct folk literature but to project a lost worldview for the modern reader.

Maureen McLane has read such "balladeering" as part of an ethnographic engagement with folk literature, drawing on the thinking of French philosopher Michel de Certeau to explore the "epistemological-technical process through which the speech or song of 'primitive' others is textualized."[81] The result is a hybrid genre that incorporates ballad and song, narrative and lyric, secured through the bibliographic apparatus and authorial presence of the editor, who collates and coordinates the collaborative labor that is presented. It is not hard to see in this triangulation between scrupulous attention to the jointure between true and fabricated history, and the guise of the scrupulous antiquarian, a model for the strange triangulation between historian, character, and editorial command that characterizes the Waverley novels.[82] The revisions of "Spens" are just one example of how the act of editorial curation evolved into a complex form of authorship intimately interlinked with the editorial function. As McLane indicates, the narrative density of the historicized ballad, with its mixture of mediums and modes, and the substantial preface and annotations, looks forward to the complex narrative histories Scott would later provide in his novels.[83] If the instances of interpolation in Scott's ballad are brief but consequential, later novels extend interpolation into a discrete narrative mode of historical re-creation.

Such practices imply that Scott sees interpolation under the sign of history as true to history, in the absence of fact.[84] In this way, Scott's editorial work alters the landscape of the Percy-Ritson debate, which was about beauty versus fidelity to the letter. Later in life, Scott would retain this warm appraisal of Percy's work, though acknowledging many of Ritson's trenchant criticisms. It is in this vein that Jonathan Oldbuck, on hearing a ballad recitation, admires "a genuine and undoubted fragment of minstrelsy! Percy would admire its simplicity—Ritson could not impugn its authenticity" (310). As the comments demonstrate, Oldbuck is an antiquarian still caught within the contending critical and aesthetic standards of late eighteenth-century criticism.

Scott is after something different. The elaborate positioning of the editorial matter is a way to hedge against a forensic antiquarianism that would deny the value of factitious history. Within the parameters of Ritson's strict protocols for accuracy, there is no language in which to value the fabrication of historical truths—that is, re-creations of the past that interpolate elements which, for various reasons, cannot be recovered. Scott's careful parrying of critics who wished to see his sources marks the fine line between creative restoration and counterfeit, a line defined, troublingly enough, by the restoration's ability to go unnoticed, its capacity to hide the "very touches of the file." Within the obscure letters of his personal archive, Scott created space for a middle ground of fidelity to a history reimagined. His editorial practice is thus marked by a confusion of identity and authenticity in service of a larger historicizing intent.

From the perspective of critical history, Ritson is judged to have won the debate with Percy because he successfully argued, as in the preceding epigraph, that such forgeries endanger historical fact in the same way that monetary forgery endangers the economy of credit.[85] Yet, if we credit the larger franchise of imaginative historicism extended by Scott, the reverse is true. Redefined in terms of interpolation, historicism demands the touch of the file. This is because time is unkind to history and its artifacts. In the 1830 ballad imitation essay, Scott describes how "the general progress of the country led to an improvement in the department of poetry [largely due to the influx of classical education], tending both to soften and melodize the language employed, and to ornament the diction beyond that of the rude minstrels, to whom such topics of composition had been originally abandoned."[86] But this "general progress" is actually the problem; the mark of the ancient poetry is its roughness, a roughness that is lost as it is passed through ages. Scott describes this as the wearing of an old coin in his 1803 introduction to the *Minstrelsy*: "Thus, undergoing from age to age a gradual process of alteration and recomposition, our popular poetry and oral minstrelsy has lost, in a great measure, its original appearance; and the strong touches by which it had been formerly characterised, have been gradually smoothed down and destroyed by a process similar to that by which a coin, passing from hand to hand, loses in circulation all the finer marks of the impress" (1:12).

Forensic antiquarianism was obsessed with the processes that "smoothed down and destroyed" the "strong touches" of the historical event, and the flip side of this antiquarian coin was that it provided a powerful way of thinking about how to restore those touches. Studies by Ian Haywood, Nick Groom, and Margaret Russett have already impressed us with the impact of forgeries on the Romantic imagination, particularly as it coordinated the problem of historical distance, authenticity, and the instabilities of the Romantic subject.[87] Russett has demonstrated how forgery and imposture brought focus to the conditions of truthful experience achieved through fiction; as she puts it: "Poetic identity, even and especially in the honorific mode called 'authenticity,' is a *fictional* construction, but this does not make it false."[88] Cast as the problem of the authentic past, a past that must be interpolated because it cannot be encountered within fact, historical fiction proved the *epistemological* value of the forged artifact, that is, how forgeries contribute knowledge through their purchase on the history they re-create. This was an essential contradiction from the perspective of antiquarianism, which viewed fiction as inimical to historical value. It was a problem for the forged artifact, too, insofar as its capacity to acknowledge fiction's historical value capitulated the loss of its own. Despite the rare exception, to disclose a forgery as such was to radically depreciate its value. By passing itself off as a piece of the past, the forged artifact precluded its capacity to mediate history and address its relation to the present.

Ironically, the careful analysis of forgery provided by forensic antiquarianism refined the possibility of creating new value through imitation of the old. As Pinkerton observes, in a passage with startling implications: "The reader must beware of looking upon all forgeries in the more precious metals as modern. On the contrary, many pieces are of ancient forgers of the public money; and are often more esteemed than the genuine coins, because plated, or otherwise executed, in a way that no modern forgers could attain to; and of consequence bearing intrinsic marks of antiquity. The ancients themselves held coins ingeniously counterfeited in such high esteem, that Pliny informs us many true denarii were often given for one false one."[89] The admission that forgeries could outstrip the value of the objects they faked, an admission that documents a real complexity in the market for antiquities, threatens the entire premise of antiquarianism. Pinkerton explains this in terms of the antiquity of the forgeries themselves, but what makes them special is their excellence in manufacturing the patina of the past, "plated, or otherwise executed, in a way that no modern forgers could attain." Forgeries were *usually* conceived as damaging to the market for antiquities, much as antiquarianism was seen as a problem for philosophic history. The most excellent forgeries, however, diagnosed a more general truth, that changes in the antiquities market, like the literary market, are not zero sum: the value forgeries add to the market is greater than the impact of depreciation.

This surplus is possible because historical re-creations—whether Scott's historical fictions or forged coins—partake of the past. We usually think of such facsimiles as mediating or reflecting the past; they substitute for true history, its events and its artifacts, even as they seek to represent it. But the value of forged coins marks the capacity of facsimiles to collapse the distance between imitation and original, to be set alongside true historical artifacts, and so to stand alongside them as objects of comparative study and as contributions to a fuller understanding. This is most clearly seen in Jobert's approval of "enriched" coins: forgeries that complete a collection series, which were produced either by coating an authentic base metal specie in more precious metal or by fabrication from scratch. While such coins still stand in for the absent artifact, they stand in immediate relation to the other coins of the series, collapsing the gap between antique and forgery and considerably augmenting the value of the collection as a whole, both for comparative study and on the antiquities market. In this way, the formal analogy a forgery bears to the object it represents—its ability to represent the shape, texture, and weight of the original, often achieved by distinct technologies of fabrication—is complemented by its harmonic analogy with the collections as a whole. Such forgeries have the capacity to participate in the larger network of economic and social customs they document, helping to collapse the distance between the present and lost events.

The value of Scott's *Minstrelsy* accrued not only in its attempt to re-create an ancient poetic tradition but as an economic agent that brought money to Scott and expanded the market for Scottish history in the British public. At this early stage in his literary career, Scott produces the *Minstrelsy* less from a desire for monetary gain than to accrue critical and social capital. Yet these secondary effects had more important consequences, securing a position in literary publishing that would ultimately become both fabulously lucrative and precarious. By these means, accurate forgery produced an expansionary historicism that enriched history. By the same falsifying token, Scott's historical fictions remodeled historical engagement and opened a new medium for historical experience.

Linguistic Anthropology in *Ivanhoe* and *Waverley*

Whoever does not place himself into the characteristic ways of thinking of such a
different people, whoever cannot judge things according to their characteristic
spirit, their secrets and education, he knows only very little.

Johann Gottfried Herder, "On the History of the Nation" (1769)

The fictive past brought clarity to a central dilemma of any historical representation that does more than collect accounts. As Georg Simmel observed, histories
rely on certain quasi-logical suppositions that are conditions of their possibility—a
class of a priori convictions specific to historical understanding. This is particularly true for the conceit that historical personages can be re-created and
encountered. For Simmel the fact that such re-created figures are, strictly
speaking, factitious, does not mean they are untrue but, rather, that the standard
of historical truth is distinct. In Simmel's view, the historian registers the satisfaction of these conditions as an "overtone or accent" of confidence, "a sense in
which necessity can be ascribed to this psychological construct." This sense of the
truthy reality of the fiction, its ability to fabricate truth, "provides the criterion
which determines whether a mental construct that has a purely subjective origin
is also objectively valid."[90] Though Simmel does not insist on the point, this sense
of necessity holds for the reader as well.

The objective reality of historical fiction is a slippery idea, but it only reformulates an older historical precept, insisted on by the German historian Johann
Gottfried Herder in the preceding epigraph, that the historian "place himself
into the characteristic ways of thinking of such a different people."[91] Herder
recognizes that it is impossible to read oneself completely into the characteristic
customs, forms of speech, and self-understanding of an ancient society. Herder's confidence that the historian can enter into history—as a condition of his
effort—is authorized by a more general confidence, an intuition, that these
efforts can grasp historical truth. C. Frederick Beiser notes that this is less a
methodology than an ambition—a sense of where historical research should
lead, without a clear sense of how to get there.[92] In fact, that is the point: to
"place" oneself in the past requires an alienation from present thinking along
axes that, by their unknown nature, cannot be anticipated. This point was taken up
by nineteenth-century historian Leopold von Ranke, who insisted the historian
must try and show "how things actually happened" (*wie es eigentlich gewesen*).[93]
Though this famous statement is often glossed as a naïve faith in the historian's
craft, it was actually a caution against reading the past as a precursor to the

present. The past is "actually" distinct from the present, and so it retains, for Ranke, an equivalent value and complexity, and a necessary degree of autonomy. Historical events are not simply waypoints to modernity. In order to "show how things actually happened," then, the historian must mediate between past and present, attempting to disclose the alterity of the past in terms we might understand without overwriting it.

For both Herder and Ranke, this understanding was intuitive as much as methodical, much as Scott and his critics struggled, early in his career as a novelist, to explain how his historical fiction worked. His novels succeeded in their own terms, persuading his contemporaries that the past could be interpolated from extant records and artifacts, in fictions that are populated by figures and complex social worlds that lend an intuitive sense of history as a real presence. Certainly, this is how Macaulay read him. Eventually, Scott developed a persuasive description of how historical fiction worked. In the "Dedicatory Epistle" to *Ivanhoe* (1820), Scott offers his most careful exploration of what historical fiction accomplishes and the means it employs. Set in medieval England rather than more contemporary Scotland, *Ivanhoe* was understood in Scott's own time as a marked departure from the earlier "Scotch novels." One consequence, among many, was that the semimodernized English of the narrative and dialogue could not be understood as a historic idiom. It is in this context that Laurence Templeton, Scott's new editorial persona, explains that "it is necessary, for exciting interest of any kind, that the subject assumed should be, as it were, translated into the manners, as well as the language, of the age we live in."[94] The thesis that historical fiction requires translation can be taken, as Templeton seems to intend, as a characteristic of deeper histories like *Ivanhoe* that distinguishes them from more contemporary historical fiction, including the "Scotch novels." But on the contrary, it is clear from the endless scenes of translation, dialectical shift, and code switching in Scott's earlier novels that translation had long served Scott as a key strategy by which to organize historical distance and cross-historical engagement.

Scott's sense of the intimate relation between historical societies and their linguistic forms was part of a much larger Romantic transformation in how translation worked. The broad foundation of this change was vernacularization, the epochal shift from Greek and Latin to vernaculars as the primary languages of the European culture of letters.[95] In classical rhetoric, translation had a transparent and substitutive logic; translation was the exchange of content between two linguistic forms. In *De oratore*, Cicero influentially formulated such translation as a kind of monetary exchange—the substitution of one currency for another equal in value.[96] The possibility that different currencies are more than equivalent forms of value—that they have distinct and incommensurable contents of their own—does not enter. Thomas Greene has characterized such theories of translation as "paradigmatic" or "metaphoric."[97] These metaphoric models become problematic when the vernacular languages are viewed as inferior accessories to Greek and Latin, as they were for much of the early modern period. In such an understanding, any translation suffers by reason of its conversion into the debased currency of an inferior language. The best translation does most to preserve the excellence of the original. In a 1528 translation from Latin, Thomas Wyatt observes that "after I had made a prose of nyne or ten Dialogues the labour began to seme tedios by superfluous often rehersyng of one thyng, which tho [perhaps] *in the latyn shalbe laudable* by plentuous diversite of the spekyng of it . . . yet for lack of suche diversyte *in our tong it shulde want a great dele of the grace*" (emphasis added).[98]

As vernacularization proceeds and the task of the translator turns from mediation between classical and modern texts to translation between vernaculars, the formulation changes. Conceived as a relationship between two different national languages, translation becomes a conduit between two equivalent social forms. This more harmonic relation between languages—the sense that they are parallel and equivalent—is conceived as an "analogy" by both philologists and translators. Hence Henry Edmundson's *Lingua linguarum* advertised the "analogy of language" as the "natural language of languages . . . improvable, and applicable to the gaining of any language."[99] An extended presentation of analogy's role in late eighteenth-century linguistics can be seen in *Of the Origin and Progress of Language*, an authoritative treatise on philology written by James Burnett, Lord Monboddo, and published in Edinburgh from 1774 to 1792. Monboddo roots his system in an analysis of the analogies of language, beginning with the internal patterns of conjugation, especially as they provide a model for standardizing the declension of similar word patterns. Like Edmundson, Monboddo also envisioned a much larger role for analogy, as a way of understanding the patterns underpinning all language.[100]

This sense of a more equitable analogy between languages—a leveling of value provided by vernacularization—changed the terms of translation by recognizing that the peculiar attributes of the original language could find an analogue in the language into which the text was rendered. Conceived as an analogy of effect and style, this theory of translation focused on the creative capacities of the translator. In an anonymous letter to the *Gentleman's Magazine* (1771), a correspondent observes that the casual reader may peruse a "translation with pleasure; but thinks the commonwealth of letters no more indebted to the person who introduced it into the language, than the printer who printed, or the bookseller who sells the book." On the contrary, good translators must be "almost as good original authors as those they translate," because, in order to "preserve[] the fire and spirit" of the original, the translator must depart from "literal translation," correct and update idioms and language, conceal defects; one may even "venture to supply" absent beauty.[101]

The Enlightenment historian Alexander Tytler, an acquaintance of both Scott and Leyden who taught history at Edinburgh and was an acute analyst of contemporary translations, recognized that creative translation created a powerful tension between accuracy and transformation. His 1797 study of the principles of translation examines the influential works of the philosophe Abbé Batteux. Tytler observes that Batteux's protocols "seem to have for their principal object the ascertainment of the analogy that one language bears to another, or the pointing out of those circumstances of construction and arrangement in which languages either agree with, or differ from each other."[102] For Batteux, translation is rooted in a comparative linguistics, one that relies upon an analysis of the analogies between languages to elucidate a commonality of form. Tytler rejects Batteux's conclusion, which argues that Greek and Latin, as the most natural "languages," should serve as the model for composition and sentence construction *even when translating between two vernaculars*. On Tytler's view, Batteux's approach to the analogy of languages is formalizing—the common "circumstances of construction and arrangement" elucidate a universal standard that holds for any content.

Tytler, however, argues that a given work documents an intimate connection between a language and the conventions of its social milieu, a relationship that holds for vernaculars as well as classical languages. As an important contributor to the "science of man" in the Scottish Enlightenment, Tytler sees different social ages as characterized by the intimate organic interconnection of social customs. A good translation must find not only analogous words but analogous formal features in the new language, seeking a middle path between freedom and strict fidelity: "That, in which the merit of the original work is so completely *transfused* into another language, as to be as distinctly apprehended, and as strongly felt, by a native of the country to which that language belongs, as it is by those who speak the language of the original work" (emphasis added).[103] The work in the original language is, to a native speaker, as meaningful and impactful as the translated work to a speaker of the new language. Tytler's contribution is to argue that, by means of this analogy, the particularity of linguistic currency can be a productive resource for translation, rather than its obstacle. The element "transfused" is no longer the content of the original work, but a complex relation between text and context.[104] Mapped out into an analogy that holds between two different social systems, rather than an incommensurate formal exchange, translation no longer offers diminishing returns.

In this way, late eighteenth-century theorists understood the act of translating between languages as movement within what Emily Apter terms a "translation zone": a point of contact between cultural formations that requires one to mediate between different meanings, ideologies, and understandings of the world.[105] And this, in turn, helps explain the close connection in Scott's novels between translation and the continual navigation of actual borders—alternatively conceived as national, international, or intranational. All of his novels explore historical understanding as both a kind of *linguistic* translation (both among the idioms of the past and between past and present dialects) and a form of physical translation (moving character, narrator, and reader across historical borders).

Scott's extensive early work as a translator, as well as his fruitful partnership with Leyden—by all accounts, a very talented philologist—along with his travels across the various cultural borders that constituted Scotland within the British Union, sensitized Scott to the importance of linguistic and social translation within zones of cultural contact. His first literary efforts were translations from German, beginning with his rendition of Gottfried August Bürger's "Leonore." Scott also translated five German dramas in the 1790s, an effort, as Michael Gamer has shown, that culminated in a failed attempt to produce his own Gothic play.[106] Scott's "Essay on the Imitation of the Ancient Ballad" gives an early explanation of this "German craze" that picks up on Tytler's transfusion metaphor: "The prevailing taste in that country might be easily employed as a formidable auxiliary to renewing the spirit of our own, upon the same system as when medical persons attempt, by the transfusion of blood, to pass into the veins of an aged and exhausted patient, the vivacity of the circulation and liveliness of sensation which distinguish a young subject."[107] Scott seems to be arguing for a transfusion of "spirit" as distinct from form, text exclusive of context. But there is a decisive difference in Scott's sense that transfusion is grounded in the historical relationship between languages in their youth and age. The same essay alludes to the historical filiation between German and dialects of English, particularly Scots: "The present author, averse to the necessary toil of grammar and its rules, was in the practice of fighting his way to the knowledge of the German by his acquaintance with the Scottish and Anglo-Saxon dialects." Though he "frequently committed blunders which were not lost on his more accurate and more studious companions," Scott observes that his method was more successful than a fellow French student, "who, with the economical purpose of learning two languages at once, was endeavouring to acquire German, of which he knew nothing, by means of English, concerning which he was nearly as ignorant."[108]

At the time, Scott was impressed by the "curious coincidence" of the intimate connection between English dialects and German. Now writing from the vantage of 1830, Scott sees that this "coincidence" is linked to Sir William Jones's contemporary analysis of the Indo-European language family, which helped explain the close kinship of what would later be termed the "Germanic" languages. Though Jones's remarks on the connection between Sanskrit and other languages were not published in Britain until 1807 in his collected *Works*, Leyden was already a follower of his studies. Scott was aware that by 1803 Leyden had extended Jones's insight into the linguistic connection between Celtic and Sanskrit as the basis of a relation between "Druidic" and "Braminical" mythology.[109] By 1808, Jones's ideas were more generally known (particularly through his widely reviewed *Works*); an analysis of Charles Wilkins's "Grammar of the Sanskrita Language" in the *Edinburgh Review* spends considerable time weighing the comparative similarities between Sanskrit, classical, and more modern languages for their historical import.[110] In the light of Jones's discoveries, comparative philology offered a way to analyze the historical relationships between societies, both in their connection and in the social transformations that distinguish them, and so recast linguistic study as social history.

Jones's insight into the historical connection between languages was filled out by new theories of translation that emphasized the interconnection between language and social custom. Crawford has provided one of the most extended analyses of how Scott's attention to the anthropology of dialect influenced later generations of sociologists and ethnographers.[111] Scott's extraordinary sensitivity to the capacity of idiom and dialect to register historical difference is only now coming fully to light. David Hewitt, series editor of the recently completed Edinburgh Edition of the Waverley Novels (a painstaking twenty-year collaboration that corrects numerous printing errors in the original editions), observes that Scott's transformative attention to dialect created challenges for his printers: "A surprising amount of what was once thought loose or unidiomatic has turned out to be textual corruption."[112] A large number of these substantive errors are dialectical terms that Scott's editors and typesetters misrecognized. These errors were exacerbated by the procedures that maintained Scott's anonymity, which obscured his handwriting by having a third person laboriously recopy the manuscripts and revisions by hand before submission for printing.[113] Idiomatic and dialectical terms suffered more from misreading, because their alienation from contemporary English increased the likelihood that amanuenses and typesetters would accidentally substitute a more familiar word.

This alienation also constituted idiomatic value. An early example comes in *Waverley*, as Rose Bradwardine tries to explain the chieftain Fergus Mac-Ivor's position within Highland society. Told that Mac-Ivor is the person "who, they all knew, could easily procure the restoration of [stolen] cattle, if he were properly propitiated," Waverley struggles to understand Mac-Ivor's role. Rose laughs when he asks, "Was [he] the chief thief-taker of the district[?] . . . Is he a magistrate, or in the commission of the peace?" (75). Raised under the settlement of post-Restoration England, Waverley thinks in terms of the civil administration of justice. In order to clarify, she explains that Mac-Ivor will intervene if they pay "black-mail," a Scotch legal term that was not then part of standard English. Waverley asks for a definition, and Flora explains: "A sort of protection-money that Low-Country gentlemen and heritors, lying near the Highlands, pay to some Highland chief, that he may neither do them harm himself, nor suffer it to be done to them by others; and then if your cattle are stolen, you have only to send him word, and he will recover them" (75–6). For Waverley, the incomprehensible thing is that Mac-Ivor can extort money in this fashion and remain a gentleman of good standing in Scottish society; Scott later added an extensive note to the "Magnum Opus" edition to clarify that "the levying of black-mail was, before 1745, practised by several chiefs of very high rank."[114] Far from a zone of indistinction, Waverley is surprised to find, along the Scottish border, a conflict between two competing systems of civil custom and penal regulation.[115] Flora's definition and Scott's note emphasize the importance of the term as a way to specify the difference between the civil constellations of contemporary Scotland and England and, more importantly, between the past and the present. As such, the scene effectively demonstrates the power of language to serve as an insight into historically differentiated social customs. The strangeness of such terms, their capacity to alienate the novel's community of language, gives them value in Scott's historical fiction.[116] The fact that both blackmail and "protection-money," thanks to *Waverley*, are now common in English indicates the clarifying power of this strategy for the contemporary reader; if they no longer read as loanwords today, it's because they so effectively lent a new understanding of what was then a foreign social custom.

Scott understood how this process of domestication could be put to account as a way to alienate the vernacular, estranging standard English to demonstrate

its historical sedimentation. A prime example is the semicomic disquisition between Wamba the fool and Gurth the swineherd at the opening of *Ivanhoe*, as
they discuss the relation between Norman and Saxon culinary terms:

> "Why, how call you those grunting brutes running about on their four legs?"
> demanded Wamba.
>
> "Swine, fool, swine," said the herd, "every fool knows that."
>
> "And swine is good Saxon," said the Jester; "but how call you the sow when she
> is flayed, and drawn, and quartered, and hung up by the heels, like a traitor?"
>
> "Pork," answered the swine-herd.
>
> "I am very glad every fool knows that too," said Wamba, "and pork, I think, is
> good Norman-French; and so when the brute lives, and is in the charge of a Saxon
> slave, she goes by her Saxon name; but becomes a Norman, and is called pork,
> when she is carried to the Castle-hall to feast among the nobles. What do'st thou
> think of this, friend Gurth, ha?"
>
> "It is but too true doctrine, friend Wamba, however it got into thy fool's
> pate." (22)

The scene, which Wamba extends to the analogous examples of ox and beef, calf
and veal, became a set piece for introductions to the history of English, because
it so effectively demonstrates the naturalization of French vocabulary after the
Norman invasion. Though the historical precision of Scott's analysis was later
questioned, it marks an extraordinarily sophisticated synthesis of contemporary
strains of social, linguistic, and material history. The important point in *Ivanhoe* is that standard English circa 1820 contains within it a buried legacy of
social violence, conquest, and resistance. Wamba, with the privileged eye of foolish insight, sees language as a chronicle of contested material practices. Wamba's observations here are not only documentary; the historical understanding
he gleans from linguistic analysis establishes a motif of resistance and potential
revolution that is settled only by the return of Richard I and his alliances with
the "true-born native[s] of England" (169)—yeoman Locksley (a.k.a. Robin Hood),
Cedric the Saxon, and Ivanhoe himself. Such moments of translation, between
"good Saxon" and "good Norman-French," mark the Romantic analogy between
linguistic and cultural translation that organizes historical understanding in
the Waverley novels, a productive alliance between past and present that exposes language as an unsettling record of a complex and sometimes brutal
history.

For Scott, such strategies of translation are multifaceted, existing between
customs and language at any level of remove, from the major national languages

to local variations in dialect. Idiom, understood as an individual's peculiar dialect and habits of speech, plays a particularly important role in Scott's fiction as a marker of historical as well as social difference. In the "linguistic realism" of novels from the previous century, idiom serves as a comic marker of economic and regional distinction. Tobias Smollett's *Roderick Random* (1771) deploys idiom as a way to domesticate picaresque and travel narrative in analysis of the microsocial complexities of London.[117] In Scott's fiction, idiom remains a carrier of such regional and social distinction, but these distinctions are reorganized as a register of historical differentiation.[118] This is particularly evident in the speech of a character like Elspeth Mucklebackit in *The Antiquary*. Once a lady in waiting to the landed aristocracy, now a fisherman's grandmother, Elspeth's language alternates constantly between the idiom and social forms of her past and the contemporary patois of the fishing village where she now lives. Here's how she describes the crisis of her life—the moment Lady Glenallen gave her a needle and commanded her to murder young Neville, the heir of the estate: "She turned away in her fury, and left me with the bodkin in my hand. Here it is; that and the ring of Miss Neville are a' I hae preserved of my ill-gotten gear—for muckle was the gear I got. And weel hae I keepit the secret, but no for the gowd or gear either" (266). The passage, in its graduation from narrative to direct speech, mapped into a shift from standard English into Scots, recapitulates Elspeth's personal history, but also a larger social history of the reduction of the landed aristocracy in the novel. Idiomatic transition functions as a kind of cultural translation, one that fills out the social meaning of the gold bodkin and ring Elspeth displays. Here, code-switching between Scots and standard English provides this larger context, an idiomatic archaeology for her antique horde.

At the same time, the sequence of translation, from standard English into idiom, reverses the order we might expect for a mode of history that is generally read to normalize the past as a story of national formation. And this in turn shows that Scott's fiction does not overwrite the past; his novels produce historical knowledge through the continuous contact between different systems of language and thought. In his study of the interaction between different scientific communities, Peter Galison has argued that, in scientific research, continuing interaction between different technical idioms produces "trading zones" that mediate between distinct areas of investigation and expertise, sites of exchange in which "partners can hammer out a *local* coordination despite vast *global* differences," using "semispecific pidgins" and "full-fledged creoles" that allow productive interaction without forcing either perspective to conform to the other.[119] This holds equally for the way Scott explores dialect, conceived as a regionally specific variant of a language family. Among endless examples, one more will have to serve—Evan Dhu's last statement in *Waverley*. At his trial, he stands accused of treason alongside his chieftain Vich Ian Vohr, a.k.a. Fergus Mac-Ivor. He argues before the court that, on his word, a host of Highlanders would report for execution in place of their Highland chief: "If your excellent honour, and the honourable court, would let Vich Ian Vohr go free just this once . . . and if you'll just let me gae down to Glennaquoich, I'll fetch them up to ye mysell, to head or hang, and you may begin wi' me the very first man" (342). It is a clarifying moment, because the incredulous laughter of the crowded court gallery marks for the reader how far our understanding and sympathies have moved over the course of the novel. For the court auditors, Evan Dhu's intermittent turns to idiom—now figured as the influence of Scottish Gaelic, a Highland dialect—are a comic marker of an incomprehensible naïveté; their laughter refuses to trade in the understanding Dhu proffers. Yet that same idiom now registers, for the reader, as a belated language authentically bound to a social system governed by feudal honor rather than penal law. Such scenes clarify the author's explanation, in the novel's "Postscript which should have been a preface," that he has attempted "to describe these persons, not by a caricatured and exaggerated use of the national dialect, but by their habits, manners, and feelings" (364). Evan Dhu's speech avoids feeling "caricatured and exaggerated" not because it is less Scottish—that is, more idiomatic—but because his statements carefully translate between standard English and Scots, and they use this translation to bear a fine-grained analysis of social difference.[120] For a creole history, a pidgin historicism.

Evan Dhu responds to the court's laughter with scorn, marking the novel's sustained interest in cultural difference across translation: "They ken neither the heart of a Hielandman nor the honour of a gentleman" (466). Historian Dipesh Chakrabarty has emphasized the "incommensurability" of such scenes of translation: "neither an absence of relationship between dominated and dominating forms of knowledge nor equivalents that successfully mediate between differences, but precisely the partially opaque relationship we call 'difference.'" In Chakrabarty's view, this allows historians to "write narratives and analyses that produce this translucence—and not transparency—in the relation between non-Western histories and European thought," but this also characterizes the peculiar translucence of cross-cultural contact within Europe (and, for Scott's novels, within Britain itself).[121] Dhu's mixed language gives voice to the thesis that what is at stake is not the translation between heart and honor, or Hielandman and gentleman, but a more complex analogy between different social systems and the vocabulary that sustains them. His phrasing seems to give national English—"the honour of a gentleman"—its final say, but the lasting impact of the analogy is to deliver the reader's sympathy and historical understanding to the "Hielandman." Taken at face value, this also affords Dhu a purchase on the "honour of a gentleman" that is superior to that of his auditors, in their role as extensions of the British state. By these means, the relation that subsists between "honour" and "gentleman" is expropriated from the community of standard English, alienated as an extension of an endangered Highland system of values.

At the close of the novel, the reader feels with the dialect. Such scenes mark how dialectical translation in Scott's fiction provides a powerful form of historical engagement, a zone of exchange between present and past that is characterized by immediate contact and investment, and generalized as a condition of modern experience. In order for this experience of alienation to work in the novel, as a kind of semantic reiving, the reader must go along. And this is possible because there is no such thing as a dead letter. Language is not a cold artifact; reuse gives it new life. Evan Dhu, as a fictional proxy for historical personages, a "mental construct" in Simmel's terms, finds the capacity within the language of the novel to shape history. In this instance, Dhu both stands *for* and stands *in for* the preclearance Highlander. Like the forged artifact, he both represents history and operates within it. As a consequence, a personage who never lived gains a power and impact he never had. This has a truthy reality for the reader, both because the reader perceives the world Dhu inhabits as historically continuous with modernity, and because the possibility of his reality and his action are a priori suppositions for the genre. The translational analogy models this configuration. In the translation of historical language, we bring the past back to life, trading with its understanding and transfusing it with our own. The reader is interpolated into the mixed nature of history, as a link between what is known and what is imaginatively made possible, a participant in historical fiction and so an agent in history.

Conclusion: "So Leyden were alive"

The Waverley novels shaped historical translation as a model for the experience of alterity in general. As Scott's contemporary Friedrich Schleiermacher put it, "Are we not often required to translate another's speech for ourselves, even if he is our equal in all respects, but possesses a different form of mind or feeling?"[122] Scott's fictions laid a broad foundation for the intimate relationship between linguistic and social translation for later historians, placing the historian and the reader in living contact with the past.[123] The unsettling implication was that, by means of the historical imagination, the past could always irrupt into life. Michelet, in his most influential description, organizes the possibility of this revenant historicism as the fusion between reading about and speaking for the dead. In *The People* (1846), he describes this as "resurrection": an "inquest *on the living*" that discloses the "holy poetry" of the "humblest and meanest of the people."[124] Michelet insists on a discomfiting intimacy between resurrectionist and what we now term revisionist history, and I highlight his example in order to draw out the counterintuitive implications of Scott's historical fiction. As a central object for literary histories of historicism, Scott's fiction has long stood as the prime example of what Franco Moretti identifies as the totalizing intent of the historical novel: "To represent internal evenness, no doubt; and then to *abolish* it. Historical novels are not just stories 'of' the border, but of its erasure, and of the incorporation of the internal periphery into the larger unit of the state."[125] The problem is that, within the logic of historical translation, what is represented can never be abolished; what is revived in the historical imagination cannot die.

I see this as one reason why the critical arguments for Scott's ideological closure (accounts that understand his novels to invoke history's radical countermovements only to promote the triumph of a more conservative nationalism) acknowledge an uneasy tension with the unresolved, counterhistorical, even radical impulses of the same novels. Ann Rigney reads Scott's fiction as an "apology for [national] progress" that nevertheless produced a resistant *"energia* through its own ambivalence and vividness," helping sustain a surprising afterlife in counternational fictions.[126] Ina Ferris similarly traces the opposed influences of Scott's fiction on the historians who followed, contrasting Macaulay's integration of Scott's imaginative mode into the teleology of Whig history (forging a "more efficient . . . disciplinary power") with Augustin Thierry's reading of Scott, from whom he derived "a sense of time as a differentiated series rather than a single line."[127] The challenge is to weigh the balance in Scott's fiction between social diversification and national consolidation, often reduced to the discrepancy between where his histories go over the course of their plot and where they end up. If readers of Scott tend to emphasize the qualified resolutions of Scott's novels, this is because they have found it easier to read them as further examples of the close relation between ideology and form. But there is no reason to assume that the dimensions of form and ideology are uniquely aligned or that they take precedence over the orthogonal concerns of affective engagement

and historical legibility (to invoke Mark Salber Phillips's thinking about the geometry of historical perspective).[128] Scott's contemporaries registered his major
contribution in terms of personal investment and a clearer sense of the past.
The political question, *To what ultimate end?* masks the more basic question of
what more immediate ends were made possible.

In place of a concern for formal closure, I am interested in the smaller formal openings that characterize moments of engagement and translation, the
various structures of historical contact from which Scott's narratives are composed. As I have already indicated, these moments devolve in various ways from
analogies of translation and interpolation, consistently rendered through multipart comparisons between paired elements and their larger contexts. These analogies are conceived as having both forensic value, as encounters with historical
truth, and imaginative extension, as fictions forged of that history. As Fredric
Jameson notes, triangulating Lukács on the historical novel, "The very structure of our reading of the historical novel involves comparison, involves a kind
of judgment of being."[129] The historical novel holds us in that zone of contact
between past and present, a running comparison between the historical self
and contemporary experience, but it does so by reflecting on comparisons within
history, a judgment of other *beings in time.*

Within this readerly comparison I find a more extended and robust comparatism, in which the tense coordination between fiction and historical content plays out along two different axes. First, there is the problem of historical
distance, read as the nonhomologous contest between the forms of the present
(genres, vocabularies, prejudices) and past analogues. This dimension emphasizes a formalizing distance from the past, insofar as some translation is necessary to make it intelligible (as *Ivanhoe*'s Templeton observes). But this sense of
distance is at an angle to the more harmonic comparisons of interpersonal encounter. When Evan Dhu translates honor for the court, he establishes a contact
between social systems, between his life and his audience, that is carried forward
by the novel into the present.[130]

Comparative historicism adds a crucially deformalizing impulse to engagement with the past, producing an understanding of the past as a network of relationships that, even when defined by vast differences in power, are evenhanded with respect to their status as historical relations. This network (in
which the reader is implicated) is defined by an inability to cohere in a unified
form, whether that form is shaped as a national consolidation or a satisfying
marriage plot.[131] Though the relationships cannot harmonize, they are reconsti-
tuted perpetually through their contact and reverberation in the novels. The

larger impact on historicism was to demonstrate how history itself, operating under the soft teleology of what did happen, depends on the often violent exclusion of alternative histories. To read Scott's novels in this fashion is to see them as counterfactual histories, not merely because they give fiction mixed with fact but, more importantly, because they read the necessity of history as a fiction in itself.

Scott's novels are often dissatisfying, particularly in their conclusion, and their counterfactual ambition explains why this must be so. Scott's readers absorbed a sense that things could have been different in the novels, and they returned to flesh out these counterfactuals. As recent studies by Murray Pittock, Ann Rigney, and others have shown, Scott's novels are an endless resource for revisionist writers in Britain and elsewhere. Pittock observes that "in societies struggling for independence against regional powers or colonial oppressors . . . the radical undertow in Scott's writing could seem more prominent than it did to a British audience."[132] This subversive undertow was an important source of investment for many British readers, too. Take the relation between Rebecca and Rowena in *Ivanhoe*. The novel effectively stages their confrontation at two levels: first, as Rebecca and Rowena compete as heroines within Ivanhoe's romance plot, and so, potentially equal objects of investment, and more generally, as representatives of Jewish and Celtic society. The reversal of fortune by which Rowena claims Ivanhoe as her lover was a constant object of revision for later readers and writers. So Maggie Tulliver comments, in George Eliot's *The Mill on the Floss*, "I'm determined to read no more books where the blond-haired women carry away all the happiness. . . . If you could give me some story, now, where the dark woman triumphs, it would restore the balance. I want to avenge Rebecca."[133] To take this as a criticism of Scott's fiction (as is common) misses Eliot's point in foregrounding how Maggie's ultimate destruction provides a further example of the conflict between given history and its preferable alternatives. Such imaginative revisions keep Scott's novel and its potentialities alive, developing possibilities suggested by his reading of history. More recently, a group of scholars at University of Virginia's Scholars Lab have revived the "Ivanhoe" game, which allows readers to rework narrative fictions in a series of interlocking "moves." In its first iteration, the Ivanhoe game consisted of an exchange of emails between Joanna Drucker and Jerome McGann, one of which discovers Rebecca's power to shape the novel's events.[134] The game emerged from the recognition that "*Ivanhoe* contained within itself many alternative narrative possibilities," and I see this as the particular organizing concern for historicism after Scott.[135]

Scott's novels resuscitate rather than lay to rest the possibility that history could be and perhaps *should* be different. Ian Duncan insists that Scott's narratives enjoin the reader to live with, as well as within, the past. The contemporaneity of the past produces instabilities in the "condition of modernity" and casts the contemporary world as a "global network of uneven, heterogeneous times and spaces . . . the dynamism of which is generated by the jagged economic and social differences of the local parts."[136] I see this as the legacy of the mode of comparative historicism, subtle and ramifying, which Scott's novels catalyzed. His novels implicate modernity in a tangled world of narratives irreducible to a single end, and present comparative historicism as the only analytic suited to that complexity. It is competent to address patterns and differences that extend over time and within the network of relations that constitute historical moments. It surrenders coherence in the service of interconnection and increased precision.

At the close of *Waverley* hangs the famous portrait of Edward and Fergus Mac-Ivor in Highland dress, carefully mounted with Waverley's musket and claymore as part of the miraculous restoration of the Tully Veolan estate and its possessions, which were previously reduced and scattered in suppression of the Scottish Rebellion or "Rising" of 1745. It is common to read this portrait and the restoration of the estate as a figure for Scott's relationship to history. The success of *Waverley* and the novels that followed famously allowed Scott to expand his estate, Abbotsford, into an eccentric neo-Baronial collage filled with relics of Scottish history, from Edinburgh's "Heart of Midlothian" gate to Rob Roy's musket. A modern visitor to the estate sees how Scott's novels serve as an imaginative provenance for his collection of such artifacts, which still hang in bewildering profusion from Abbotsford's walls.

When Scott finished *Waverley* in 1814, Abbotsford was still a small farmhouse, but it was already cluttered with his collection of books and baubles. Among them hung a large Malaysian dagger, sent by Leyden several years before, a gift that marked his emigration to India, as a functionary of the East India Company. It is also a token that anticipated what both assumed would be Leyden's successful return, laden with translations, philologies, and literary extracts—the spoils of orientalist research that would make his fortune (figure 6). The dagger had already served as the subject of Leyden's "Ode to a Malay Cris," which described a narrow escape from a French privateer on his perilous voyage out, and it would eventually find its way into Scott's *Surgeon's Daughter* (1827), as part of the equipage of the "Queen of Sheba," daughter of a deceased Scottish emigrant. When Leyden himself died unexpectedly in 1811 (he

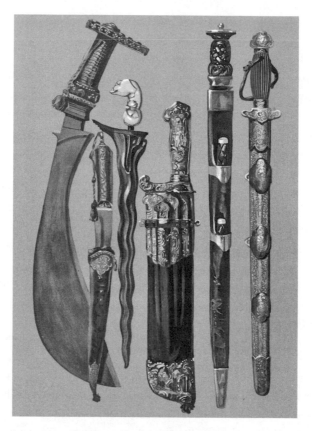

Figure 6. "Malay Kris" *(third from left)*. Watercolored engraving by William Gibb. From *Abbotsford: from the Personal Relics and Antiquarian Treasures of Sir Walter Scott* (1893), 128858, pl. 13. Courtesy of The Huntington Library, San Marino, California

fell ill after entering an ancient library during the invasion of Java, an actual victim of archive fever), the event transformed the dagger into Scott's last material connection to his friend. In the following years Scott was closely involved with unsuccessful efforts to collect and publish Leyden's works and wrote the extended "Biographical Memoir" that fronts the second volume of the *Edinburgh Annual Register for 1811*, an annual account of events significant to the population and environs of Scotland. In that account, Leyden's quixotic search for imaginative literature is the center point of his mixed character, "formed between the lettered scholar and the wild rude borderer," and his death is read as an extension of his enthusiastic passion for ancient literature and romance, which began early in life. As Scott describes it, this zeal extended through his extensive tour of the Highlands, which (like Scott's own later tour) "diverg[ed] from the common and more commodious route, visited what are called the *rough bounds* of the Highlands, and investigated the decaying traditions and Celtic manners and story which are yet preserved in the wild districts."[137]

Rather than insist on the extensive connections between Scott's description of Leyden and the portrait he would draw when he resumed the manuscript of *Waverley*, I want to underline the connection between the recuperative mode of the biographical memoir, which pulls together fragments scattered across the globe, and the imaginative historicism of the novels that followed. As a biographer, Scott is careful to maintain a critical balance between description and judgment. He is careful to acknowledge both Leyden's perceived faults and Leyden's status as an "intimate in the family of Mr Walter Scott, where a congenial taste for ballad, romance, and border antiquities, as well as a sincere admiration of Leyden's high talents, extensive knowledge, and excellent heart, secured him a welcome reception."[138] The description is curiously constrained by the distance between biographer and subject; if Leyden can be "intimate," Scott, his biographer in the *Register*, cannot. Scott's feelings are far more frank in an allusion to Leyden within the first volume of the *Register*, in a description of the military invasion of Java (where Leyden died), concluding: "Unhappily our conquest cost us the life of one who, had his days been prolonged, would probably have added more to our knowledge of eastern literature and antiquities than all his predecessors . . . whose early death may be considered a loss so great—so irreparable,— . . . that I will not refrain from expressing a wish that Java had remained in the hands of the enemy, so Leyden were alive."[139] Leyden's death, like Vich Ian Vohr's and the violence of history in general, is indeed "irreparable." The sweep of imperial history cannot justify the personal loss. Scott wants 1811 other than it was. And within the imaginative space of historical fiction, such victims of history can be given new life and reencountered. The *Register* cannot imagine that Leyden still lives, but the historical novel can, because comparative historicism, with its investment in encounter, places us in contact with the past.

In the following chapter, I explore how Alfred Tennyson's *In Memoriam* profits from this comparative and collaborative engagement with the dead. Here, I see this restorative historicism at work in the peculiar way Waverley disappears into his portrait at the close of that novel; his depiction alongside Fergus MacIvor in the "large and spirited painting" being, strictly speaking, his last appearance (361). Baron Bradwardine, Emily Bradwardine cum Waverley, Mr. Morton, Mr. Rubrick, even Colonel Talbot feature at the dinner that follows, but not

Waverley. It is as if Waverley has been flattened into the history depicted in his portrait. This effective absence allows him to live on with Vich Ian Vohr, reoccupying a past we have just quit. For Waverley to live, the reader must keep the novel alive, both in memory and in later readings. More generally, Waverley seems to persist at the close of the novel as the narrative ligature that holds together, by means of his history, the artifacts that situate the painting: from the "arms" that hang alongside, to the restored estate of Tully Veolan itself, to the baron's loved armorial cup (returned as a favor to Waverley). The final scene does not really stage Waverly's history so much as his role in aligning the various histories that constitute the novel, beginning with Fergus Mac-Ivor's own. In the complicated relation between the partisans of the portrait, in their commonality and their difference, in their various distances from the restorative collectors who assemble and explain the artifacts that reconstitute history, I find Scott's own complicated relationship to the novel: part history and part fiction, part artifact and part forgery, part affectionate tribute to a friend and partner and part self-interested pecuniary gambit. Here *Waverley* broaches the basic formal problem of elegy (a thread I will pick up in the next chapter): how to profit from the dead. The portrait "draws tears," but its viewers must move on, even if they will return. Scott apologizes: "Men must however eat, in spite of both sentiment and vertu."

CHAPTER THREE

Spooky Action in Alfred Tennyson's
In Memoriam A. H. H.

> . . . I see in part
> That all, as in some piece of art,
> Is toil cöoperant to an end.
>
> Tennyson, *In Memoriam* (1850)

The premise of elegy is that, by incorporating the dead into the life of the living, poetry will revivify our experience of life and its loss.[1] For *In Memoriam* to do its work, art must do more than reflect life. In this chapter's epigraph, a conjunction of labor—the "toil" already done by the dead and yet to be accomplished by the living—offers a larger analogy to social history. Tennyson's qualified belief that human means have larger ends, immediately following his references to the "red fool-fury of the Seine" and "vast eddies in the [social] flood," marks both his concern for contemporary social upheaval (especially the revolutionary year of 1848) and his hard-won optimism that history tends toward the good.[2] So, approaching the close of the series of 133 verse fragments that constitute *In Memoriam A. H. H.*, Tennyson resuscitates a qualified faith, abandoned in the first section of the poem, "That men may rise on stepping-stones / Of their dead selves to higher things" (1).

This chapter turns to Tennyson and *In Memoriam* in order to reexamine the relation, discussed in my first chapter, between poetic form and comparative historicism. Earlier, I argued that Erasmus Darwin's epic verse fails to supply the narrative flexibility needed for a comparative approach to history—an approach that juxtaposes alternative accounts of history and compares the disparate trajectories of individuals and societies. Tennyson's elegy, with its oscillations between disparate memories and locations, and its reflexive movements between the present and the past, offers a wider ground for analogy and analysis. As we will see, the verse form of *In Memoriam*, with its well-known abba rhyme scheme, weaves a comparative impulse into the texture of this movement, as stanzas continually fall back, review, and revise their own understanding. Moreover, Tennyson's marked interest in science stands in sharp contrast to Scott; while my previous chapter explores Scott's engagement with history in the context of the comparative sciences, Scott and his novels are notably quiet on the status of those sciences, and their purchase on human history. *In Memoriam*, by contrast, gives extended attention to the significant implications of contemporary geology, anatomy, and astronomy for how we understand our place within history. For these reasons, *In Memoriam* belongs at the center of this study, as it focuses our attention on analogy as a strategy of historical interpretation important to both the sciences and the humanities.

In Memoriam serves two very different ends. On the one hand, as an elegy for Arthur Henry Hallam, Tennyson's dear friend and almost family member, the poem refuses, for most of its duration, to let Hallam go. Through the poet's recursive visit to the sites of their old haunts, through his troubled handling of both his own memories and the relics of Arthur's life, and, above all, through analogical verses that insist on the possibility of contact across social, historical, and metaphysical divides, the poem accrues as a web of interconnected experiences, a mosaic of encounters, a pattern of diffraction, rather than as a stately narrative that moves its subject through coherent stages of grief and acceptance. On the other hand, the poem struggles to assemble a wide canvas of political, scientific, and religious perspectives into a coherent vision of "toil coöperant *to an end*." It struggles to reach a vision of social history that is both progressive and benedictory, sanctifying historical crisis in a longer view of ultimate progress. Ostensibly, this tension is resolved in the epilogue of the poem, which tries to weave Hallam back into the family through later marriages and namesakes, even as it asserts that social evolution allows us to read Hallam into history as the "type" of "the crowning race" to come. By these means the poem discharges the burden of elegy, exchanging the serial encounter of difference for a formal understanding of collective purpose, translating rupture into coherence.

The sense of coordination between social progress, a benevolent nature, and natural development seems to have been an enormous comfort to Tennyson's contemporaries, divided, as they generally were, between conflicting political, theological, and scientific views. Though it took Tennyson nearly twenty years to work, rework, arrange, and polish the poem, its publication in 1850 was timely. Albert, both prince consort and a student of contemporary social and scientific debates, was so impressed with the poem that he threw his weight behind Tennyson's selection as the new poet laureate, succeeding William Wordsworth. And shortly after Albert's death a decade later, the grieving Queen Victoria told Tennyson, "Next to the Bible 'In Memoriam' is my comfort."[3]

The poem's extraordinary capacity to reach across Victorian intellectual and social divides cemented Tennyson's mediatory role and made him celebrated by

the leading scientists and theologians of his day. When he was buried in Westminster Abbey between the graves of poets John Dryden and Robert Browning,
the attending president and officers of the Royal Society testified that, as the
scientific journal *Nature* observed, "above all Others who have ever lived, [Tenny-
son] combined the love and knowledge of Nature with the unceasing study of
the causes of things and of Nature's laws."[4] In a gesture that would be almost
inconceivable today, contemporary scientists accorded Tennyson and his poetry
unmeasured respect for confronting the unsettling implications of contemporary
scientific discoveries. The physicist John Tyndall considered his scientific
curiosity "profound" and recorded Tennyson's assiduous attention to both the
discoveries of contemporary naturalists and the scientific accuracy of his own
descriptions.[5] *In Memoriam* calibrates this address to "nature red in tooth and
claw" with a searching religious inquiry in which faith accommodates nature,
rather than triumphing over it. In this way, the poem is as an agent of *secularization*, understood in the more capacious sense that Charles Taylor gives that
word: conceived not as the collapse of Christian belief but, rather, as the growing recognition that Christianity, like all religions, is part of a larger community
of sustainable convictions, variously religious as well as materialist, scientific,
and atheist. The poem works by flattening metaphysical and moral hierarchies,
placing discordant perspectives on a common plane where their respective merits and implications can be explored.

If scientists embraced Tennyson's "toil coöperant to an end," that resonant
phrase, with its emphasis on collaborative labor, gives another way of reading
the formal charge of his elegy. Over its 133 verse lyrics, each ranging from a few
to dozens of stanzas in length, *In Memoriam* addresses not merely the death of
an extremely dear friend, Arthur Henry Hallam, but also the end of a kindred
poet with whom Tennyson had closely cooperated.[6] The devastating news of
Hallam's death forestalled both a future of collaborative verse and a planned
marriage between Hallam and Tennyson's sister Emily that would have con-
joined their families. In taking up Hallam's memorial verses, Tennyson engaged a tradition governed by the towering examples of John Milton's "Lycidas"
and Percy Shelley's "Adonais," both also a "poet's elegy" that mourned the loss
of a fellow crafter of verse. This chapter makes the case that, within *In Memoriam*, Tennyson did not let Hallam die but rather worked with Hallam to give
new life to the tradition of the "poet's elegy." Extending Isobel Armstrong's case
for Hallam's influence on Tennyson's early poetry, I argue that Tennyson used
In Memoriam to collaborate with his friend long after he died, engaging Hallam's views on religion, natural science, and philosophy, and especially his

verse. In this way, Arthur Henry Hallam (the A. H. H. of its title) became an author of the elegy that bears his name.[7]

As an elegy, a poem that attempts to incorporate the dead into the life of the living, *In Memoriam* engages a radical possibility inherent in what I have termed *harmonic analogy*, understood as a heuristic that allows elements of the world to interact and intervene in the production of meaning. As I explained in the prelude, I see this as one implication of Bruno Latour's "actor-network theory," which interprets the objects of scientific study, alongside the scientists themselves, as actors in the chains of signification that produce scientific research.[8] *In Memoriam*, I argue, is constructed as a kind of sounding box, an instrument that allows its dead poetic subject to intervene in the act of poetic composition. By means of this "spooky action," *In Memoriam* examines how elegies might allow the dead, as the poem puts it, to "reach a hand through time" and interact with the living: in this sense, the dead can be "coöperant," operating within the elegy itself.

In part, this possibility is a feature of elegy as a genre marked by intense historical awareness, a point we can underline through contrast to the (conventionally) ahistorical character of lyric verse. Studies by Sharon Cameron and Monique Morgan demonstrate that, even as the lyric operated in dialogue with longer narrative forms in the nineteenth century, it served to organize an experience that transcends time.[9] On this view, lyric experience lifts the poet and reader outside history. As we will see, in examining Tennyson's own lyric experiments within *In Memoriam*, this experience of transcendence gains further power when set within and in contrast to longer verse forms that directly confront history. Nineteenth-century British poetry abounds in experiments with longer modes of narrative verse—novel poems like Elizabeth Barrett Browning's *Aurora Leigh* (1856), the dramatic monologues favored by Robert Browning, and even, as Herbert Tucker reminds us, failed experiments with the epic—that sought to historicize poetic experience and place it in relation to past and present forms of life.[10] This chapter reads *In Memoriam* as an attempt to explore elegy (understood as a form invested in history as it is lived, reflected upon, and memorialized for the future) as a deeply interpersonal genre of history writing. As a "poet's elegy," *In Memoriam* seeks to understand Hallam's place in Tennyson's life, and within human history, but also to secure Hallam's place within the literary canon.

Feeling the weight of Hallam's hand within *In Memoriam*, I take up a concern implicit in discussions of Tennyson's poetry at least since Hallam Tennyson (named for Arthur) published the first influential biography of his father.

Almost from its first publication, Tennyson's readers have acknowledged the importance of *In Memoriam* in constructing a place for Arthur Hallam in posterity.[11] Yet, for many critics, Hallam's heavy biographical and (to a lesser degree) intellectual influence on Tennyson has been overemphasized by readers. Herbert Tucker's continually rewarding *Tennyson and the Doom of Romanticism* is an extended attempt to free Tennyson from Hallam's anxious influence. In Tucker's view, Tennyson freed himself by estranging the material of Hallam's life and death, objectifying Hallam as figure for Tennyson's "ambivalent fascination with the mutual antagonism, or dependence, between creative agency and contextual determination."[12] To read *In Memoriam* this way is to underline a distinction between Hallam as object and the poet as subject that the poem refuses to accept. The sympathy it expects (as Hallam himself once argued) makes "the object the same as the subject."[13] From the rapturous moment when "his living soul was flash'd on mine" to the continual blurring of tense and clausal dependence that joins Hallam and the poet in a coordinate syntax, *In Memoriam* seems to insist on the possibility, though not the certainty, of Hallam's continued action in verse.[14]

To see those hands within *In Memoriam* is to recognize that the "tumult of acclaim" (LXXV) that the poem eventually received was directed at Hallam as well as Tennyson. The first section of this chapter examines the enfolded rhyme scheme of the *In Memoriam* stanza as a system of metrical analogy (a formal pattern that structures the warp and woof of correspondences within and between lines). In the section that follows, I explore Hallam's contribution to *In Memoriam*, providing a model for both poetic collaboration and the collective possibilities of the sonnet sequence. In particular, I argue that Hallam furnished a way of using enfolded rhymes to ingrain reflection, recursion, and contact within verse. The third movement of this chapter explores the logic of analogy by which Hallam's hand in shaping the poem is understood as a corollary to the work his hands perform "out of human view," in the afterlife. In moving back and forth between modes of harmonic contact and formal reorganization, the poem tests the relation between keeping the past alive and memorializing it in the present. In order to understand this dynamic, I confront its "analogic" head-on, framing it in terms of contemporary debates between William Whewell and John Stuart Mill over the logical status of induction. By these means, I raise the question of *In Memoriam*'s epistemology. In my view, the poem engages a larger mid-nineteenth-century argument about what comparative inquiry adds up to when recognized as an uncertain empiricism. The nineteenth century percolated with discussions of knowledge that was not strictly logical, particularly (as this book shows) the insights furnished by analogy. What is not often noted is how these debates spilled over into arguments on the plurality of worlds—the possibility of life on other planets—with which Whewell was concerned and Tennyson fascinated.

Ultimately, I argue, *In Memoriam* must be understood as part of a much larger midcentury debate over the status of analogy and the uncertainty of the knowledge it furnishes. As we will see, Richard Owen gave Tennyson his most powerful model for imagining the fruit of comparative historicism as a unified theory of "homologies" within nature, governing archetypes that could reconcile the plurality of forms within a higher category. Tennyson was vibrantly in tune with these discussions of analogy—both of its limits and of its surprising power—and the way they amplified the problem of the afterlife. Analogy furnished a way of figuring the dead in both their continued existence and possible dialogue with the living. But poems must end, and Tennyson fashioned *In Memoriam* to draw these possibilities to a close, organizing them around a single unitary vision. Whereas Scott read history as a network of unresolved possibilities, Tennyson recast historical complexity as a product of the modern subject's uncertain struggle toward coherent faith—a faith in both progress and God. In this way, we can read Tennyson as engaging comparative historicism, and the uneven history it disclosed, as a formal test of the poet's capacity to fashion the perfect line.

Analogical Verses

Rhyme has been said to contain in itself a constant appeal to memory and hope. This is true of all verse, of all harmonized sound; but it is certainly made more palpable by the recurrence of termination.

Arthur Henry Hallam, "The Influence of Italian Works of Imagination" (1834)

The possibility of a sustained analogy between the living and the dead is opened by the enclosed rhyme scheme of the *In Memoriam* stanza itself, which serves as a kind of epistemological engine, designed to gather information from both previous experience and new inspiration. I will get to Hallam's influence on *In Memoriam* later, but now I want to call attention to the peculiarly comparative texture of its stanzas. In order to draw out these implications, it helps to consider the alternatives—other, less comparative possibilities for *In Memoriam*'s verse form. Let us first examine the poem's alternative four-line stanzaic schemes. Delivered as blank verse, the four-line stanzas would more closely approximate natural language, but in setting aside the firm structure of end rhyme, blank verse acknowledges the imprecision of language and recuses rhyme's capacity to disclose unsensed resonances between words and the world they disclose. As *In Memoriam* puts this, "words, like Nature, half reveal / And half conceal the Soul within" (V). Though offered in a moment of doubt, this insight asserts a productive analogy between poetic form and natural inquiry in their mutual relation to discovery and loss. Though the *In Memoriam* stanza is generally characterized in terms of progress and regress, efflux and reflux, it more often moves through cycles of assertion and review, statement and critique. In one of many crises of doubt, the poet's despair provides a key example:

> Vague words! but ah, how hard to frame
> >In matter-moulded forms of speech,
> >>Or ev'n for intellect to reach
> Thro' memory that which I became . . .   (XCV)

Here, the interior rhyme (b/b) immediately takes up and revises the assertion of the first two lines—the problem of capturing what he "became" in poetic form—and recognizes within it a more basic struggle to remember an experience that has no adequate language. (Ultimately, as we will see, the poem also finds opportunity here: the analogy between poetic composition and memory asserts their basic interconnection and suggests how, in the struggle to compose, the poet might find the capacity to breathe new life into old memories.) Hence, what Erik Gray terms the "willing regression" of the enfolded rhyme, the "tendency to end where it began" marks a commitment to persist until things are truly worked through.[15]

Christopher Ricks has argued that the "ebb and flow" of the *In Memoriam* stanza leaves it "utterly unsuitable for sustained argument," but it also demonstrates how important argument, to be sustained, depends on variation, the turns and detours that support its evolution.[16] This gets at the epistemological charge of *In Memoriam*'s enveloped rhymes, which fold the second half of the stanza back as perspective on the first, in a commitment to find new life in what has just been laid to rest. Even as the enclosed rhymes lend coherence to the individual movements of the poem, they commit the poem to a mode of ongoing critical extension that moves the poet and reader beyond first (and second, and third) take. Its stanzas return in form, yet never to the same place. This affords the hermeneutic quality that Sarah Gates identifies in the stanza, described in terms of an evolving "spiral": a "vacillation (a to bb, and back to a), of gesturing backward (a<---), and of leading beyond (bb--->a)."[17] If the stanzas were instead organized by the alternate rhymes of the ballad stanza or hymn (abab or abcb), the verses would be more colloquial but also more directive, advancing a sense of movement, driving along the narrative of the story. Gates's searching analysis articulates the importance of both extension *and* return; the relation between the rhymes moves the stanza in both directions, between the distant "a"s (which look forward and back) and proximate "b"s that "clinch" at the center. As Hallam puts it, in this section's epigraph, rhyme is both a "memory and a hope," both recalling what came before and advancing beyond.[18]

Most analysts of the *In Memoriam* stanza overlook the important fact that its primary formal gathering is delivered at the level of syntax instead of rhyme, in the strong caesura or stop (marked by a comma in the stanza just quoted) that usually divides the first pair of lines from the second. I find it helpful to think of the *In Memoriam* stanza as two long "sixteener" lines, lines of octameter that each enwrap the end rhyme of the other. If the four lines of its stanzas were organized as couplets (aabb), the rhyme would emphasize that divided syntax, asserting confidence in unilateral advance. This is a key reason why, as Ricks has observed, the *In Memoriam* stanza is "so unlike the disputatious sequences of the heroic couplet."[19] Interlinked by the enclosed rhyme, the stanza asserts a parallel movement between its two syntactical groups, a mirroring that infers a deeper semantic relationship, an echo of sound that suggests a harmonics of sense.

As a form that unsettles preconception (as Tucker has put it, "unfixed 'to form'"), the *In Memoriam* stanza stands out in one of the poem's most important turning points, an unsummoned optimism that appears for the first time in section LXV.[20] It is a passage that, as the poem's first great commentator, A. C. Bradley, noted, is "perhaps the first section of *In Memoriam* that can be described as cheerful or happy":[21]

> And in that solace can I sing,
>> Till out of painful phrases wrought
>> There flutters up a happy thought,
> Self-balanced on a lightsome wing:
>
> Since we deserved the name of friends,
>> And thine effect so lives in me,
>> A part of mine may live in thee
> And move thee on to noble ends.

In section LXV, communion produces a tentative analogy between Tennyson and his dead friend, namely, that Hallam's continued influence in his friend (and, ultimately, their poem) indicates the possibility that Tennyson likewise influences Hallam and his ongoing work. This possibility rests on some significant uncertainties: it is no simple matter to claim that Hallam continues in a contemporaneous afterlife, that he continues to work "noble ends" there, that Tennyson's influence is still with him. Shortly we will see that this continued influence is based, in part, on the continuing inspiration that *In Memoriam* takes from Hallam's verses. In such moments of contact, the poem naturalizes the afterlife, drawing it into dialogue as a mirror of the material world, compressing (without collapsing) the metaphysical strata that constitute the Christian understanding of the distinct spheres of existence within God's kingdom. The formal analogy presumed to persist between these strata takes an increasingly harmonic character; they seem to function as parallel worlds.

In both stanzas, the syntactic and semantic divisions between the first and second paired lines is sharp, part of the syntactic revolution, from independent to dependent clause, and then back. As the same time, the strong rhymes insist on a strong semantic mirroring, between what the poet can "sing" and what then takes "wing," between what is "wrought" and then, unbidden, "thought," between intimate "friends" and analogous "ends," most importantly, between "me" and "thee." Such passages crystallize as moments of extraordinary "self-balance" because they snap the systematic doubling of the formal elements—four lines of four beats, paired lines with paired rhymes—into focus, as a filiated network of syntactic, semantic, and sonic analogies.

In Memoriam inscribes the analogy by which, as Susan Wolfson puts it, "choices of form and the way it is managed often signify as much as, and as part of, words themselves."[22] In the first chapter, I explored how analogies drawn between words, meanings, and sounds sustain Erasmus Darwin's theory of natural knowledge and the uncanny sufficiency of verse to capture it. In lines that anticipate *In Memoriam*, Darwin set this analogous theory of poetics to the language of astronomic motion, as "Orbs wheel in orbs, round centres centres roll / And form, self-balanced, one revolving whole." In that analysis, I noted a syntactic pairing that allows Darwin's lines to enfold their meaning and gravitate toward the accentual stress of a four-beat line. Tennyson's poem, with its mania for close repetition, is filled with twinning moments like this, but more important is how *In Memoriam* fully realizes the reflective possibility of the four-beat line.

For both Darwin and Tennyson, this collusion between meaning and pattern is an example of the "self-balance" of both their verses and the patterns of the larger world. For Darwin, this spontaneous form is a vital impulse toward order and pattern, observed in the self-organizing processes of physical, biological, and mental development; order is an emergent impulse that patterns the natural world. Yet, for Tennyson, the source of "self-balance" is not so clear. In the second and third lines drawn from LXV ("Till out of painful phrases wrought, / There flutters up a happy thought"), this balance "flutters up" as if from below the "painful phrases," which establishes the conditions, but not the motive of its action. It erupts in the phonetic acceleration that draws the passage's third line away from its second. The "thought" must come from below, buried as it is below the line above, but its impulse evidently rises. In this way, the *In Memoriam* stanza catches at uncertain inspiration.

The poem leaves the source of that "happy thought" undetermined. Usually such moments in the poem, which stage momentary contact with a higher wisdom, or actual converse with Tennyson's dead friend, are cast into doubt. It is only a few sections later that Tennyson will open up a more explicit "commerce with the dead," and that assumption is immediately qualified by uncertainty, "Or so methinks the dead would say; / Or so shall grief with symbols play" (LXXXV). This later crisis of doubt prepares the way for the rapture of communion that immediately follows. In section LXV, however, the problem of doubt does not enter, in part because the poet does not go so far as to assert that the unbidden "thought" is from the dead, and in part because the section takes form in a way that so fully catches and completes the "self-balance" of the *In Memoriam* line. Doubt is engrained within the line itself, in the act of writing, an uncertainty that lies in the question "What will come next?" as much as "What do I know?"[23] Such moments of inspiration seem to provide a sufficient answer.

Hallam's Perfect Danäe

The gallery is grand and I longed for you. . . . oh Alfred such Titians! by Heaven, that man could paint! I wish you could see his Danaë. Do you just write as perfect a Danaë! . . . Titian's imagination and style are more analogous to your own than those of Rubens or of any other school.

Arthur Henry Hallam to Tennyson, September 6, 1833[24]

It took some time for Hallam's letter, sent from Italy on September 6, to reach Tennyson. On October 6, he received a second letter communicating Hallam's death.[25] By the end of the week he had written the first lyric that would find its way, substantially altered, into *In Memoriam* as section IX: "Fair ship, that from the Italian shore."[26] In the months that followed, Tennyson began to write ad-

ditional lyrics, including a group of poems that described Hallam's slow journey home, a series dealing with the experience of the first and subsequent
Christmases after Hallam's death (Hallam had spent Christmas 1832 with the
Tennysons), and a lyric on Lazarus that would eventually expand to a group of
meditations on death and resurrection (X).

Tennyson's Trinity and Huntington notebooks document early attempts to
group these verses into a larger form. The former has a long initial draft of section LXXXV, which begins with perhaps the most famous lines of the poem,
"This truth came borne with bier and pall, . . . 'Tis better to have loved and
lost, / Than never to have loved at all." It is a comforting thought, and in 1835
Tennyson shared an early version of it with Hallam's father, who recorded it in a
private notebook, as did Queen Victoria after Frederick's death in 1861. The initiating lines of section LXXXV also mark one of many moments where verse
is "borne" to Tennyson rather than forged by him. The poet imagines he holds
"commerce with the dead," though he worries the resulting conversation on the
nature of the afterlife, on sympathy between the dead and the living, is only
"fancy-fed." Conversation proves inadequate, and the poet "look[s] to some
settled end, / That these things pass, [where he] shall prove / A meeting somewhere, love with love." He "apologize[s]," but he must have substance as well as
words. "I woo your love!" the poet cries in a demand that was repeated both in
the middle and at the close of the lyric in its earliest versions.[27]

In the section that immediately follows, one of the most beautiful lyrics in the
poem, the poet receives that love:

Sweet after showers, ambrosial air,
   That rollest from the gorgeous gloom
   Of evening over brake and bloom
And meadow, slowly breathing bare

The round of space, and rapt below
   Thro' all the dewy-tassell'd wood,
   And shadowing down the horned flood
In ripples, fan my brows and blow

The fever from my cheek, and sigh
   The full new life that feeds thy breath
   Throughout my frame . . .   (LXXXVI)

It is a difficult lyric to excerpt, because its four stanzas constitute one sonorous
flowing sentence. Surprisingly, it does not appear that anyone has noted that it

depicts Tennyson's possession, as descending from the "gorgeous gloom," the "showers . . . rapt [those] below," including the poet. The clausal structure of the passage, with its serial predicates (rapt, fan, blow, sigh), makes it clear that "rapt," rather than an adjective, is implicitly transitive, though the transitive objects—the earth and its contents—are absent, displaced by "brows," "cheek," and "frame" of the poet.

The poet could not refuse Hallam's demand, effectively a dying wish, that he paint in words as "perfect" a Danaë. In the Viennese example of Titian's famous depiction of Jupiter showering Danaë with gold (which so inspired Hallam to "long" for Tennyson), the god, dimly seen in a murky cloud, appears to blow the coins from his hand. Cornelia Pearsall has given the most extended analysis of Tennyson's mode of rapture, reminding us that the experience of rapture comprises both the sense of being physically seized but also the metaphysical sense of being caught up, lifted into some larger self.[28] Here, even as the poet is rapt by that "full new life" within his frame, he lets his "fancy fly" (LXXXVI). In Tennyson's Danaë, the breeze mints a new "commerce with the dead," just as Hallam breathes a "full new life" found after death but capable of entering into the life (and frame) of the living.

Later in this chapter I take up the trajectory of that fancy, soaring "To where in yonder orient star / A hundred spirits whisper Peace," as the site of an intense investment in the productive capacities of alienation. Here it is more important to note how the allusion to Danaë enhances the erotic saturation of the lyric. This saturation draws substantially from Tennyson's study of Hallam's verse and from a complex amatory metaphysics, figured in Hallam's sonnets, that connects sensual experience with the divine. Section LXXXVI is a major example of the intricate texture of desire, friendship, homosociality, and heterosexual union that has teased readers and critics since its publication. An anonymous review attributed to Manley Hopkins, father of the poet, complained of this "defect" of "amatory tenderness."[29] For Hopkins, this purported deficiency derives from Tennyson's unwitting participation in a poetic tradition that is sexually confused. Hopkins excerpts a passage from LXXIV and identifies its progenitor: "Shakespeare may be considered the founder of this style in English. . . . His mysterious sonnets present the startling peculiarity of transferring every epithet of womanly endearment to a masculine friend,—his master-mistress, as he calls him by a compound epithet, harsh as it is agreeable." Hopkins follows this with an extract of one of Shakespeare's sonnets and notes that they, too, have their precursor: "Is it Petrarch whispering to Laura?

We really think that floating remembrances of Shakespeare's sonnets have beguiled Mr. Tennyson." Hopkins's recoil at this sexual and historical triangulation, his desire to drown this floating "tenderness," did not override his admiration of the poem. A few years later, on the death of his son Felix, he penned a short elegy that used the *In Memoriam* stanza without the enclosed rhyme (although he also pointedly insisted that "no faithless fear / Troubled our heart about his rest").[30]

Construed as the place where Hallam and Tennyson meet, *In Memoriam* becomes a vehicle for Hallam's ability to "reach a hand through time and catch / The far off interest of tears," but also a way to revive his poetic legacy as a presence in the poem's long afterlife. Hopkins, in tying Tennyson to Shakespeare—and, through him, Petrarch—is unwittingly astute in summoning this potential. It is unlikely that Hopkins knew Hallam's "Oration on the Influence of Italian Works of Imagination on the Same Class of Compositions in England," a work that was published in the posthumous *Remains of Arthur Henry Hallam, in Prose and Verse* (1834), and which argues that Petrarch, as well as Dante, had impacted Shakespeare directly, revitalizing "amatory" language as a vocabulary of Christian revelation. "I cannot help considering the sonnets of Shakespeare as a sort of homage to the Genius of Christian Europe," Hallam explains.[31] In his view, the poets served as a conduit to the mediatory love of Christ described in Hallam's "Theodicaea Novissima," or "Hints for an effectual construction of the higher philosophy on the basis of revelation," an essay that Hallam's father also included in the first edition of the *Remains*, at Tennyson's particular urging.[32] The intimate connection between Dante, Petrarch, Shakespeare, and Christian belief in Hallam's theology helps frame their close relation in *In Memoriam*—which Tennyson sometimes cryptically described as "a sort of divine comedy."[33]

Hopkin's review, in raising the nature of Tennyson's relation to Hallam, and the sexual economy of the poem, opened a source of continuing interest to scholars and biographers. Gerhard Joseph has provided one of the most extensive analyses of Tennyson's "feminization" within the erotic staging of the poem, while Christopher Craft reads the sublimation of Hallam as an attempt to "relieve the speaker's desperate erotic distress, a distress that is indistinguishable from grief."[34] I think the energetic desire of the poem works because it loads the tension between physical and interpersonal longing, so magnifying our sense of the latter's amplitude. "The gallery is grand and I longed for you," Hallam wrote, in a mirrored desire for shared experience and the ability to reflect on that experience together. It was a relationship constituted through "longing" for reunion—a longing to belong together. The figure of rapture is strictly insufficient to describe this experience, because its implicitly transitive nature renders the poetic subject passive, captured by some larger power. For this reason, even as rapture provides an experience of being lifted up into some larger self, it is intensely personal rather than shared; it provides no indication that the larger spirit reciprocates, in a communion that preserves what it brings together. Seized by rapture, the subject is imperiled. Section LXXXVI bolsters the ground of that imperiled subject through the generous extension of the "rapt" object across brake and bloom, wood and flood, while also loosening the bonds between the subject "frame" and the freed "fancy" of the spirit. Some part of the subject must remain grounded, retaining contact with the world, while liberating its complement to seek a reciprocal communion.[35]

For Hallam, this "amatory" tradition was inseparable from poetry's capacity as a vehicle for Christian intuition.[36] The formative motivation for *In Memoriam*'s extensive engagement with the sonnet was not the legacy of the sonnet but that of Hallam himself, who was a close student of the sonnet tradition, particularly in the examples of Shakespeare and Petrarch. The Petrarchan sonnet is generally understood to consist of fourteen lines of iambic pentameter divided into an eight-line octave and a concluding six-line sestet, while Shakespeare preferred to divide the fourteen lines between three four-line quatrains and a concluding couplet. Yet, in Hallam's view, these traditional distinctions were not as important as we suppose. In his essay on the influence of the Italian poets, he insists that "the structure of [Shakespeare's] sonnets is perfectly Tuscan [i.e., Petrarchan], except in the particular of the rhymes." As Hallam uses it, "structure" intends more than the meter, comprising also the larger metaphysical movement of the sonnet from the earthly problem of the initial lines to the transcendent reversal of the volta. The ebb and flow of the sonnet, its movement into dilemma and then above it, constituted its central current in either form.

The sonnet (often in sequence) was Hallam's chief poetic outlet in his Cambridge years and the form with which he was most closely identified. In 1835 W. H. Brookfield wrote to Tennyson to beg him to "make a sonnet for me as Hallam once did. I could not value it more, and should not less, than his."[37] Hallam favored the Petrarchan form and was an accomplished imitator of his verse in Italian; before publishing his son's Tuscan sonnets in the *Remains*, Henry Hallam consulted Antonio Panizzi, a naturalized Italian scholar and British Museum librarian, who offered enthusiastic admiration.[38] From his work with Petrarchan sonnets, Hallam came to heavily favor envelope rhymes, both in initial quatrains and, in various enclosed configurations, in the sestet, shaping the abba rhyme scheme that would later constitute *In Memoriam*.

Later in life Tennyson expressed a preference for the Shakespearean sonnet, but experimented with these Petrarchan strategies while with Hallam. Tennyson's first serial publication, the sonnet "Check every outflash" (which Hallam submitted to the *Gentleman's Magazine* without his friend's permission) adopts the envelope rhyme scheme though it retains the concluding couplet of Shakespeare.[39] And the conclusion of the Petrarchan sonnet Hallam and Tennyson wrote together while at Cambridge, "To Poesy," resolves on a complexly enclosed sestet (efgegf).[40] Tennyson's "The Kraken," which dates from the same period, is also strongly Petrarchan and experiments extensively with intervolved rhyme in the concluding septet.

Hallam's sonnets are most innovative when they explore enclosed rhyme as a way to interpolate the larger vaulting movement of the poem within the closer structures of its stanzas. There are many examples in the "Somersby Sonnets," as he titled them, a sequence that captures Hallam's courtship and successful proposal to Emily Tennyson during his 1831 stay with the family. Hallam's own "sort of divine comedy," it consists of eight sonnets, numbered sequentially in Roman numerals in his notebook, which carry the reader from the first line of the sequence, in which Emily "put[s Hallam's] question by," to the concluding phrase, which looks forward to Emily's state as "an English maiden and an English wife." Henry Hallam drew three of these poems, apparently from Arthur's private notebook, for publication in the *Remains*. These are V, VI, and VIII ("Why throbbest amain, my heart," "Still here—thou hast not faded from my sight," and "Lady, I bid thee to a sunny dome").[41]

"Still here" is the most lyrical, and its initial lines display a control and clarity that are unusual for Arthur Hallam:

Still here—thou hast not faded from my sight,
  Nor all the music round thee from mine ear:
  Still grace flows from thee to the brightening year,
And all the birds laugh out in wealthier light.   (78)

The envelope rhyme allows the poem to fold Emily into the poet's memory even as it encapsulates, at the opening of the sonnet, the larger movement by which such self-enclosure fortifies the poet for his concluding foresight of a time "When I may shape the dark, but vainly bid / True light restore that form, those looks, that smile." The immediate concern is their impending separation, as Hallam returns to Cambridge for the summer term. A Shakespearean reflection on mortality gathers in this closing vision, opening a new confidence that the poet's memory, embalmed in these lines, will transcend, despite the inevitability of death.

The opening quatrain of "Still here" also acknowledges the immediate kinship between Hallam's verses and the *In Memoriam* stanza. Tennyson described his approach to poetry in terms of precise economy—the attempt to launch poems in "a small vessel on fine lines" instead of a "great raft."[42] It is tempting to speculate that Tennyson, an obsessive editor, would have recognized an opportunity to tone Hallam's slightly overballast lines. I take that liberty here:

> Thou hast not faded from my sight,
>> Nor all thy music from mine ear:
>> Still grace flows from thee to the year
> And all the birds laugh out in light.

Though Hallam worried that it would be "vain" to "bid / True light restore that form," Tennyson is willing to try the experiment. XCI demands that Hallam "Come, wear the form by which I know / Thy spirit in time among thy peers," while LXXXIX describes that "form," in terms reminiscent of Hallam's own, as "beauteous in thine after form, / And like a finer light in light."[43] Such passages indicate how the *In Memoriam* stanza evolved Hallam's sonnet form into verses more lithe, less formal, but equally concerned with the intervolved nature of experience and memory, life and death. *In Memoriam*'s verse is a "form" that draws from the known Hallam (the examples of his earthly verse) to elucidate the possibility of his continued influence (his "finer light in light").

In Memoriam was not Tennyson's first experiment with enclosed rhyming tetrameter; his long political poem, "Hail, Briton!," which was probably composed between 1831 and 1833, uses the *In Memoriam* stanza to advance a doctrinaire Whig thesis of gradual political improvement. It resembles several other political poems that deploy the *In Memoriam* stanza, all dating to the period of Hallam's death, including "You ask me, why," "Love thou thy land," and "Of old sat Freedom on the heights." An additional poem, "The Statesman," which Hallam Tennyson later published in the *Memoir*, adapts several of its stanzas into a portrait that seems, recognizably, to imagine Arthur Hallam's potential political career. Tennyson developed this idea more explicitly in section LXIV of *In Memoriam*, which asks if Hallam "look[s] back on what hath been, / As some divinely gifted man" occupied with affairs of state, long separated from "his earliest mate; / Who ploughs with pain his native lea."

Criticism of *In Memoriam* is divided on the question of whether the tetrameter of these verses and *In Memoriam* itself, particularly in its departure from the pentameter line, seek a disarming mode of more frank political address or a depersonalized voice that alludes to the ballad tradition. In contrast to the staid political formulations of "Hail, Briton!," the personalizing turn of section LXIV explores the wider potential of the form, as it modulates from the oratory of the statesman who can "mould a mighty state's decrees" to the disarming need to know "Does my old friend remember me?"[44] The key difference lies in the use of the enclosed stanza. "Hail Briton!" strings them together as ringing theses and rhetorical questions that rarely divide within themselves. Section LXIV, by contrast, uses the enclosed rhyme to turn the social question back on itself. The last three stanzas break cleanly into a format of call and complicating response: the statesman "feels . . . A distant dearness in the hill" of his childhood; the "vocal springs" of the local waters remember his earliest games of "counsellors and kings"; the poet who still works there, and "reaps the labour of his hands," looks back, "musing" on his "old friend." At stake is a need for connection more powerful than the dialogue, established through pastoral, eclogue, and idyll, between urbanity and the rural scenes of intimate memory.

Hallam's most important contribution to *In Memoriam* was this feel for envelope rhyme as a vehicle of interconnection, a strategy that enfolds the largest religious, ethical, and political questions within the closer formal embrace of the reversing stanza. In essence, it allows the poet to work the evolving thought of the verse sections into the revolution of its constituent parts. It also affords the stanzas a coherent movement that makes the larger boundaries of the section less important. Often, *In Memoriam*'s epigrammatic stanzas encapsulate sentiments and images that are later revived, revised, and extended, as in the "fall'n leaves" that interleave LXXIV and XCV, or the famous statement of love and loss that closes XXVII only to return as the opening of LXXXV. I noted earlier that the first groupings of *In Memoriam* verses were organized thematically. David Goslee notes that they tend to operate "on the edge of Tennyson's central grief. . . . they never focus directly on Hallam, on his loss, or on the personal and philosophical consequences of it."[45] As the poem evolved into a longer series, Tennyson distributed these clusters throughout, leaving these closures ajar and opening up space for more direct engagement. As Joseph Sendry first observed, the thematic gatherings, notable in the Trinity and Huntington notebooks, were an important early staging for resonances that Tennyson would later disseminate throughout the larger poem, giving the series the flexibility to "absorb a greater quantity and diversity of material."[46]

Anyone who has read *In Memoriam* more than once has noted its aleatory qualities; the way it encourages the reader to skip forward and back, to explore how these patterns ripple across its verses. The coherence of *In Memoriam* emerges in these passing alignments between sections dispersed throughout the poem, elements of changing constellations. The evolution of the *In Memoriam* notebooks, the way thematic clusters are gathered in the Huntington notebook, only to be dispersed in later versions of the poem, document Tennyson's work as he feels his way into this effect. I see this dispersed quality as part of the seriality of *In Memoriam*, taking "serial" here in the more general sense of the distribution of a set. Though there are larger waypoints in the poem's movement, most notably, the three Christmas sections that mark its movement, the poem actively resists a pattern of consistent development in favor of a more dispersed network of thematic reflection and interaction. The serial extension of *In Memoriam*, in turn, flattens the metaphysical hierarchy that a stronger formal coherence would assert. In chapter 1, I observed that the epic, as a verse form predicated on a confidence in a larger metaphysical order, was endangered by an age of increasing political, scientific, and religious upheaval. *In Memoriam* explicitly addresses that new world, both in its insistent doubt about larger providential order and in its refusal to adopt a strong form. Charles Kingsley saw this clearly in his reading of *In Memoriam*: "Without faith there can be no real art, for art is the outward expression of firm coherent belief. And a poetry of doubt, even a skeptical poetry, in its true sense, can never possess clear and sound form."[47] Along these lines, and yet with admiration, James Kincaid observes it is "more proper . . . to speak of [*In Memoriam*]'s organization than its structure. . . . The poem explores how man can live in a world that has denied easily perceived structure, one where human experience can no longer be explained by analogy with linear mathematics."[48] The unevenness of this larger and looser "organization" stands in uneasy tension with the precise metrics of its stanza structure.

The tension between formal structure and flexibility underlines the complicated relation between faith and doubt within the poem, as Tennyson attempts to structure a verse form sensitive enough to register the inspiration of his friend and, in this way, to cooperate with him. But it also marks the poem's sensitivity to contemporary debates over the sources of knowledge and the security of knowledge derived from the natural world, particularly the relation between observation in general and scientific induction. Kingsley (like Tennyson) felt that knowledge of nature provided an important perspective on human experience, but he was confident (in contrast to Tennyson) that it bolstered Christian faith. In Kingsley's view, the solidity of Baconian induction provides a storehouse of "fresh analogy" for Christian belief, deploying the fortifying stability of science to shore up faith.[49] Tennyson, well versed in the uncertainties of scientific knowledge, saw nature as equally unsure. Within *In Memoriam* this is most evident in the extinction lyrics, LIV–LVI, which weigh and then discard the assertion that nature, "red in tooth and claw," is so "careless of the single life" and yet "so careful of the type" (LV–LVI). Tennyson was both a passionate student of scientific discoveries and an astute judge of the uncertainty woven into its results; "he asserted," as Hallam Tennyson put it, "that 'Nothing worthy proving can be proven,' and that even as to the great laws which are the basis of Science, 'We have but faith, we cannot know.'"[50] This generalizes the problem of doubt within *In Memoriam* as a basic condition of knowledge in general. The important question is how to derive knowledge absent surety.

The Logic of Analogy and the Plurality of Worlds

The publication of *In Memoriam* in 1850 coincided with the high point of a mid-Victorian debate over the status and nature of scientific inference. Tennyson's poem, deeply concerned with what can be known about other possible realms of existence, takes up its central questions. Though this debate had a much longer history (and, indeed, continues today), the two main antagonists in Tennyson's day were William Whewell and John Stuart Mill. They disagreed over the nature of language and the mind; whether words were merely useful labels for factual observations and generalizations (Mill) or vehicles for new conceptions of the world (Whewell).[51] Whewell's *Philosophy of the Inductive Sciences* (1840) used his earlier study of scientific history to derive a "logic of induction" that explained how specific scientific theories are subsumed, over time, into more general scientific laws. Mill, in his *System of Logic* (1843), attacked Whewell's account by arguing that most scientific examples of what Whewell considered "induction" were really just more accurate descriptions, reserving the term *induction* for cases where concrete, observed facts are generalized to apply to all similar cases. Mill's attack upon Whewell is extensive and runs over several sections and both volumes of his treatise; Whewell responded at length in the second edition of his *Philosophy* (1847), attempting (though, by most accounts, failing) to refute Mill point by point. The debate necessarily piqued Tennyson's interest; not only was Whewell the tutor of Tennyson and Hallam while at Cambridge, but astronomy, one of Tennyson's passionate interests, was the central example in their tug of war.[52]

This debate over induction can sharpen our sense for how analogy shapes knowledge within *In Memoriam*, and for this reason, it's worth taking some time to examine the dispute between Whewell and Mill. There are many ways one might choose to analyze their argument: as a disagreement over the utility of logic in science; a battle over the practical and theoretical tenets of positivism; or even a precursor to the conflict between logical positivism and more socio-logical theories of scientific practice. The important point, really, is that Whewell uses the language of logic casually, because he is more interested in providing a practical understanding of both scientific progress and its operation as a pro-cess. As Laura J. Snyder has argued, Whewell had a strategic interest in the logi-cal features of induction. He hoped to salvage induction from a classical syllo-gistic account (like that given by Richard Whatley in his *Elements of Logic* [1826]) and show that, in practice, science developed through the formulation of new "ideas" from direct empirical study. By these means, he hoped to show that sci-ence did not operate by reasoning from first principles. For this reason, while Mill is concerned with a precise logical dissection of the distinct functions (comparison, generalization, induction) that constitute ratiocination, Whewell is more interested in how models, like the financial ledger, give form to scientific discovery. In Whewell's analysis, as over against Mill's, the mediation between the formal system and the event matters because it is here that the scientist calls forth the "*new conception* [that characterizes induction], a principle of connexion and unity, supplied by the mind, and superinduced upon the particulars."[53]

The "mind" here is evidently that of the individual scientist, but in the larger view—supplied by the scientific history from which Whewell derives his account—one can see that this "new conception" has an important social com-ponent, forged of the interaction between contemporary and past networks of scientists and the world they engage. In this way, the Whewell-Mill argument looks forward to the later debate between Thomas Kuhn and Karl Popper over the contribution of social practices and logical content to scientific theory. But my interest here is to use Whewell's account of induction to cast fresh light on Tennyson's effort to draw Hallam out on the page. Tennyson, like Whewell, un-derstood form as an *epistemological* as well as *organizational* tool; the analogical verse form and the distributed gatherings of *In Memoriam*'s sections help ex-plore Hallam's legacy and make space for fresh discovery. At the same time, there is no straightforward way to perform, for verse, the kind of "summation of the results" that Whewell ascribes to the inductive method. Verses cannot be induced into a larger generalization, on either Whewell's or Mill's account, because their complexity, as a network of semantic, syntactic, and sonic correspondences, resists generalization.

The analogical form of the *In Memoriam* stanza, the way it interrogates and revises its propositional content, makes this point inevitable, as it continually threatens to undermine, even as it amends and extends, what has just been found. To return to the closing stanza of LXV, the poet adopts the form, if not the content, of one of Mill's syllogisms when he supposes that:

> Since we deserved the name of friends,
>> And thine effect so lives in me,
>> A part of mine may live in thee
> And move thee on to noble ends

Compare this to a classic example from Mill's *Logic* (preserving Mill's format):

> All men are mortal,
> Socrates is a man,
>> therefore
> Socrates is mortal.[54]

Mill's analysis of the syllogism helps here to draw out the difference between these distinct logical and what I would call *analogical* "styles of reasoning" (as Ian Hacking has put it).[55] In both, the first line serves as the major premise—it specifies that the class (alternatively "we" or "men") has a specific attribute ("friends" or "mortality"). Their distinctions emerge in the following lines. The second line, in Mill, constitutes the minor premise, which asserts that a new individual or group (here, "Socrates") belongs to the class described. The second line from *In Memoriam* fails Mill's test, because membership in the class is only implied (that "thine effect," because it still "lives in me," is a part of the friendly class "we"). The problem with all syllogisms is that their conclusions are uninteresting (in part, because they are inherent in the premises). Mill's is no exception. Only slightly more intriguing would be a more strictly syllogistic conclusion to Tennyson's verse, an assertion that Hallam's "effect" also deserves the name "friend." If we follow Mill's analysis, Tennyson commits any of a number of logical fallacies here (e.g., equating unequal premises, changing the premise, using ambiguous terms). But the closing lines of the stanza give an analogy, not a conclusion, and the tension between the quasi-logical format of the premises and the tentative extension of their import is what gives these lines the strength to cast knowledge beyond what can be justified by logical means.

Comparison between Tennyson's verse and Mill's syllogism indicates a formal congruity between the *In Memoriam* stanza and logical deduction, insofar as the first line must end with a statement that it then takes up, in modified form, at the close of the last.[56] Mill, in fact, acknowledged the capacity of analogy to move beyond logical categories. In the *Logic*, he analyzes it as a kind of incomplete induction, the extension of an attribute or observation to a larger posited class that is defined by degree of resemblance. In the vocabulary we have been using, Mill's critique of analogy recognizes the harmonic property of analogy, insofar as such analogies sense without fully specifying a shared but distributed pattern.

For Mill, the degree of likeness between cases increases the probability that the analogy is true. Analogy is really an inchoate form of logic in this account; its value derives from the possibility that it will evolve into a complete induction, providing a solid axiom that can be used in deductive reasoning. The incomplete form of analogical thinking, as over against true induction, is central to Mill's critique of Whewell and the latter's claim that science works by discovering new ideas that subsume previously unrelated phenomena. In a longer view, analogy in fact played an important role in models of scientific discovery, constituting, for example, a central feature of Baconian induction.[57] In fact, Mill argues, such developments begin with "the first confused feeling of analogy" between different phenomena and proceed through a Baconian process of "comparison and abstraction," which in turn is formalized as the "conception" on which Whewell places so much weight. This conception constitutes induction only if the resemblances that constitute its basis are completely specified.[58] For Mill, the important point is that such conceptions, viewed as mental phenomena, are simply generalized from analogy and the comparison of things in the world; they do not represent new ideas furnished by the mind, as argued by Whewell. As Mill puts it, "The conception becomes the type of comparison."[59] The difference between comparison and analogy, on this view, operates at their degree of formalization; as the analogy becomes more clearly specified, it becomes amenable to comparative generalization as a concept. On this account (and here Mill basically agrees with Whewell), "conception" provides what I have been calling a *formal analogy*, the governing model that collects together particular cases.

But this account does not adequately address the uncertainty and risk of Tennyson's assertion that, if "thine effect so lives in me / A part of mine may live in thee." From one view, this seems to evidently constitute a formal analogy, mapping a pattern from one case to another. Both Mill and Whewell would likely object that it is not constituted between two phenomena in the world (this world, anyway), and yet the proposition remains so. *In Memoriam* cannot rest with this formal analogy (at least not immediately); it reaches for something beyond the assertion that Tennyson's life, after Hallam's death, might model Hallam's afterlife. There are many reasons to believe this. Beyond the syntactic equipoise of the lines themselves, they are necessarily interdependent. It cannot simply be the case that the initial lines assert a fact that "may" be extended to Hallam; given the possibility that Hallam does, in fact, continue and remember Tennyson, the real question is whether they do "deserve[] the name of friends," and it depends, in part, on Hallam's current opinion of his friend, and whether Tennyson's effect does in fact "live" in Hallam. In order for the harmonic analogy between Hallam and Tennyson to hold, so must the analogy between friendship and reciprocity.

More consequential is the ambiguous source of the sentiment; it is here that unbiddenness, the unmarked source of the thought that "flutters up," takes importance. The degree that this source is beyond the doubtful poet, is precisely the degree to which the statement flattens the metaphysical divide between the living and the dead and produces a point of contact between two distinct but interlinked positions. This is true at the level of Hallam's influence on the verse form but also in the content of the assertion. Above all, "thine effect" is an effect of love, and as Hallam argues in his *Theodicaea Novissima*, the eternal reciprocity of love is its defining (and distinctly Christian) feature: "Is it not reasonable therefore to conclude that the love of the Eternal Being will require similarity in the object that excites it, and a proportionable return of it, when once excited?"[60] At the very least, Hallam's theology suggests the distinctly Christian reading he would give of LXV and the important place of divine mediation in a harmonic love shared between the living and the dead.

Tennyson was not content with an afterlife of abstraction and deferred interest, a relationship with his friend that depended on the mediation of God rather than on immediate contact. Strikingly, *In Memoriam* turns instead to the possibility of other worlds—the idea that the stars have other populated planets— as a way to image the afterlife as a physical place, a world, like ours, where they could potentially meet. The importance of astronomy for Tennyson and for *In Memoriam* in particular has long been known, and has received a sweeping new treatment in Anna Henchman's recent study.[61] A. Dwight Culler was first to call critical attention to the importance of the plural world hypothesis as a way to give the afterlife a material claim in the poem.[62] Astronomy, as a science that depended on inference across vast, inaccessible distances, also provided the central example over which Mill and Whewell argued in their accounts of induction. In many sections, *In Memoriam* rises to a perspective that is extraterrestrial: viewing the earth from without (XXIV, XXXIV), imagining a transit through the stars (LXXVI), imagining the purgatory of souls as a kind of Kuiper belt of undifferentiated spirit (epilogue). But the more general purchase of the possibility of other worlds is given in the specific material configuration of lines that analogize the possibility of Hallam's current experience to Tennyson's own, as in section XL: "My paths are in the fields I know. / And thine in undiscover'd lands." Whereas, in Hamlet's famous soliloquy, the "undiscovered Country" limns the knowledge of death and draws the prince, turned "coward," back to the embrace of life, Tennyson's energies are directed to knowing more of what is undiscovered. As his poet puts it, this knowledge is crucial to lend purpose to his lasting life; it is vital that "I shall know him when we meet," that Hallam, rather than merged into "the general Soul," waits upon "Some landing-place, to clasp and say, / 'Farewell! We lose ourselves in light'" (XLVII).

Among a host of contemporary theorists of possible worlds, Whewell was most important in helping Tennyson find this place among the stars. Though his 1853 study "Of the Plurality of Worlds" gets most attention, it was his eighth Bridgewater Treatise, *Astronomy and General Physics with Reference to Natural Theology* (1833), that shaped the terms of the debate for *In Memoriam*.[63] Published soon after Tennyson left Cambridge, on a topic that had fascinated him since childhood, the treatise was filled with intriguing speculations on the import of astronomical knowledge. Whewell backed into the theory of plural worlds in an attempt to humanize the universe revealed through the telescope, as a way to refute the charge that the vastness of space weakens the astronomer's faith that God would be concerned with individual humanity: "We then find, that a few of the shining points which we see scattered on the face of the sky in such profusion, appear to be of the same nature as the earth, and may perhaps, as analogy would suggest, be like the earth, the habitations of organized beings;—that the rest of 'the host of heaven' may, by a like analogy, be conjectured to be the centres of similar systems of revolving worlds."[64] Whewell continues by reasoning that this inhabited universe gives more evidence of divine care, not less, much as the microscope revealed an earth populated with infinitely more living complexity and intent that previously imagined. It was a radical though long-standing idea. The astronomer William Herschel—who, as I noted in my first chapter, explored a vitalist theory of nebular creation—also used his astronomical studies to promote the existence of life on the moon and other planets.[65] With Whewell's imprimatur, others quickly joined the cause. Robert Chambers saw the plurality of life as a clear implication of the nebular hypothesis in his *Vestiges of Creation* (1844), which Tennyson read eagerly: "the whole of which form but one portion of an apparently infinite globe-peopled space, where all seems analogous."[66] For most Christian thinkers, however, the inhabited universe was a threat to the anthropocentric vision sketched out in the Bible, and in 1853 Whewell reversed himself (though anonymously), writing against the existence of other life, in an argument Tennyson judged "anything but satisfactory."[67]

Tennyson needed the satisfaction of a larger, living vision of the cosmos, and he turned to the theory of plural worlds because it satisfied his desire for an afterlife consistent with both material science and Christian belief. The possibility of life on other planets rehabilitated analogy as a kind of projective knowledge rooted in scientific observation and opened the possibility that the "spheres" of the afterlife might be actual planets. If, as Whewell argued, astronomy relies on the extrapolation from what "appears" to what "analogy would suggest," this both argues that other planets are like our own *and* raises the possibility that they harbor life. In working through the analogic of this extension by analogy, Tennyson's verse is able to circumvent Mill's demand that we base knowledge exclusively in connections between observed facts. Analogy, as organized by the verse form of *In Memoriam*, works to communicate knowledge directly between cases, even when one of those cases is alien, and so operates outside the formal procedures of induction (generalization from experience) or deduction (application of a governing axiom). I accept Mill's analysis of the important role that comparative analysis can play in both of these processes (arguing, as he does, that all deduction depends upon previous induction), but his entire *System of Logic* is designed to undermine "intuitionism"—the belief that knowledge can be drawn from sources outside direct experience.[68]

If anything, Tennyson uses analogy in a way that looks more like Charles Saunders Peirce's "abduction," which Peirce locates at the "opposite pole[] of reason" from induction. "Abduction makes its start from the facts," Peirce explains, in contrast to induction, "without, at the outset, having any particular theory in view, though it is motivated by the feeling that a theory is needed to explain the surprising facts."[69] This clarifies how new hypotheses enter into science, in advance of their formalization. As Peirce puts it elsewhere, "Abduction is Originary in respect to being the only kind of argument which starts a new idea."[70] This account of abduction shadow's Mill's description of analogy, insofar as it is a kind of pattern recognition that "suggest[s] the hypothesis by resemblance—the resemblance of the facts to the consequence of the hypothesis," and therefore precedes induction and deduction.[71] But unlike Mill, and like Whewell, Peirce believes that abduction generates new intellectual content, a way to theorize what is observed that is not simply an inductive generalization of fact. This is why abduction is needed in scientific discovery.

For this apprehension of new knowledge to work, Peirce emphasizes the "pure play" of the abductive perception—the need to observe with an eye that does not yet grasp what it sees.[72] This is of a piece with the more general epistemology of disinterest so crucial to eighteenth- and nineteenth-century theories of objectivity, the belief that the observer must discipline or otherwise abnegate the self in order to understand.[73] Tennyson also emphasizes the "pure play" of receptive sensitivity, and *In Memoriam* puts great weight on the "calm" that must precede accurate insight. Erik Gray has given extensive study to this "poetry of indifference" in the poem, which is driven by elaborate recusals of poetic ambition. Such recusal is necessary to the elegy, insofar as its success threatens to profit from the death of its subject (Joseph likens elegy to an act of cannibalism for this reason).[74] Peirce's analysis suggests a less violent purpose of this indifference to both tradition and legacy, as a precondition to discovery within *In Memoriam*. Insofar as Tennyson is after a knowledge that extends beyond the frontiers of scientific advance, he aims at a mode of perception that outstrips preconceived notions.

Tennyson tests astronomy's potential, within the poem, to furnish knowledge of the connection between material experience and Hallam's afterlife. As an analogic, abduction engages the abstruse, even if such cases do not produce the kind of knowledge that Peirce (or Mill, or even Whewell) would accept. Astronomy, a science calibrated to capture the faintest glimmering of light, gives this mode of analysis widest scope. As Henchman explains, the elegy had a long-standing interest in the astronomy of loss: "The elegiac convention of stellification grows out of stars' unique relation to the perceptual and conceptual. Stars . . . provide a reassuring image for the dead: no longer physically present, the deceased remains just accessible to the senses."[75] The plural world debate gives this analogy epistemological bite, offering a way to imagine how we might learn about the continued life of the dead, and even gain the ability to communicate with them. In the long and attenuated view of astronomy, "Where all the starry heavens of space / Are sharpen'd to a needle's end" (LXXVI), our ability to see other planets means that we exchange some bit of light with alien worlds. What we can see may see us. As Culler observes, *In Memoriam* "is concerned with how, consistent with intellectual integrity and the dictates of reason, [the poet] can believe that they *do* communicate" (emphasis added).[76] Tennyson's sidereal analogy does not use astronomy to "construct a science of immortality" (as Culler puts it), so much as provide a technology, assembled at the limit of scientific understanding, which can bridge the mortal divide.

Despite *In Memoriam*'s ambition to connect earthbound experience to life in other spheres—in the afterlife and on other planets—the poem ultimately finds its solace in those living back here on earth. Shortly before publication in 1850, Tennyson added a verse prologue and epilogue that both frame the poem as an expression of Christian faith and document a marriage that, in the poet's view, provides solace and makes Hallam's grave "bright." The continuing challenge of *In Memoriam*'s prologue and epilogue, which insist on the larger Christian framework of the poem, is not only that they assert a greater confidence in providential order than the poem elsewhere accepts but also that they insist on a determinate beginning and end to the poem's argumentative line. By that measure, the tendentious codas reinforce the important sense that *In Memoriam*'s endpoints are arbitrary. The poem cannot treat Hallam's death as its discrete beginning if Hallam is to continue to live within it. The form of *In Memoriam* has to be open to his intercession, his ability to reach through time and shape its movement.

This is one way of saying the poem's codas are not so closed as they seem. This is evident in the felicity of the epilogue, a benedictory wedding song (or epithalamion) for the marriage between Edward Lushington and Tennyson's sister, Cecilia. Criticism is roughly divided between those who find the song inconsistent with the larger poem and those who see the epilogue as a culmination of the poem's evolution. *In Memoriam* suffers no illusion that the marriage between Lushington and Cecilia finds a gain that matches the loss of Hallam. At the same time, Hallam's extensive engagement within the poem, his influence on its lines and in its commitments, offers a different way of reading the epithalamion (and the poem at large). If we consider this "tactful complement" (as Kincaid puts it)[77] to the marriage as equally Hallam's valediction, then it both serves to free the family from mourning Hallam's legacy and allows the marriage to stand in the public place of Hallam's private but ongoing success *within* the poem, through his poetic marriage with Tennyson. This is not a metaphoric logic of substitution—Lushington for Hallam, Cecilia for Emily— but a logic of analogy, an analogic that can formalize an ongoing interaction between the present and the past, between the living and the dead. The past lives on, both through the Lushingtons and through the life of the poem in the world.

Tennyson's desire to prove that thought and feeling are more than the "magnetic mockeries" of the brain (CXX) is organized through a verse form that insists on equating the material world with its alternate possibilities, sustaining the spiritual through dialogue with material experience. Jason Rudy has emphasized the "spiritualist" dimension of a poetry that "exemplifies the Victorian desire for connectedness"—the "flip side," he notes, of contemporary interest in the science and technology of communication.[78] And Jesse Hoffman has shown how Tennyson reflects within his poetry on his encounter with a possible "spirit photograph" of Hallam, a convergence between elegiac aims and photographic technologies that promised "contact with the dead."[79] Within *In Memoriam*, the material purchase of the metaphysical strikes when,

> . . . word by word, and line by line,
>   The dead man touch'd me from the past,
>   And all at once it seem'd at last
> His living soul was flash'd on mine.   (XCV)

Much has been made of Tennyson's decision to revise "The living soul" to give it more general assignation (exchanging "His" for "The" in 1872). The ambiguous object is complemented by the passive construction and the subjunctive cast of the closing lines. These uncertainties emerge as possibilities only through the lasting and ongoing engagement of the previous lines. Verse provides a vehicle for this transition, a place to ground those Swedenborgian "disembodied spirits" in which Tennyson is reported to have believed, as they go "flying hither & thither . . . their outlet into human existence . . . by contact with life in human beings."[80] Rapt, the poet creates the conditions by which Hallam may "strike a sudden hand" in Tennyson's, and so collapse the metaphysical distance between them.

Section XCV is one of several key moments that feature the poet's alienation from material bounds (another being the Danaë lyric), as the "strange" features, "silent-speaking words," and "dumb cry" of the dead become suddenly familiar, launching the poet "about [the] empyreal heights of thought." A central object for this chapter has been to explain how the formal commitments of *In Memoriam*, and particularly, the analogic of its stanzas, make possible what Tucker terms "the poet's indecisive flight to and from the world, his alternation between cultivating and suppressing an alien self."[81] This is the reverse of the poem's metaphysical coin: in flattening the relation between life and death, the poem not only draws the dead closer to the living but raises lived experience into contact with the afterlife. This alienation of the poetic subject is a necessary consequence of a poem that tries to do more than incorporate the dead into the life of the living; it gives the dead poet substantial power to constitute the poetic subject and the hand that writes it, "word by word, and line by line."

Comparative Anatomy and the Archetype

Would there not also be some glory for a man to know how to burst the limits of time, and, by some observations, to recover the history of the world, and the succession of events that preceded the birth of the human species? The astronomers have, without a doubt progressed more rapidly than the naturalists.

Georges Cuvier, *On the Study of Fossil Bones* (1812)

Yet an unsettled metaphysics—one that elides the distance between the living and the dead—puts at risk much of what we know about the world, under either a scientific or a metaphysical view. I am not a fan of conclusive readings that grieve the loss of possibility, but it's hard to avoid the epilogue's effort to defuse the formative tensions of the poem as it closely aligns a terrestrial marriage with the divine economy, a theory of evolutionary progress, and the memory of Hallam. The tense divisions that have furnished the substantial energy of the poem are suddenly ranked and marshaled under the banner of "One God, one law, one element, / And one far-off divine event" (epilogue). In the last lines, the pluralizing spirit of the poem is exorcised; its ability to sustain multiple possibilities and views pressed into a singular, strident vision. Hallam himself becomes merely a "noble type," a figure, devoid of significance in his particularity, for Christ and the "crowning race" to be. This is a regressive hermeneutics, which sets aside the substantial power of analogy to expand horizons and the interconnected patterns of experience in favor of a unitary typology that formalizes narrative. Though intent on comfort, there is much to grieve for in these lines, even if we take up the palliative ambiguities of the language, appearing at the close of a poem obsessed with the productive possibilities of words that "half reveal" and "half conceal."

This movement from diversity to unity closely follows the efforts of contemporary naturalists, who responded to the plurality of natural forms by arguing for transcendent principles of order and development. Isobel Armstrong has given one of the post powerful readings of this influence within the poem, arguing that the opposed geological perspectives of Charles Lyell and Richard Chenevix Trench structure the differences between the sections of the poem and its epilogue, and drive the poem's effort to "fix and stabilise . . . the authority of the Type."[82] And in a wider view, the asserted stability of "type" marks *In Memoriam*'s concern for a much larger argument among contemporary naturalists over the status of analogy in the study of both fossil and living forms. The typological strategy of the epilogue (which worries the "link" that makes us "half-akin to brute" and an indication of the "crowning race" to come, and asserts coherence under the direction of "One god, one law, one element") can't be understood without attention to these epoch-making debates over the comparative study of natural history. Analogy was the central analytical principle of comparative anatomy—the premier biological science of the early nineteenth century—and so analogy was at the center of hot debates over the meaning of geology, the antiquity of the human race, and our relation to animal species. It is worth canvassing these arguments over the natural order, and their disputes over the status of analogy (known to Tennyson through his friendship with Richard Owen), in order to place *In Memoriam*'s typology in relief. But this discussion also will fill out our understanding of analogy's importance to the sciences of comparative historicism. If, as I argued in chapter 2, forensic antiquarianism was a science that furnished Walter Scott with comparative approaches to *human* history, then comparative anatomy had a far-reaching influence, marked within *In Memoriam*, on the study of humanity's place in *natural* history.

The power of comparative anatomy was rooted in its productive use of analogies between physical specimens and among the different structural patterns these analogies disclosed. The explosion of insights gleaned through comparison, magnified by the exponential growth of specimen collection in the eighteenth and nineteenth centuries, created a demand for formalizing strategies that could give coherence to these rapidly expanding webs of relation. After the pioneering work of earlier explorers and naturalists like Sir Hans Sloane (whose collections helped found the British Museum), Alexander von Humboldt, and Georges-Louis Leclerc (Comte Buffon), naturalists saw specimen collection as continuous with the imperial project of territorial exploration.

This explosion in the collection of biological specimens—living, dead, and fossilized—reshaped science at the dawn of the nineteenth century.[83] While Jean-Baptiste Lamarck is most generally remembered for his theory of adaptation through the inheritance of acquired characteristics, it is important to remember that these narratives of adaptation were secondary adjustments to a larger progressive analogy he postulated between all organisms (a theory he developed for the task of organizing the specimens of the Muséum national d'histoire naturelle in his role as zoology professor): "Among living bodies the name affinity has been given to features of analogy or resemblance between two objects, that are compared in their totality, but with special stress on the most essential parts. The closer and more extensive the resemblance, the greater the affinities. They indicate a sort of kinship between living bodies which exhibit them; and oblige us in our classification to place these bodies in proximity proportional to their affinities."[84] The primary organization of creatures for Lamarck was rooted in their structural "affinities," an analogy of structure that advertised the broader organizational plan of a creator. The central force of Lamarck's argument was to emphasize natural investigation as a comparative process, driven by examination of the relations between specific elements, rather than driven by the blinkered investigation of individual species. Such investigation, Lamarck insisted, will reveal a broad system of analogical order: "Those who have gone in exclusively for the study of species," he asserts, "find it very difficult to grasp the general affinities among objects; they do not in the least appreciate nature's true plan, and they perceive hardly any of her laws" (14). "Laws" are not recognized through the exclusive focus on individual elements of nature but through the formalization of their common patterns.

In contrast to these laws of natural progression, his famous argument for the inheritance of acquired characteristics was a hedge. When Lamarck turned to the natural world, his idealized gradation of animal analogies—from nearly vegetative animals like sponges to the advanced state of sentient man—appeared broken, full of gaps. Lamarck notes these discontinuities, pointing to holes in his schema between different clusters of closely related organisms: "I do not mean that existing animals form a very simple series, regularly graded throughout; but I do mean that they form a branching series, irregularly graded and free from discontinuity, or at least once free from it" (37). These flaws, "irregularly graded" but "once free" from these rifts, point to the tension between the competing influence of system and particular experience. In Lamarck's view, the transcendent analogies are primary, a "simple series" that subsequently diverged through Lamarckian inheritance. But in practice his approach reverses this sequence, reading through the "irregularly graded" series to recuperate the order it distorts. Lamarckian evolution is a narrative justification for his formalized developmental tree; a formal model, rooted in the vital patterns of nature, created in an attempt to bring coherence to the harmonic analogies of comparative anatomy.

Lamarck's great antagonist, Georges Cuvier, decried Lamarck's narrative model of biological change, arguing instead that the precise gradation of similar structures argued for a mathematically precise order characterized by "measure [*règle*] and direction" rather than contingency—a system removed from incidental history, like astronomy: "Genius and science have burst the limits of space, and some observations developed by reason have unveiled the mechanism of the world. Would there not also be some glory for a man to know how to burst the limits of time, and, by some observations, to recover the history of the world, and the succession of events that preceded the birth of the human species? The astronomers have, without a doubt progressed more rapidly than the naturalists."[85] Cuvier, in looking to formalize the observations of comparative anatomy, turns to the Newtonian model of classical mechanics. In the "Discours préleminaire," Cuvier casts this work as a new kind of historicism: "As a new species of antiquarian, I have had to learn to decipher and restore these monuments, and to recognize and reassemble in their original order the scattered and mutilated fragments of which they are composed; to reconstruct the ancient beings to which these fragments belonged" (183). As I noted in chapter 2, Martin Rudwick and Noah Heringman have extensively studied the productive analogies between early paleontology and the contemporary science of antiquarianism.[86] This disciplinary analogy elevated comparative historicism as a collective paradigm that both Lamarck and Cuvier could leverage to support widely divergent understandings of history. Pattern, in itself, does not produce history—it is the formalization of that pattern, whether to some basic narrative, or as an expression of basic formal laws, that gives history shape.

My first two chapters explore two different perspectives on the narrative order of comparative history: in choosing epic as a progressive narrative that consolidates a theory of cosmic, natural, and social development, Erasmus Darwin overwrote the substantial complexity of human and natural history, while Walter Scott crafted a mode of historical fiction that employed comparisons drawn between multiple differentiated narratives to give substantial texture to historical accounts. The debate between Lamarck and Cuvier can similarly be understood as an argument over the nature of "original order": whether it consists in a single narrative of progress (however fragmented), or a universal law of interrelation that stands outside of history and its incidental narratives.

The formalizations of analogy offered by Lamarck and Cuvier remained closely disputed by midcentury, as naturalists from Lyell to T. H. Huxley challenged Lamarckian inheritance, and the "geometrical laws" of Cuvier's taxonomy failed to materialize. This is the interpretive problem Geoffroy Saint-Hilaire pointed to when he complained that analogy's many senses left it "far from having the power to furnish precise indications," and in response, Geoffroy proposed that there should be two classes of comparison: direct analogies between specific limbs or organs, and the larger pattern or "homology" that characterized these patterns globally.[87] Analogy's long history of use in natural theology and biblical hermeneutics left it rife with implications.[88] Homology, by contrast, was a more precise mathematical neologism, recently developed to describe the common geometric properties of various "homologous" figures and their reflexive transformations.[89]

It was not until Richard Owen set out, in the 1840s, to formalize the relationship between analogical comparison and interpretation that naturalists began to adopt this analogy-homology distinction. The relation Owen set out between analogy and homology is often mischaracterized, particularly in the longer view of Darwinian theory, and bears unpacking. Owen was the preeminent anatomist of Tennyson's era (dubbed, to his chagrin,[90] the Cuvier of England), and he and Tennyson were good friends and regular visitors.[91] As early as 1843, in his *Lectures on the Comparative Anatomy and Physiology of the Invertebrate Animals* (1843), Owen had defined a "homologue" as "the same organ in different animals under every variety of form and function" in contrast with an "analogue," a "part or organ in one animal which has the same function as another part or organ in a different animal."[92] The key terms of Owen's mature thinking are already evident here; analogous organs are similar because they have the same *function*, homologues are the *same* organs in different animals. This distinction between function and identity was only later reworked, in light of Charles Darwin's theories, as a distinction between analogy through convergent evolution and homology through common descent.

Owen's distinction between homology and analogy, which has continued to evolve, is just one of the many ways, and many vocabularies, that nineteenth-century thinkers used to organize the disciplinary practices of analogy.[93] Owen's innovation was to tie the systematic relation of homology to a universal "archetype," the common, idealized form that expressed the common features of a group of homological organs, much as Johann Wolfgang von Goethe had once speculated on an "ur-plant" or primary plant.[94] As Owen makes clear in *On the Archetype and Homologies of the Vertebrate Skeleton* (1848), the turn to archetype is powerful, not because it fixes a meaning to the larger patterns of analogy, but precisely because archetypal homology is ahistorical and conventional in character.[95] In other words, Owen made homology an explicit formalism capable of dealing with the varieties of analogical similarity offered by comparative anatomy. Hence, in Owen's taxonomy, homology captures a kind of *formal analogy*, in contrast to seeing analogy as a harmonic relation without higher model. Owen's homology helped translate a methodological problem into a representational challenge. In defining an "arbitrary" method, an archetypical system of terms and forms, Owen established a conventional framework for interpretation and representation. Owen's homologies, though they are explicitly mental fictions, are the peculiarly powerful fictions of scientific modeling.[96]

At the same time, Owen argued that these "arbitrary" models captured something real about the ingrained pattern of nature, and he believed that this basic plan was evidence of a higher design. In his critical review of Darwin's *On the Origin of Species*, Owen holds up Cuvier as an exemplar of the empirical "principles" of modern naturalism including "daily observation, comparison, and reflection, on recent and extinct organisms. . . . These principles," he continues, "based on rigorous and extended observation . . . have tended to impress upon the minds of the most exact reasoners in biology the conviction of a constantly operating secondary creational law . . . [as demonstrated by] the law of unity of plan or relations to an archetype" (though, it should be noted, Owen admits the uncertain mechanism of that "law").[97] As a term of art, "special creation" helps recognize the substantial interaction between natural theological speculation and scientific naturalism in the nineteenth century, a topic I pursue in chapter 5. But it also demonstrates that, while Owen advanced his theory of archetype on the grounds that such conventionalism could bracket the larger problem of how natural patterns arise, he believed these conventions verified his own Christian belief.

Conclusion: The Higher Type

They wrought a work which time reveres
A precedent to all the lands
And an example reaching hands
For ever into coming years.

Tennyson, "Hail, Briton!," Heath MS[98]

Tennyson's *In Memoriam* proves the suppleness of verse in mapping the divide between the ramification of pattern and a desire—shared with Owen—for higher order. Contemporary readers as well as subsequent critics were sensitive to Tennyson's engagement with the implications of contemporary geology and comparative anatomy. The poem is famous for its struggle with the implications of a nature "red in tooth and claw," both "careless of the single life" and of the "type" that it belongs to. If within the geological evidence of mass extinction, nature screams "A thousand types are gone: / I care for nothing, all shall go" (LVI), Tennyson's poem yet searches for a higher pattern in which such extinction and destruction are meaningful, part of a larger purpose. This ambi-

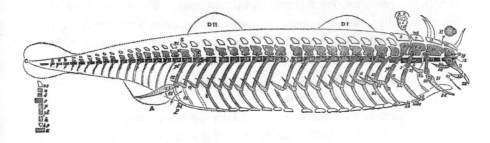

Figure 7. "Archetype Vertebrate Skeleton." Woodcut. From Richard Owen, *The Principle Forms of the Skeleton and the Teeth* (1856), 712501, fig. 7. Courtesy of The Huntington Library, San Marino, California

tion resonates powerfully with recent critical discussions of the "anthropocene"— the period (which we inhabit with the Victorians) that marks humanity's catastrophic intervention in the climate and the geological record.[99] For his part, Owen found confidence in higher purpose through his study of the homologous patterns of living structure—the internal echoes and marked physical similarities that indicate the interconnection of all life.

For Owen, such patterns demonstrated (as he put it in his personal seal) "The One in the Manifold."[100] Something extraordinary happens if you look closely at Owen's archetypes with *In Memoriam* in mind (figure 7) and explore the archetypal patterns that govern both the "serial homology" between the constituent vertebral columns and their coordination within the ultimate archetype of all vertebrates (conveniently for Owen's Christian belief, a fish). In the relation between typical relations and ultimate form, I see a crucial model for how Hallam is incorporated into the larger history imagined by *In Memoriam*. The poem ultimately turns toward such a homology, assembling its variations, through the serial interrelation of its sections and the reflecting pattern of its stanzas, in order to draw them together as the body of Tennyson's visionary historicism. The interrelated sections of the poem, like the "stepping stones" of history, are gathered as evidence of the connection between the "noble type" and the "race to come," under the governing pattern of "One god, one law, one element."

My interest here is in how, by means of an ultimate faith in progress to the noble type figured by Hallam (a progress that justifies the loss of both Hallam and the other victims of natural cause), the poem overwrites the differential implications of comparative historicism with a single, capacious history, a vision of total incorporation. For this reason, I do not read *In Memoriam* as a work of comparative historicism, though it is intimately concerned with comparisons of form, of understanding, and, above all, of past and present experience. Comparative historicism reflects a commitment to the plural past, the sustained difference not only of experience but of ultimate trajectory, in a centrifugal analysis of individuals and of larger social formations. The point of Scott's *Waverley* is that Edward's future excludes that of Vich Ian Vohr, even as his past lives alongside the Highland chief (as figured in the closing portrait). While the power of the *In Memoriam* stanza is rooted in its formal capacity to explore experience and memory as a field of cooperant action—populated with the histories of actors who act together, even as they act apart—the ultimate belief that this toil is "coöperant to an end" resists the threatening possibility that their ends are different. I am tempted to speculate that this is the inherent limitation of elegy, as a mode of engagement that, like epic, moves emphatically from rupture to integration.

The problem is not, strictly speaking, that the teleology of a common end lacks the flexibility for a plural vision; in my final chapter, I explore how Charles Darwin reformulates a heuristic of intent that was intrinsic to natural theology, in order to better explore the coordination of divergent purposes. The gamble of *In Memoriam* is that the plural past can be recast through a shared future, through a marriage of poetic labor that cannot be sundered by the contingency of events, in a powerful centripetal pull that "close[s] with all we loved, / And all we flow from, soul in soul" (CXXXI). This is the trick of *In Memoriam*'s extensive serialization of experience. The extended series of events, memories, recollections, and visions of *In Memoriam* evoke a governing pull toward not just reunion but unification. On this reading, Hallam's "noble type" is equally his experiment in intervolved rhyme, fulfilled through the ripening of *In Memoriam* itself as it is finally set to type. Even as its stanzas, by falling back, emphasize the balance and inconclusion of specific formal efforts—as formal operations are destabilized in a return to serial encounters that resent consolidation—the repeated instances of this movement set the pattern, consolidating the eponymous "In Memoriam Stanza."

In this fashion, *In Memoriam* remains a cooperant form, constituted through networks of collaboration and continued engagement. The poet's insistent failures—that is, his inability to comprehensively master his own intent through form—is a condition of his ability to work with others, the living and the dead, as much as a condition of his ability to reach for a perception beyond subjective experience. The success of this formal coordination is reflected in the fact that later poets have so rarely taken up its stanzaic form despite (or because of) its prominence after 1850. *In Memoriam* tests the limit of analogy as a strategy of pattern analysis. Pursued with enough intensity and extension, the comparison of experience threatens to wash out the particularities that make such comparisons valuable. In the chapter that follows, I explore how George Eliot's fiction takes up the reverse movement: her novels argue that *disanalogy*, as the mark of valuable distinction, constitutes the opportunity for fresh understanding. For Eliot, cooperation toward proximate ends succeeds only when subsumed in the larger dialectical movement toward a better understanding through difference.

CHAPTER FOUR

Falsifying George Eliot

"How very ugly Mr. Casaubon is!"

"Celia! He is one of the most distinguished-looking men I ever saw. He is re-markably like the portrait of Locke. He has the same deep eye-sockets."

"Had Locke those two white moles with hairs on them?"

"Oh, I dare say! When people of a certain sort looked at him," said Dorothea, walking away a little.

"Mr. Casaubon is so sallow."

"All the better. I suppose you admire a man with the complexion of a cochon de lait."

"Dodo!" exclaimed Celia, looking after her in surprise. "I never heard you make such a comparison before."

"Why should I make it before the occasion came? It is a good comparison: the match is perfect."

Miss Brooke was clearly forgetting herself, and Celia thought so.

George Eliot, *Middlemarch* (1871–2)

Eliot loved a "good comparison." Her novels feature bifurcated plotlines that prompt readers to draw comparisons between characters and incidents, across social and imaginative distances. The longest novels (*Felix Holt* [1866], *Middle-march* [1871–2], *Daniel Deronda* [1876]) are divided almost evenly between two primary stories that describe the fortunes of different principal characters. This is true even of *The Mill on the Floss* (1860), though, as Eliot once complained, we overlook Tom Tulliver's significance. F. R. Leavis, in his canonical study of Eliot's place in the "great tradition," notoriously saved *Daniel Deronda* for that tradition by excising one of its halves as a novel he titles "*Gwendolen Harleth*."[1] *Middlemarch*, a novel that is famous for its social portrait of the organic inter-connection of a single community, in fact began as two different manuscripts: "Middlemarch," which was to explore Tertius Lydgate's semitragic story, includ-ing his search for a "primary tissue" from which all organs are formed, and "Miss Brooke," which followed Dorothea, in her equally unhappy marriage to Isaac Casaubon, and his search for the "key to all mythologies."[2] The point of internal division is to emphasize a common pattern; meaning accrues not within isolated parts but in the relationships that exist between them.

In chapter 3, I argued that Tennyson uses the formal organization of *In Memoriam*—particularly, the enclosed rhyme scheme of that elegy's stanzas—to elucidate analogies drawn between memory and experience, the present and past, the living and the dead. In a similar fashion, Eliot's novels deploy an array of formal strategies, in addition to their double plots, to foster a comparative reading of character and experience, continuously drawing focused comparisons within and between the chapters themselves, both by the explicit pairing of characters and by the echoing patterns of her epigraphs, chapter, and book titles. Whereas, in my reading of *In Memoriam*, Tennyson's commitment to comparatism ultimately submits to his ambition to discern the consonance of "one God, one law, one element" within the complexity of experience, Eliot's works demonstrate a sustained commitment to relative dissonance and uncertainty. The current chapter takes an unusually long view of Eliot's career, from her translations of German biblical scholarship in the 1840s and 1850s, to her tenure as de facto editor of the *Westminster Review* from 1852 to 1854, to "The Legend of Jubal" (1870)—one of her later experiments with epic verse—and even her posthumously published essays and notes, in order to explore the grounds of Eliot's comparative imagination and her profound commitment to the comparative study of social history. Above all, I argue, it is in Eliot's novels, particularly *Middlemarch*, where we find the most powerful statement of her belief that such comparisons, particularly in their ability to diagnose previous errors, produce new knowledge.

Eliot uses cognates of "compare" seven times in the first three chapters of *Middlemarch*. In the first chapter Celia Brooke serves primarily to establish, by way of contrast, that her sister Dorothea is an unusual heroine; Mr. Brooke, that Casaubon is a deep thinker; Sir James Chettham, that Casaubon is an awkward lover. Dorothea seems to compare everything with everything else, but especially her own ambition and understanding to Casaubon's, "seeing reflected there in vague labyrinthine extension every quality she herself brought."[3] Nevertheless, even as Dorothea insists in this chapter's epigraph that she makes a "good comparison" (20), it is evident to the reader, and virtually all of the other characters, that her comparisons lead her astray. When she looks at Casaubon, as the narrator reminds us repeatedly, she sees herself; she does not, at the outset, recognize his independence as an "equivalent center of self" (211). At the same time, her belief in this reflection establishes expectations for Casaubon and his actions, hypotheses about their future together. When Casaubon violates those expectations, Dorothea is primed to take note. This is possible because, at the outset, the heroine does not have the rest of the novel to work with. She compares first characterizations and first incidents; she does not have the experience to compare them over the long haul, as a more complex relationship between lives, narratives, and their respective conditions. She does not have the experience to move from a formal to shared understanding that respects difference. Her problem (and here Celia gets it wrong) is that Dorothea is not "forgetting herself" effectively. The challenge she faces in *Middlemarch* is to learn how to bracket one's self in the right way, as one node within the larger network of the novel. If we (as Neil Hertz has put it) feel nearly "submerged in the flood of visionary comparisons" that interlink Eliot's characters and place them within the "world-history" of her novels, Dorothea ultimately helps us to find our footing by gaining perspective on the complex pattern of social history.[4] Her evolving engagement with Casaubon and their world narrates this evolution from isolated comparisons to the texture of comparative historicism.

Ultimately *Middlemarch* is about the central "march" or "boundary" between self and other, and between the present and the past, the middle ground where patterns of similarity and difference subsist. It studies the interstitial histories we weave between lives, cultures, and understandings. The novel's primary concern is the source and limitation of knowledge as it is constructed between moments and between individuals. Walter Scott's novels find these zones of contact in physical borders, often organized geographically in the division between Highlands and Lowlands, Scotland and England, Norman France and Saxon England, but in Eliot's narratives the system of contact is much more dispersed. So, in a novel like *Daniel Deronda*, contact operates between the largest (and most problematic) social and historical forms—London and Zion, occident and orient—but it is also woven into the fabric of mundane experience.[5]

These expanded border zones accommodate the moderating generic impulse of realism, which, as George Levine explains, "belongs, almost provincially, to a 'middling' condition and defines itself against excesses, both stylistic and narrative, of various kinds of romantic, exotic, or sensational literatures."[6] For Christopher Herbert, Andrew Miller, and David Kurnick, it is hard to miss the alliance between the self-government of this " 'middling' condition" and the systems of self-mastery, discipline, and libidinal control we now generally see as important features of nineteenth-century institutions, both religious and secular.[7] Indeed, Eliot's novels are deeply concerned with the mechanisms of restraint that are both imposed upon and cultivated by her characters, particularly her heroines. As her narrator puts this in a particularly excruciating simile from *Felix Holt*: "The finest threads, such as no eye sees, if bound cunningly about the sensitive flesh, so that the movement to break them would bring torture, may make a worse bondage than any fetters."[8] Here the simile is applied to Mrs. Transome and the intolerable cross-pressures of familial loyalty, social censure, and desire that leave her unable to change the semitragic course of her son's passage through the novel.

Yet I join Kurnick in seeing such "threads" as constituting both social pressures and the foundation for social engagement; it is also Mrs. Transome who refers to her remaining connection to her son as her "last poor threads." These threads are both fragile and precious to Eliot's characters. They emphasize the importance of a different kind of self-quietism in Eliot's fiction: the sensitization that allows one to reach toward, if not grasp, what the narrator of *Middlemarch* famously describes as that "keen vision and feeling of all ordinary human life . . . [which is] like hearing the grass grow and the squirrel's heart beat" (194). Though we are cautioned that, could we hear it fully, we would die of that "roar which lies on the other side of silence," Eliot's novels argue that we can learn to be quiet enough to hear some portion of its discords and harmonies. The trick is to sensitize us, as readers, to relations that subsist between individuals and between experiences but which, as Barbara Hardy observed, are often left implicit in Eliot's novels.[9]

When we encounter this more sensitive mode of discovery in Eliot's fiction, we find, to our surprise, that it depends not upon a quieting self-mastery, but the stunning shock of new encounter and, especially, error. Dorothea Brooke is important among Eliot's characters for the way she internalizes larger comparisons across the character system of the novel as a method of interpersonal understanding.[10] But the movement in this understanding is constituted through her many mistakes. The first three chapters of *Middlemarch* set up the narrative of Dorothea and Casaubon as what happens when "good comparison" goes bad. Ultimately, the novel is determined to show how these errors are productive, as we begin to understand, along with Dorothea, where her comparisons went wrong, and where a more productive similarity (and sympathy) lies. By these means Dorothea gives comparatism purchase on the relation between alterity and sympathetic understanding central to Eliot's realism. In the context of this book, Eliot's fiction demonstrates the critical place of *disanalogy* in comparative historicism's effort to better understand historical and social difference. If, as I argued in my discussion of Walter Scott's novels, translation offers a way of understanding the analogy between present and past in historical fiction, disanalogy is a form that furnishes Eliot's social realism with "an approach to literary history," as Emily Apter has recently put it, "that recognizes the importance of non-translation, mistranslation, incomparability, and untranslatability."[11]

This chapter takes up the complicated relation between analogy and comparison that Eliot deployed to study the relational dimensions of characters and their understandings, exploring in particular how such comparisons subtend Eliot's insight into the sedimentary nature of historical language. For this reason, I emphasize Eliot's interest in history writ large over natural history. Eliot is often credited as the Victorian novelist most invested in contemporary scientific discoveries and in their ability to shed light on the fictional worlds she explores. *Adam Bede* (1859) is often given exemplary status as a novel that, in carefully elaborating the idiomatic, religious, and social customs of a Western Midlands farming district at the beginning of the Napoleonic era, gives a natural history of daily life. And yet, even *Adam Bede*'s most overt engagements with contemporary research into anthropology—for example, the way the narrator first explains the eponymous hero's physiognomy as a joint Saxon-Celtic heritage—are counterbalanced by a comparative sense of its limitations. Eliot's handling of this ethnological vocabulary is distrustful; we are immediately informed that, in his brother Seth Bede, "the strength of the family likeness seems only to render more conspicuous the remarkable difference of expression both in form and face" and, we might add, personality.[12] By contrast, the idiomatic peculiarities that characterize the social and historical locations of her different characters receive far more focused and extensive concern in the novel. So, at the opening, the contrast between Seth Bede's "There! I've finished my door to-day, anyhow" and Sandy Jim's crusty retort, "What! Dost think thee'st finished the door?" (11) carefully registers the historical sedimentation of the rural community within a single group of laborers.

The contrast between Seth and Sandy Jim does more than mark a gap between the known community (implicitly modern) and the incompletely grasped knowable community of the past; it sets a tense engagement between present and past that underlines historical concern.[13] Idiom served Eliot, as it had Scott, as a way to bring historical complexity into contact with the mediating idiom of her narrator. Moreover, Eliot understood that idioms, operating within historical and social coordinates, made a master language (a key to all philologies) impossible. Eliot rejected the total knowability of communities, of individuals, and of empirical knowledge generally as anything more than a notional goal.

This strict limitation is particularly clear in her skeptical engagement with contemporary science. Gillian Beer has urged the relationship between contemporary scientific methods and the "series of structural comparisons" that organize *Middlemarch*, a novel that takes the comparative sciences as confirmation that relational analysis provides fresh historical understanding.[14] Virtually all of the famous "parables" Eliot draws from science are concerned with the irreducibility of distortion, the need to settle with approximation, and the inability of science to gain comprehensive access to the object of concern. This is true whether we speak of the optics of the "pier glass" in *Middlemarch*, in which the scratched circles of light surrounding a reflected candle appear to change as the light source moves; or the "make-believe of a beginning," described in *Daniel Deronda*, that astronomers use to fix celestial time; or the "personal equation" referenced in *Impressions of Theophrastus Such* (1879), which corrects for the bias of the individual scientific observer by comparing the aggregate of their observations to those of others.[15] All demonstrate Eliot's sense that empirical science confronts the same limitations as any description or historical account. She uses these examples to emphasize the continuity of scientific research and imaginative history in their dependence on a careful comparative analysis that discriminates between provisional truth and error. Eliot used scientific examples (alongside examples from mythology, literary history, art criticism, and various other fields) to pose her more general concern for the imaginative possibilities and limitations of comparative historicism.

For this reason (and even though it is generally read as exemplifying a close alliance between scientific and social description in Victorian Britain), I don't think *Middlemarch* is really interested in science per se but rather in how scientific knowledge can figure more basic problems of epistemology, social understanding, and the texture of history. Among all of the writers I examine in this book, Eliot had the broadest sense for comparative historicism as a methodology shared across various scientific, historical, and social disciplines. In one of her notebooks, not published until 1980, Eliot makes this clear. In a series of speculations about social evolution, the notebook critiques the ongoing "discussion of origins"—including those of Charles Darwin—for reading common patterns as evidence of common descent.[16] Her immediate object is the study of social customs and mythology, as explored by comparatists like Herbert Spencer, Max Müller, and Edward Burnett Tylor. As I will argue at the close of this chapter, Eliot's larger point is that we cannot map theories of natural change onto the social world. Human societies are characterized by "increasing complexity" and change rapidly; a common pattern is as likely to be the result of independent responses to similar conditions as evidence of common descent (390). It is an argument that articulates Richard Owen's distinction between homology and analogy (discussed in chapter 3) to the study of social forms. The problem only gets worse when we try to think about origins, which are shrouded in permanent obscurity: "And discovery? That too must end somewhere & under the name of knowledge has long been recognized as a mere parenthesis in a context of irremovable darkness" (387).

This chapter begins by surveying Eliot's editorial work for the *Westminster Review* in order to establish a contrast between her explicit interest in the problem of historiographic method and the more general tendency to read her realism through the lens of contemporary science and objectivity. I then turn to conflicting models of history explored in *Middlemarch*, particularly the way it invokes distinct perspectives on the history of Rome, in order to flesh out the deeply comparative impulse in that novel's historiographic imagination, with its history of histories. Turning to the "Legend of Jubal," I then suggest how Eliot used musical theory to explore what I have termed *harmonic analogy* as a social form that patterns sympathetic understanding.[17] Rather than naïvely asserting fellow feeling, Eliot calibrates this sympathy with *disanalogy*, as both a critical feature of harmonic analogy and a structure of falsifiability that aligns the realism of a novel like *Middlemarch* with the uncertainty of historical interpretation. In the section that follows, I take up Eliot's theory of form and explore how her grasp of the historical contingency of language emerged from close study of the German "higher critics" and their comparative investigation of Christian history. If Eliot's formal understanding grew from her comparative studies of religious and social history—rather than natural history—this undermines the seeming naturalism of her theory of organic form. The realist novel, as handled by Eliot, precipitates out of the historical novel and turns comparatism to account in the interrogation of the grounds of experience, our knowledge of others and of the world in general. Virtually all of Eliot's novels are historical, and it is precisely this displacement into the past that liberates the comparative energies of the novel and its critical interrogation not only of genre but also of more basic forms of life: customs, languages, daily experience.

Along the way I will be arguing that what Harry E. Shaw terms a "realist habit of mind" and Rae Greiner "sympathetic realism" turns on moments of interpersonal and historical comparison, usually organized through explicit analogies that place both reader and narrator within a network of similarity and difference.[18] Eliot understood comparative historicism to be based in analogies that mediate between (as Catherine Gallagher puts it) "the competing needs to adhere to type and to deviate, to mean and to be, to have significance and to become real."[19] In order to make this case, I find it helpful to set the traditional categories of realist fiction (typification and particularity, metaphor and metonymy, judgment and sympathy, observation and participation, narrative and event) in dialogue with the relation between *formal* and *harmonic* analogy that this book examines. Comparatism is a methodology, after all, and so a method organized by its characteristic forms. But as Eliot observed with regard to poetry, forms emerge through application: "Form was not begotten by thinking it out or framing it as a shell which should hold emotional expression, any more than the shell of an animal arises before the living creature; but emotion, by its tendency to repetition . . . creates a form by the recurrence of its elements in adjustment with certain given conditions."[20] Eliot's routine use of such organic models is commonly read to naturalize social fashions, whether habits of speech, local customs, or the novel itself. As I argue at greater length, however, her emphasis on the recurrence of pattern points to a more contingent and unstable relation between history and conditions, for both forms in general and analogies in particular. Just as Eliot's mollusk analogy serves to characterize form as a similarity of pattern within varying context, this chapter works out to a better understanding of Eliot's comparative mode by working through a series of its more prominent instances.

The *Westminster Review* and the "Historic Imagination"

[By the "historic imagination"] I mean the working out in detail of the various steps by which a political or social change was reached, using all extant evidence and supplying deficiencies by careful analogical creation. . . . There has been abundant writing on such turning points [as the conversion of Constantine I], but not such as serves to instruct the imagination in true comparison.

George Eliot, *Leaves from a Notebook* (1885)

Less a manifesto than a protocol for effective history, Eliot's brief note on the "Historic Imagination" carefully balances historical distance and language's place within it. Eliot emphasizes the value and importance of such histories, including their "analogical creations," in establishing an understanding of both the substantial similarities and crucial differences between the present and the past. As she explains, retracing an argument made, in some version, in each of her novels, "A false kind of idealization dulls our perception of the meaning in words when they relate to past events which have had a glorious issue: for lack of comparison no warning image rises to check scorn of the very phrases which in other associations are consecrated."[21] Eliot insists that the chief object of the historic imagination is to address the complexity of history head-on. I have already argued that the "false idealization" which closes *In Memoriam* is in tension with the richer canvas of conflicting perspectives entertained throughout Tennyson's poem. This tension is a key source of the continuing interest and power of that elegy, as the poem insists on the pull between looking forward and looking back, between faith and doubt. Here Eliot imagines how these impulses converge. Analogical creation is "careful" because it looks forward and back at the same time, setting the present and past in a dialogue that captures the complex relation of similarity and difference within the moment, of continuity and differentiation over time.

In Eliot's view, the significant dissonance of history is amplified and distorted by two opposing tendencies: the false idealization that does not recognize historical differences between what words mean now and meant then, and a scorn that alienates other time-bound "phrases" that, in a different context, have continuity with the present. In both cases, comparison has failed to do its work; either too pat or absent entirely, it has failed to be "true," to provide its "warning check." For both author and reader, true comparison sifts similarity and difference for substantive pattern, while "analogical creation" mediates between them, translating that pattern into historical fictions that fill gaps in the historical record with plausible truth.[22] In a novel (as opposed to an elegy), there is ample space for the "warning check" of comparisons that provide precision, granularity, and a sense of how much is yet unknown. The extended network of dialogue and reflection in Eliot's fiction makes the historical density of communication a continual object of comparative analysis. Eliot's fiction explores this history as the elaborate network of analogies, parallels, and allusions that interrelate characters, incidents, plots, and—most especially—historical perspectives.

This is part of my more general thesis, that Eliot helped shape comparative historicism as a broadly shared approach to finding patterns among people and things within time. Eliot was well positioned to see this common pattern; before becoming a novelist, she spent ten years translating the German higher critics (an effort I'll take up later) and six years closely engaged with the *Westminster Review*, an influential quarterly journal that canvassed books on a wide variety of subjects, especially literature, science, history, and religion. This extraordinarily wide reading gave her an extensive prospect of the various fields that explored the past comparatively. For three years, from 1851 to 1854, she helped the publisher John Chapman by serving as the *Westminster's* effective (though unacknowledged) editor. During these years, the *Westminster* published an omnibus review of "The Future of Geology"; reviews of recent novels by Charlotte Brontë, William Makepeace Thackeray, and Elizabeth Gaskell; an assault on the "metaphysics" of Arthur Schopenhauer and Johann Gottlieb Fichte; a review of Honoré de Balzac's (largely historical) novels; an extended study of philosophical skepticism since Hume; a demographic analysis of the 1851 census; as well as George Henry Lewes's study of "Goethe as a Man of Science," which examined his lasting importance to comparative anatomy and botany.[23] During Eliot's tenure, the journal also showed a keen interest in the works of the comparative anatomist Geoffroy Saint-Hilaire, publishing reviews of his individual works and a more comprehensive study of his "Life and Doctrine."[24]

On reading the pages of the *Westminster* in this period, it is striking how important comparatism became, both as a technical term within specific sciences and as a more general analytical procedure. This convergence encouraged Eliot to reimagine the format of the quarterly review—which collects and juxtaposes essays on a range of common subjects—as itself a genre of comparative study. The April 1854 issue of the *Westminster* (the last that Eliot edited) exchanged the national subdivisions of the "Contemporary Literature" section for specific subjects like "Theology, Philosophy, Sociology, and Politics" and "Belles Lettres," promoting reviews that read these topics across national and linguistic divisions. In addition to allowing more books to be treated, the editor explained that "it is now intended by a careful analysis and grouping of each quarter's productions at once to exhibit the characteristics of the individual works reviewed, and to supply a connected and *comparative history* of Contemporary Literature" (emphasis added).[25]

Instead of a review, the new *Westminster* imagines itself as a series of comparative studies. In modern histories of comparative literature, the 1877 launch of the *Comparationis Litterarum Universarum* by Hungarian scholar Hugo Meltzl is recognized as the first journal of comparative literature. The *Westminster's* emphasis twenty years earlier upon the "comparative history of Contemporary Literature" shows an emergent concern for what Meltzl and his collaborators imagined as a *vergleichende Literaturgeschichte*—a comparative literary history. It was modeled on comparative science, especially comparative philology, with its close focus on differences in regional language and idiom.[26] But in contrast to Meltzl and an understanding of "world literature" founded by Johann Wolfgang von Goethe, the *Westminster's* reorganization was a move *away* from national categories and discrete literary traditions—an imagination of comparative literature that did not take the nation-state as its ordering principle. The *Westminster*

Review is one important waypoint in an alternative genealogy for comparative literature, one that includes Scott, and which focused on the literary differentia of comparative historicism rather than the unities of nation or ethnic group.

Fionnuala Dillane has warned us to remember that this editorial work came *before* Mary Ann Evans chose "George Eliot" as her pen name, when she was instead hard at work writing and revising articles, soliciting essays, and editing, with seemingly no end, the submissions of a wide range of prominent scientists, historians, and philosophers.[27] Her work reading, writing, and editing across such varied disciplines furnished a deep understanding of the interconnection of their historical methodologies and their relation to the comparative hermeneutic of the sciences and of German scholarship. Her experiment with the *Westminster*'s format suggests an attempt to imagine a form better suited to the comparative analysis of perspectives.

In emphasizing her commitment to comparative historicism as a model for the "comparative history of Contemporary literature," I want to reconsider the relation between social and natural history in Eliot's writing. If Eliot sometimes described her novels as "experiments," they were experiments organized to extend this capacious understanding of comparative historicism's scope into her fiction. On the one hand, Eliot's deep investment in contemporary scientific thought has received extensive attention, particularly her use of scientific allegories to examine problems of perspective and social form in *Middlemarch* (characterized by George Levine as "that most scientifically knowledgeable of novels").[28] So, for Gillian Beer, Eliot's concern for the experimental procedures of Charles Darwin and other scientists makes "the analogy with scientific procedures essential" to *Middlemarch*.[29] For her part, Sally Shuttleworth has emphasized the impact of contemporary science on a theory of organic form that serves Eliot as a governing "model" for both social and natural interrelation.[30]

On the other hand, as I have already indicated, I think that *Middlemarch* is more interested in the *history* of natural history than its *nature*. For one thing, the novel resolutely refuses to adjudicate matters of fact in scientific terms. How and where does Fred Vincy get sick? Does the final dose of brandy kill John Raffles? What procedures are used by Tertius Lydgate in his search for the "primary tissue"? Repeatedly, the novel either insists that such questions cannot be resolved scientifically or simply ignores them, turning our attention instead to consider how the nonscientific community evaluates those problems and the impact of these issues on that larger community. This shift historicizes science itself, militating against science's capacity to adjudicate truth and error after the fact. Raffles's death is complicated, *both* because (as the novel notes) the medical community disagreed at the time over how to treat alcohol poisoning *and* because the question is irrelevant to the *Middlemarch* community, which assigns a shared responsibility to Nicholas Bulstrode (who cared for Raffles) and Dr. Lydgate (who treated him). These differing perspectives are, in the sense used by philosopher Thomas Kuhn, "incommensurable"—they configure the problem in incompatible terms.[31] In my own terminology, the incident illustrates a critical *disanalogy* of perspectives. By these means, the conflict between scientific and lay perspectives elucidates, for the reader, the uneven texture of experience within historical time.

For Eliot, this interest in interpreting the past was inseparable from her life-long engagement with historical fiction. As George Levine observes, Eliot's fascination with historicism grew from her early immersion in historical novels. In his account, the novels of Walter Scott "revealed to her that goodness had little to do with belief in any particular religion" and helped turn a more critical eye on her fervent Anglican background.[32] They sounded a keynote for the consistent relational analysis of her later studies of social life. Other scholars have taken historicism itself as a key focus of Eliot's fiction. Harry E. Shaw reframes Eliot's narrative realism through a close engagement with the historicism of Walter Scott, while Neil McCaw finds in Eliot's fiction a struggle with the master narratives of progressive history.[33] Each sees progressive history as the crucial element in Eliot's historicism, with one foot in the Whig political history promulgated by Thomas Babington Macaulay and the other in a theory of positive advance developed by Auguste Comte and his followers.

Yet, as I argue in chapter 2, Scott's more important contribution to historicism was to turn the imagination of the past away from unitary narratives and toward a more complex and more comparative feel for historical texture. Certainly, Eliot considered Scott (in the words of George Henry Lewes) both the "longest-venerated and best-loved Romanticist."[34] Eliot possessed more works by Scott than any other author; the collection she amassed with George Henry Lewes contained more than a hundred volumes—roughly the same as their entire collection of histories.[35] Of course, this was representative for the Victorian imagination of the past, which was profoundly conditioned by Scott's fiction in both its popular and scholarly forms. Levine sees Scott's influence as evidence that his "historical particularity . . . was for Eliot importantly ahistorical because it allowed her to see beyond local events into the essence of human nature itself."[36] I think Eliot found greater possibility, however, in the comparative investment of Scott's fiction, which places its analysis of social forms and individual actions within (rather than outside of) a larger, relational, and inconsistent historical canvas. For Eliot, as for Scott, an ahistorical perspective would be use-less, were it possible, because it would stand outside the systems of meaning that constitute history. As the narrator puts the problem in *Middlemarch*, "To Uriel watching the progress of planetary history from the sun . . . one result would be just as much of a coincidence as the other" (412). Rather than deriving an Archimedean historical vantage from Scott, Eliot adapted a model of histo-riographic comparison that emphasized difference as well as continuity, both within the past and between past and present.

This sense for the complex texture of historical perspectives receives its most extended treatment in *Middlemarch* during Dorothea and Casaubon's honey-moon in Rome. Those chapters, which move between ancient architectural spaces, famous museums, and the private studios of historical painters, dem-onstrate Eliot's commitment to a mode of historiographic comparison that plays alternative perspectives on the past against each other, emphasizes disjunction, and unsettles the conceit of a more traditional and more coherent history. The "city of visible history" stands in the novel as the quintessential artifact of his-torical complexity, a touchstone for the capacity of individual characters to make sense of its limitless narratives of change. As a mirror of imperial history, clas-sical Rome was an object of fascination for Victorian Britain.[37] Edward Bulwer Lytton's *Rienzi: The Last of the Roman Tribunes* (1835) was only the most popular of a range of dramatic and prose fictions that explored Rome's past as an ana-logue to contemporary debates. Victorians found, within Rome, a reflection on the relation between popular radicalism, demagoguery, and republicanism; the imperial ambitions of France and Prussia; the status of mercenary forces in modern warfare; and, most importantly, contemporary efforts to unify the Ital-ian states. Within *Middlemarch*, however, Roman history is defined by its resis-tance to such analogies. Its historical confusion makes Dorothea nauseous, and the narrator explains its effect by drawing a contrast to "those who have looked at Rome with the quickening power of a knowledge which breathes a growing soul into all historic shapes, and traces out the suppressed transitions which unite all contrasts, [for whom] Rome may still be the spiritual centre and interpreter of the world" (193). The city's profuse architectural styles, art, and customs are bewildering because she sees only "contrast," not connection and "transition." The important contrast here is really to Will Ladislaw, who within a few pages describes his "enjoyment" of the "miscellaneousness of Rome, which made the mind flexible with constant comparison, and saved you from seeing the world's ages as a set of box-like partitions without vital connection. . . . [H]e confessed that Rome had given him quite a new sense of history as a whole" (212).

Will's "history as a whole" remains tantalizingly unrealized but seems to consist in a sense of interconnection produced through comparisons that unsettle fixed notions of stadial or national history as a "set of box-like partitions." By organizing a more even and distributed network of analogies between individual artifacts and monuments, "constant comparison" provides Will with a "new" understanding, because these differentiated perspectives mark a productive movement away from the strong focus of a more traditional history. Dorothea's contrasting problem is that her "toy-box history of the world" (86) and "meagre Protestant histories" (193) give no purchase on the complex social history of Rome; she knows virtually nothing about the city, not even Edward Gibbon's influential description of its classical republican decline.[38]

The fact that much of this analysis of Rome's history is routed through artwork underlines the close coordination between historical fiction and contemporary aesthetic debates about historical painting and realism.[39] In *Middlemarch*, the problem of making sense of Roman history stands in for a more general problem with finding the "vital connection" that constitutes the real meaning of history—whether at the level of the individual, the political body, or the school of art. For this reason, the contrast between how Dorothea and Will encounter Rome's art underscores the value of comparison to the historical understanding. The differences between these experiences reflect a question of aesthetic engagement recently analyzed by Dehn Gilmore in her study of Victorian exhibition culture and its impact on the literary market. Gilmore suggests a contrast between the "familiar looking" of the connoisseur who evaluates the individual artwork within the framework of previous knowledge and the "repeated looking" of less experienced viewers who draw inferences across the serial experience of fresh example.[40] In *Middlemarch*, "familiar looking" is organized in historical terms, whether at the level of the individual experience or the nation. Will's "constant comparison" argues that repetition and familiarity stand in dialogue; repeated comparison can destabilize what is familiar and intuit alternative ways to understand the past.

In the Roman chapters, the art object stands as the consummate focus for our insight into the nature of historical perception and realism, as well as the importance of comparison in navigating between them. When Dorothea first encounters Will, a Roman gallery serves as a canvas for testing alternative theories of the real. Ladislaw and his painting companion, Adolf Naumann, spot Dorothea first in the gallery, and the narrator carefully sets the scene by arranging her in an ekphrastic tableau: she leans against a column among the sculptures, "one beautiful ungloved hand pillowed her cheek, pushing somewhat backward the white beaver bonnet which made a sort of halo to her face around the simply braided dark-brown hair, . . . her eyes were fixed dreamily on a streak of sunlight which fell across the floor" (189). Naumann, our guide, gushes about her "antique beauty . . . arrested in the complete contentment of its sensuous perfection." His speech is stocked with the formal language of German Romanticism and Hellenist aesthetics, but he uses it to make a case for historical representation, insisting that Dorothea's form reflects her intense "consciousness of Christian centuries." The passage tries on the features of what Jonah Siegel has termed the "art romance" in nineteenth-century fiction, a genre that interprets desire through the formal protocols of art criticism.[41] By this point in the novel, we are prepared to guess that this romance will be subject to some version of realist critique.

Art was central to the vocabulary of realist aesthetics.[42] Kate Flint and Ruth Yeazell have explored the importance of art criticism to discussions of literary realism in midcentury. In the larger view, writers found within art criticism an appealing model for the materiality of art's depiction of the world, an example of what Flint terms "the potential of material objects to bear witness to the process of social history that underpin the world of the text."[43] More particularly, as Yeazell notes, literary realism, particularly in Britain, looked to central currents in contemporary art criticism, especially the contemporary vogue for the domestic verisimilitude of Dutch realism, for a realist vocabulary.[44] Naumann identifies Dorothea in the gallery as a public instance of the relation between particular and type, and her status as both object and vehicle for historical consciousness seems to point in the direction of a mimetic realism. He articulates a way of seeing, in the "complete contentment of [Dorothea's] sensuous perfection," an answer to realism's dual need to both particularize and typify character as a token of reality.

Naumann's confidence, as well as Dorothea's silence, persuades us to accept his judgment, setting us up for the next chapter, which begins with Dorothea in her private apartment, "sobbing bitterly" (192). The deflation of Naumann's theory of "contentment" exposes the inability of his aesthetic vocabulary to get at Dorothea's interior view, while opening a new insight into Dorothea's life. More important, the collapse establishes an analogy between our misunderstanding and Dorothea's own. As always in Eliot's fiction, the emphasis is on the movement from type to incommensurate token; from formal model to a particular instance that does not fit. In marrying Casaubon, Dorothea had hoped to develop a union of mind; her essential failure was to imagine that Casaubon was another example of herself, that he, too, desired a communion of scholarly labor.

The misunderstanding is reciprocal. Dorothea assumes that Casaubon shares her desire for scholarly communion, while Casaubon mistakes Dorothea's eager questions for the first jabs of an invidious scholarly quibbler after his own dry heart. For Casaubon, there is no hope for this failure through identification—he will never discover how to span the different perspectives that separate them, and he ultimately dies isolated by a deep misunderstanding. For Dorothea, another path opens. At the close of the following chapter, after Dorothea asks that he forgive her curiosity, and after Casaubon expresses his moldy satisfaction, the narrator continues: "Dorothea remembered it to the last with the vividness with which we all remember epochs in our experience when some dear expectation dies, or some new motive is born. To-day she had begun to see that she had been under a wild illusion in expecting a response to her feeling from Mr Casaubon, and she felt the waking of a presentiment that there might be a sad consciousness in his life which made as great a need on his side as on her own" (211). At this level of focus, the macrohistorical unit of the "epoch" finds meaning in the important transitions of personal history, punctuated by a shift in understanding. The personal epoch is constituted by an alignment between a series of disjunctions; between Naumann (with the reader) and Dorothea in the gallery; between Dorothea and Casaubon in their early marriage. The weight of the three chapters we've been discussing, bursting with narrative digressions, revisions, and shifts in historical perspective, is distilled into this single precious analogy between the independent configurations of self and other.

Eliot elsewhere states her avowed program succinctly as the "extension of our sympathies," and the rich narrative texture gathered so carefully in these passages, organized around this single troubled instance of sympathetic identification, registers the delicate formation of that sympathy in time.[45] The point is that Dorothea lacked the background to work through such comparative discovery in art history, but a few months of married life has given her sufficient (and sufficiently painful) experience to work through personal history in the comparative mode. The narrator draws the reader's experience, as well as the narrator's own, into contact with Dorothea's discovery, comparing Dorothea's reflections to "the vividness with which we all remember epochs in our experience when some dear expectation dies, or some new motive is born" (211). An alignment of error generates new possibilities. Here, the coordinating "or" is causal: it is the death of the "dear expectation" that quickens the birth of "some new motive" in a movement from convention, to alienation, to new recognition. By recognizing that Casaubon is *not* like her, Dorothea begins to perceive how he *is*. His "sad consciousness" makes "*as* great a need on *his* side as her *own*" (emphasis added). The grammar of analogy is rarely so explicit. Produced through a negative epistemology, a way of learning through error, this analogy of need is a precious and unstable insight. Its tenuous nature is triply marked by its "wakening" state, its status as a "*pre*" sentiment, the hedge that it "might" hold true. It is a hypothesis—an abductive insight—rather than secure knowledge.[46]

This is the only knowledge available in a novel that commits to the social nature of history and meaning; knowledge is not the province of individual insight but, rather, something that must be shared. Dorothea's discovery is marked as an "epoch" in relation to her own life, but its parallel connection to the lives of the readers and the narrator—and, more immediately, to Casaubon's own—discloses the social nature of the experience. Rather than a determinate point in the linear narrative of her history, it is a point of true contact between her own history and Casaubon's. The structure of analogy that sustains the insight marks this new and better-founded proximity between these two narratives; it is fair to say that, in this instance, Dorothea and Casaubon have never been closer. To put this differently, comparative historicism, as practiced in *Middlemarch*, transitions from the successive order of the individual narrative to the synchronous contact between narratives. Hence the moments that J. Hillis Miller considers antihistorical and Levine transhistorical are really movements into comparatism; we are not outside of history but engaged in a space that opens between histories. I belabor this point because it provides a distinct way of answering what Harry E. Shaw terms the "grand hermeneutical problem facing historicist realism: How is it that someone in the stream of history can have sufficient distance on other historical moments to give an account of them that achieves more than a simple reflection of his own historical moment?"[47] Seen as a way to achieve sufficient *contact with* rather than *distance from* "other historical moments," comparative historicism provides a particularly compelling solution.

Formally my argument here brushes into what Shaw calls the "historicist metonymy" of realism, a sense of the part-to-whole relation between historical narratives and the "ontological claim" of the realist novel: that it "tells us what our world is really like."[48] Shaw's formulation gives a powerful historicist bent to the link between realism and the figure of metonymy (in contrast to metaphor), first asserted by the linguist Roman Jakobson, and later popularized by the literary theorist David Lodge. At the same time, Shaw's careful work to reshape "metonymy" as meaning something more than part-to-whole connection, suggests metonymy's inadequate depiction of realism's compelling and complex features.[49] I take this as a tacit admission that any single figure will struggle to account for the intricate relation between realism and historicism. In chapter 2, I explored how Scott's historical novels interrogate major narratives of eighteenth-century historicism, particularly Whig progress, Christian eschatology, and stadial history, through a more complex and conflicted study of historical sedimentation, differentiation, and alternative possibilities. J. Hillis Miller and Neil McCaw see a similar tension in Eliot's novels between history's "master narratives" and their critique.[50] The ability of comparative historicism to function without a strong thesis of history's pattern is important for a novelist like Eliot, who rejected strident progressivism explicitly, comparing it to the millennial fervor of her youth, a fantasy that eventually "superstition will vanish, and statistics will reign for ever and ever."[51] Instead, a novel like *Middlemarch* links realism and history by means of comparative analysis, through which patterns emerge between histories, supporting the sense of a "real" order, even if its true nature can only ever be partially glimpsed. As Eliot puts this, at the close of the novel, "the fragment of a life, however typical, is not the sample of an even web" (832)—though it is a real sample.

Analogy is key to this mode because, unlike metaphor or metonymy, it has an explicit syntax that can assert a serial or "harmonic" equivalence between the objects compared (even if, in practice, it often works to map a formal model instead). Metaphor and metonymy are syntactically reticent; in substituting one term for another—vehicle for tenor, part for whole—they appear to make a semantic error, forcing the reader to reimagine their semantics and capitulating a different meaning. This play of semantics against syntax is why Paul Ricœur describes metaphor as an "impertinent predicate" or, with reference to Gilbert Ryle, a "category mistake."[52] Analogy, by contrast, emphasizes the pertinence of pattern to its grammar, using an *explicit* syntax to elucidate and reinforce a set of semantic relations. So, in the preceding passage—"there might be a sad consciousness in his life which made *as great* a need *on* his side *as on* her own" (emphasis added)—the adverbs and prepositions structure a comparison drawn between lives and their respective desires. Eliot's realism invests in comparative historical analysis, leveraging analogy's power to flatten formal modes of historical characterization (including both metaphor and metonymy) into opportunities for contact, engagement, and recognition. Realism, as I understand it here, is about relationships, between things, between people, between people and things, both within the novel and beyond. Analogies allow such similarity to extend through sentences (and in time), resonating between individuals, incidents, objects, and, most importantly, narratives.[53]

"Higher Criticism" and the Natural History of Social Life

What the science of Bible-criticism, like all other science, needs, is a very wide experience from comparative observation in many directions, and a very slowly acquired habit of mind.

Matthew Arnold, *Literature and Dogma* (1873)

I have already noted that Eliot's work for the *Westminster Review* gave her an extraordinarily broad sense of the comparative method as it operated within various disciplines of historical, social, and scientific study. This long engagement with comparatism, particularly in studying how habits of thought change over time, began earlier, as part of her first experiments with biblical study and Christian history. Her deep concern for history, and how language documents its changing contours, initially developed through her effort to understand the religious tradition in which she was raised. We often forget that, before Eliot's climactic falling away from her Evangelical faith, she was a devoted practitioner of biblical prophecy—the art of using internal dating and typological analysis to make predictions about the coming apocalypse and its significant events. A lack of primary documents has left the young Eliot's early millennial efforts tantalizingly undeveloped.[54] And while she eventually turned from Christianity, Eliot retained a strong sense, informed by biblical hermeneutics, of language's complex signification.

This set the condition for her translations of David Friedrich Strauss and Ludwig Feuerbach, two major proponents of the German "higher criticism" who explored how Christian accounts changed over time. The "higher criticism" was really a host of associated practices that responded to the insight, generated by the "lower" critics and their comparative studies of individual biblical accounts, that the Bible was a composite of different historical moments and authors. The earlier generation of critical comparatists had used contrastual comparison to reject false emendations—part of an ambitious effort to restore the unitary original text—but they only succeeded in producing a sense of biblical incoherence. The higher critics turned this effort around, working to restore a sense of general unity by isolating the common patterns that comparatism disclosed, especially in their exploration of the impact of historical changes upon the ideology and worldview of biblical accounts.

Eliot's translation of Strauss's *Life of Jesus* (1846) began amidst the liberal-theological foment of the Rosehill Circle and contemporary efforts at Anglican reform.[55] Unitarian theologians and freethinkers drafted Strauss in their effort to displace the Trinitarian dogma of the Anglican Church. As Suzy Anger and Susan Hill have shown, these new approaches to biblical interpretation, which furnished a robust hermeneutic understanding, shaped Eliot's sense for how we read through common experience and language to achieve social insight.[56] Eliot's attention to German higher criticism, and the challenges of linguistic and cultural translation, drew the close relationship between sympathy, objectivity, and history into sharp resolution. It also afforded her a sharp sense of how linguistic forms changed over time. Strauss emphasized an approach to the Bible that was "formal" rather than "genetic"; rather than asking what events *really* happened, he explored how the accounts changed over time, eventually taking the "form" we find them in today.[57] Strauss drew on early work in comparative mythology and, by these means, placed Christianity in serial relation to other systems of belief. As he put it in Eliot's translation, his argument was "supported by the analogy of all antiquity, political and religious, since the closest resemblance exists between many of the narratives of the Old and New Testament, and the mythi of profane antiquity" (1:32–3).[58] Feuerbach followed Strauss's formal method closely, particularly in his emphasis on the close relation between linguistic form and abstraction. One of his key arguments was that the sentence-level distinction between subject and predicate induces a false distinction between divine predication and being: "The true sense of Theology is Anthropology. . . . [T]here is no distinction between the *predicates* of the divine and human nature, and, consequently, no distinction between the divine and human *subject*."[59] Feuerbach's "anthropological" analysis argued that central elements of Christian doctrine took their significance as resonant celebrations of central elements of the human experience (in the case of the Eucharist, the need for food and drink). In this way, Strauss and Feuerbach were central to higher criticism's study of language as an element of the changing (as well as unchanging) features of human experience over time.

As we have seen, Eliot generalized Strauss's "formal analysis" as a model for the historical inflection of form in general, but she never forgot this deep connection to the comparative analysis of historical accounts. In a late critique of comparative mythology, she argued that "in whatever concerns the life of man there is a constantly increasing complexity or interaction of various conditions, so that similar phenomena may often be produced by a different concurrence of facts & events & that you can hardly find any phenomenon which is not a highly mixed product. Surely it is unreasonable to explain mythologies exclusively by the personification of natural forces as by the deification of kings & heroes or the symbolical exoteric teaching of priests" (390). As a product of history, form is always mixed and changing. Accounts that characterize Eliot's critical analysis of form as "self-defeating" fail to see how that formal analysis deploys comparatism to attend to historical complexity.[60] As I will shortly argue, this historicization of form makes it peculiarly responsive to human experience and adaptive to new conditions.

Eliot's review of Wilhelm Heinrich Riehl's *Natural History of the German People*, generally seen as an important statement of the relation between scientific naturalism and literary realism, explores the capacity of Strauss's "formal" method to analyze colloquial language as a record of human experience. Her review of Riehl remains "an oblique piece of writing," as Fionnuala Dillane notes, and we should hesitate to read it as a statement of her intent as a novelist. For Dillane, this is further evidence of the important differences between Eliot's early career as editor and essayist and her later work as author.[61] Yet this careful revision of Riehl contributed to her ongoing effort to shape a mode of comparative historicism suited to her interest in the idiosyncrasy and importance of everyday life, a historical realism that contributes substantially to her fiction.

Riehl represents himself as a natural scientist, examining the particularities of culture and locale in a manner explicitly "analogous" to the naturalist in the field. The journal of Eliot's experience reading Riehl during her seaside retreat, as she helped her lover, George Henry Lewes, look for new fauna in the rocky shoals, is often adduced to show the impact on Eliot's naturalistic brand of description: "When one sees a house stuck on the side of a great hill, and still more a number of houses looking like a few barnacles clustered on the side of a great rock, we begin to think of the strong family likeness between ourselves and all other building, burrowing, house-appropriating, and shell-secreting animals."[62] We have already seen an example of how Eliot uses this organic metaphor to think about form. Here, it is evident that Eliot responds to Riehl as she takes up his evocative speculations on the interrelationship between local culture and local building materials. Riehl proposed that "a comprehensive analysis of the influences that building materials exert on a national character would be a major task for the cultural historian."[63]

Eliot's description of Ilfracombe, and particularly Lantern Hill, is seen as illustrating an essential commitment to a style of scientific observation that emphasizes the organic unity between the individual and local conditions, with a concomitant conservative gradualism.[64] But the passage departs significantly from Riehl by putting equal weight on the division between the natural world and the human. For the bivalve, the "fact of individual existence" and the completion of its house are closely linked in a naturalized coordination of environment, individual, and home. Eliot emphasizes that for human habitations, in contrast, there are many more "steps and phenomena," an unnaturally labored effect that makes humanity seem "parasitic." Those steps register a fissure that marks the distance between "the individual [human] existence" and the complex forces of society and environment. The analogy between house and mollusk shell calls attention to the inorganic relation between individual and culture. This tension between the individual, nature, and culture is reflected throughout Eliot's writing, which overflows with characters physically and mentally out of step with their world.

In the review, Eliot creates space for a freer interpretation of Riehl by cautioning that she "must be understood not as quoting . . . but as interpreting and illustrating."[65] In fact, the review translates Riehl's insights freely into the historical and philological dimensions of her distinct perspective on social history, placing Riehl in close dialogue with Strauss and Feuerbach. Conceiving of himself as a naturalist, Riehl sees language as a largely unmarked and unproblematic medium through which he recuperates a catalog of local customs and tradition in order to produce a composite analysis of the correspondence between social groups and their environment.[66] Language is not an object of complexity in Riehl's view. Eliot, on the contrary, argues that "the language of cultivated nations is in anything but a rational state; the great sections of the civilized world are only approximately intelligible to each other, and even that only at the cost of long study" (164). This is a major problem for Riehl's approach to ethnographic analysis but not for the view of social history that Eliot seeks to locate within his study. She emphasizes the value of these "subtle shades of meaning, and still subtler echoes of association" precisely for their opacity, their social and historical complexity.

In foregrounding the virtue of our common implication in historical language, Eliot provides an important qualification to the discourse of detachment that Amanda Anderson identifies in Eliot's argument; a qualification that Anderson terms "both subtl[e] and ambigu[ous] when considering the purpose and consequences of cultivated distance."[67] Eliot's corrective is directed at Riehl, who forcefully advocates an objective vocabulary that overlooks the vicissitudes of "historical language." By way of illustration, Eliot sets up a counterexample, a truly "detached" and scientific language:

> Suppose, then, that the effect which has been again and again made to construct
> a universal language on a rational basis has at length succeeded, and that you
> have a language which has no uncertainty, no whims of idiom, no cumbrous

forms, no fitful simmer of many-hued significance, no hoary archaisms "familiar with forgotten years"—a patent deodorized and non-resonant language, which effects language as perfectly and rapidly as algebraic signs. Your language may be a perfect medium of expression to science, but will never express *life*, which is a great deal more than science. With the anomalies and inconveniences of historical language you will have parted with its music and its passions, and its vital qualities as an expression of individual character. (165)

The allusive passage, with flourishes of both Wordsworth and commercialese, emphasizes the contrast between the mechanics of a formal language like mathematics and the musical qualities of "historical languages."[68] The coordination between a critique of the delusive "universal language" and the illusive transhistoricism of "universal history" is implicit in the value she finds in language's historical function. Instead, Eliot emphasizes both language's deeply sedimentary nature as a repository of social tradition and our necessary implication within it; as she puts it, "There is an analogous relation between the moral tendencies of men and the social conditions they have inherited." The review grafts a substantive historical understanding of language's evolution onto Riehl's ethnography. Eliot undertakes all this in an attempt to characterize what she describes as the "conception of incarnate history" that serves as "the fundamental idea of Riehl's books" (166). The particularity of this "incarnate" history, as formulated by Eliot, cuts across the grain of Riehl's effort to build a systematic natural history of the German folk; "uncertainty," "idiom," and "archaisms" do not contribute unity.

In emphasizing the particularity of incarnate history over systematic ethnography, Eliot advances a more sophisticated reading of Walter Scott's fictions than Riehl can provide. Hence, if Riehl lauds Scott's historical novels for capturing the "social core" of the "English," Eliot understands how Scott's novels, particularly in their deep comparative investment, are predicated on the *lack* of this "social core," a point repeatedly emphasized in his careful use of idiom and dialect. This is what I have termed the "translational historicism" of Scott's fiction, the ability to translate between different moments rather than assuming a common language. As Eliot puts this in her 1854 review on Ruskin's Edinburgh lectures: "The aim of Art, in depicting any natural [or historical] object, is to produce in the mind analogous emotions to those produced by the object itself; but as with all our skill and care we cannot imitate it exactly, this aim is not attained by *transcribing*, but by *translating* it into the language of Art."[69] Eliot's own translations of Strauss and Feuerbach afforded this fine-tuned sense for linguistic translation as way to mediate between distinct cultures and historical periods. Recalling Eliot's essay on the "Historic Imagination," we may recognize Eliot's purpose in reviewing Riehl's study as a careful and differentiated attempt to reframe the natural history of social life in more comparative terms, emphasizing complexity and differentiation over an organic social "whole."

Harmonic Sympathy in *Middlemarch*

Then, as the metal shapes more various grew,
And, hurled upon each other, resonance drew,
Each gave new tones, the revelations dim
Of some external soul that spoke for him.

George Eliot, "The Legend of Jubal" (1867)

The characters in Eliot's novels, in their struggles to understand each other, allegorize this concern to interpret historical difference. Consistently, they suggest that the struggle to make sense of others—whether contemporaries or actors in the past—is characterized by long periods of misunderstanding punctuated by moments of discovery and contact. It is clear that Eliot's fiction imagines such moments of contact between various figures: between author and reader, between characters, between reader and character. As Eliot famously puts this, in her review of Riehl, "The greatest benefit we owe to the artist, whether painter, poet or novelist, is the extension of our sympathies," and even if this comes as she imagines a natural history of social life, her emphasis upon artistic professions, as well as their fictional worlds and characters, underlines her belief both that we can feel with fictional actors and that this sympathy furnishes a mode of understanding that applies equally to the living. If, in chapter 3, I argued that Tennyson's *In Memoriam* turns substantially on the possibility of collaboration with the dead, the question I ask now is how novels might allow us to feel with people who never lived.

Part of the answer lies in Eliot's emphasis on the experience and validation of feeling itself. Eliot drew her formative conception of interpersonal contact, the "extension of sympathies," from a closer consideration of how understanding could extend beyond the self, how moments of contact are distributed in time and space. Foremost among such models, I argue, was musical theory. From its consequential effect on Maggie Tulliver in *The Mill on the Floss* to Herr Klesmer's long-winded disquisitions in *Daniel Deronda*, Eliot consistently registered her interest in musical theory and the properties of musical experience. Eliot's long interest in music is well documented, and my reading of this engagement follows Peter J. Capuano, who compellingly argues that music serves Eliot to adapt German phenomenology, and particularly the work of Arthur Schopenhauer, as a theory for the immediate grounds of experience.[70] According to Schopenhauer, music "never expresses phenomena, but only the inner nature, the in-itself of all phenomena," in contrast to the objects of visual perception, which "contain particulars only as the first forms abstracted from perception, as it were, the separated shell of things."[71] If visual experience encourages us to see the distinct forms or "shells" of things, it also encourages us to abstract those forms as self-sufficient entities. On this view, Eliot cautions against "false idealization" as an artifact of discrete visual experience—hence our tendency to take forms as finished rather than contingent and historically evolving features. Music, in contrast, demonstrates how we are tangled up in the life of history. For one, music is experience in a spatial distribution alien to the presentation of objects to sight; for another, it can be shared, not only between listeners but in the sympathetic vibrations between listener and instrument, shared in a way that finds few analogues in visual experience.

Nick Dames has recognized Eliot's interest in music as part of a larger Victorian concern for the physiology of novel reading, an attempt to use contemporary developments in brain science to understand the mechanics of novels and their effect on readers. Taking up Catherine Gallagher's call for a "concept of novelistic length that includes analytic insights into the temporal nature of narrative," Dames argues that novels, which are characterized by their long, seemingly amorphous shape, "must be defined temporally, as a rhythm or time signature, rather than a synchronic structure."[72] This account recognizes a central feature of Eliot's concept of form itself, as it responds to the "simple rhythmic conditions of . . . life." As over against Dame's case for a theory of the novel "interested in horizontal, or rhythmic-melodic, analysis, rather than vertical, or harmonic analysis," I think Eliot deployed musical theory precisely to understand how harmonic instances of fellow feeling and reflection could persist over the course of her novels and between her characters.

In the "Legend of Jubal," a short biblical epic published in her 1867 collection of poems, Eliot wove these different musical perspectives together as a historical allegory in which music is invented by the sons of Cain as one way (along with agriculture and mechanical industry) for the individual to triumph over death. It is a moving but strange poem and refracts many of Eliot's concerns, among them the multivalence of Christian allegory in humanist meaning and the capacity of "analogical creation" to produce new historical understanding.

The poem traces Cain's expulsion after Abel's murder and the continuing modulation of social unity into dramas of anguish and loss.[73]

Eliot follows Genesis, assigning to each of three brothers, Jabal, Tubal-Cain, and Jubal, the invention of husbandry, metalwork, and polyphonic music. These discoveries are additive; Jabal "learned to tame the lowing kine," after which model Tubal-Cain "caught and yoked the fire," producing his bellows.[74] Strictly speaking, Jubal does not invent music; his brother, Jabal, while soothing his herds, discovers the "friendly tone" (12), while Tubal-Cain, his hammer ringing at the forge, teases out the cadences of its rhythm. Jubal, for his part,

Watched [Tubal-Cain's] hammer, till his eyes,
No longer following its fall or rise,
Seemed glad with something that they could not see,
But only listened to—some melody,
Wherein dumb longings inward speech had found,
Won from the common store of struggling sound.
Then, as the metal shapes more various grew,
And, hurled upon each other, resonance drew,
Each gave new tones, the revelations dim
Of some external soul that spoke for him.   (16–7)

On the basis of his brother's ringing hammer, and the reverberating ring that fills his home, Jubal discerns "Concords and discords, cadences and cries / That seemed from some world-shrouded soul to rise . . . , / Some living sea that burst the bounds of man's brief age" (17). It will take Jubal the rest of the poem to learn the full import of these words, as he invents the harp and travels the earth, teaching its harmonies to the people he finds, before finally returning home. Jubal dies an outcast, and initially grieves to learn that, despite the "the myriad worlds that people space, / And make the heavens one joy-diffusing quire," he will die alone (38). But Jubal finds comfort in the discovery that his harmonies live on, and, like Tennyson's *In Memoriam*, Eliot holds out hope that they transcend earthly experience. In his final rapture, Jubal is raised by an angel, "floating him the heavenly space along, / Where mighty harmonies all gently fell / Through veiling vastness" (45).

Eliot's account in the "Legend of Jubal" gives a mythical interpretation of what she knew of musical history. As the musician George Alexander Macfarren described it, in his *Six Lectures on Harmony* (1867), which Eliot and Lewes possessed: "Harmony—in the musical sense of symphony, accordance, combination—was unknown to the Greeks. . . . The art of music, therefore . . . dates only within the last six or seven centuries." Macfarren explains that Greek music consisted only of melody: "In music, the word harmony expressly defines a *combination* of notes, in contra-distinction to melody, which means a *succession* of notes."[75] In Eliot's account, Jubal does not use his lyre for melody alone (as did the Greeks); he explicitly calls himself the "sire of harmony." It is this turn toward harmony that gives Jubal's music the ability to transcend the finitude of death: whereas melody is about succession, harmony is about simultaneity within succession, the ability to resonate within the moment, and by these means, a figure also for resonance across time.

It is clear why this musical theory resonated with Eliot. Not only did Macfarren provide a surprising gauge for how much the experience of music had changed over time, but he saw harmony as a feature of social experience in general: the "Greek original [of 'harmony'] defines the fitness, propriety, accordance of things; so that we use the word in a primitive rather than a figurative sense, when we speak of harmony among the members of a society—of a harmonious whole, comprising the diverse elements in a work of art."[76] Macfarren's theory of harmony offered a way to recognize the potential of both social and musical experiences to produce communion, a way of being together in a "fitness" that preserves the "diverse elements" that constitute our parts. Harmony provides a model of sympathetic fellow feeling that is insistently shared between subjects, rather than within a single organic body, a way of thinking about sympathetic analogy—the belief that one person's feeling can reflect how another feels—as a shared experience that hangs between its terms as much as within them.

Though it is common to read sympathy as a central feature of Eliot's fiction, Rae Greiner has recently provided fresh interest in the power of the "sympathetic realism" that characterizes Eliot's novels and much of nineteenth-century fiction.[77] In Greiner's view, "Thinking of me thinking of you, thinking of you thinking of me," produces a "form of thinking geared toward others," and so "guarantees the reader's fullest contact with realism's most sought-after cognitive and emotional effects."[78] Greiner's analysis provides a compelling account of how sympathetic engagement helps apprehend the differences and complexities that sustain realism in the novel. For Greiner, working from the tradition of spectacular sympathy developed by Adam Smith, the analogy between your "thinking" and mine is formal; it is *me* "thinking of me thinking of you" that gives a model for how *I* am to start "thinking of you thinking of me." But Jubal makes it clear that, for Eliot, sympathy is a social system; as much as a cognitive or rhetorical effect, it is a mode of social harmony that exists between reader and character (and narrator), and not within the reader alone.[79]

Near the close of *Middlemarch*, the complicated harmonics of this sympathetic engagement are made startlingly clear in a scene that constitutes the emotional climax of Dorothea's plot. When the codicil to Casaubon's will becomes known, Will Ladislaw says goodbye to the now-widowed Dorothea and leaves town. A year passes, taken up with striking events: a blackmail, a murder, a communal scandal. Lydgate, who is connected to the scandal, finds himself unable to explain his largely innocent role to his wife Rosamond, and he accepts Dorothea's offer to intercede on his behalf. But when Dorothea first arrives at the Lydgate home, she is ushered in to find Will Ladislaw attending upon Rosamond. There is a misunderstanding—Dorothea believes Will is Rosamond's lover, and she leaves, heartbroken—only to return the next day, *and on the same errand.*

The sequence once seemed superfluous to me—does the narrative demand yet one more check to the potential union between Dorothea and Will? Dorothea returns to her chamber, much as she did after her visit in Rome, and is racked by anguish and jealousy. Significantly, she is torn between "two images"—one that pictures Will as a "bright creature whom she had trusted," the other, as "a changed belief exhausted of hope." But the point of the passage is how Dorothea tries to work past these "images" and recognize in Will a "living man." And in the wee hours of the morning, Dorothea begins to recover from her grief. The passage explains: "[T]hat base prompting which makes a woman more cruel to a rival than to a faithless lover, could have no strength of recurrence in Dorothea.... All the active thought with which she had before been representing to herself the trials of Lydgate's lot, and this young marriage union which, like her own, seemed to have its hidden as well as evident troubles—all this vivid sympathetic experience returned to her now as a power: it asserted itself as acquired knowledge asserts itself" (788). Here the tension between individual perspective and larger recognition is reproduced as Dorothea's struggle between conflicting desires for Will's love and her ambition to give Rosamond her understanding. Dorothea must disentangle herself from the general plot of the "woman more cruel to a rival than to a faithless lover," in order to see Rosamond as a particular person. This activates her "vivid sympathetic experience." For Dorothea, this is only the most difficult of a series of trials in which, as Gallagher puts it, she must "learn to realize others by imagining their particularity instead of pressing them into categories."[80]

The peculiar thing about Eliot's fiction is that while such particularity *is* realized, it is not through an exclusive focus on differences (the endless distinctions that isolate Rosamond from Dorothea) but rather through the network of both similarity and difference that constitutes character and allows it to resonate in society. Here, a static triangulation between three lovers is replaced by Dorothea's "acquired knowledge"—a knowledge produced by an analogy that extends beyond this triangle. Specifically, Dorothea remembers that this "young marriage union" between Lydgate and Rosamond is "like her own"—it too has "hidden as well as evident troubles." It is worth noting that this analogy has a compound object—the unions possess "hidden" *and* "evident" troubles. And this compound works to extend the analogy, mapping its structure onto a more complex network of social relationships.

There are profound implications for the way Dorothea uses analogy to work through the pattern of relations that hang between her own experience and Rosamond's, and I wish to pursue them by taking up the considerable critical back-and-forth that has developed around the novel's engagement with alterity, understood as the apprehension of otherness. Many critics have asked how Dorothea, given her sheltered and privileged existence, has the capacity to engage with Rosamond and the larger world. George Levine takes up the problem in the last chapter of his *Realistic Imagination*, where he confronts Edward Said, whose foundational study *Orientalism* continues to shape our thinking about novels and their struggle to grasp alterity.[81] Understood as a fantasy of Western knowledge, the critique of orientalism jeopardizes realism itself, given realism's ambition to understand other lives and so to provide knowledge of the "real." As literary theorist Harry E. Shaw observes, this is equally problematic for historicism and its ambition to understand the past.[82]

Orientalism in fact cites Casaubon's "key to all mythologies" as both a criticism of this fantasy and a marker of Eliot's ambition to know more. Elaborating the conditions of such fantasies, Said returns to this problem in a later study of realism, arguing that realism depends on the knowability of origins that are "divine, mythical and privileged," rather than "secular, humanly produced, and ceaselessly re-examined."[83] Eliot's rejection of "origins" is in fact closely aligned with Said's (as I will shortly argue), and this helps us understand why the "Legend of Jubal," a poem about the origin of music, explores myth as a vehicle for secular insight. The problem is how one can be both grounded and in contact with unfamiliar territory, a problem that Said recognizes as a larger problem for comparative criticism in general. His intriguing response, developed in *Culture and Imperialism*, turns, like Eliot, to musical models: "A comparative, or better, a contrapuntal perspective is required. . . . we must be able to think and interpret together experiences that are discrepant, each with its own particular agenda and pace of development, its own internal formations."[84] Contrapuntal music is characterized by the presence of two or more melodies that, though they sustain their own "agenda and pace of development," produce in harmony and co-incidence of time a sense of emergent alignment, a collective pattern. By these means, Said's contrapuntalism pluralizes the objects of comparative analysis, preserving independence and equanimity while establishing a common ground for emergent patterns. In the key terms of these discussions—"continuities and discontinuities," "comparative" versus "contrapuntal"—I read an attempt to locate the formal characteristics of what I have characterized as *harmonic analogy*: a "relation between relations" that elucidates common pattern by establishing conditions for a productive and evenhanded resonance between two systems of understanding.

Perhaps the best statement of this thesis is provided by the sequence of events that succeeds Dorothea's "vivid sympathetic experience" of the hidden troubles of the Lydgate marriage.[85] Firmly regrounded in a revised analogy between the Lydgate marriage and her own, Dorothea returns to the Lydgate household. Rosamond is disarmed when Dorothea begins by explaining *why* she could understand Lydgate—their shared experiences of troubled marriage. In effect, Dorothea reconstitutes her private analogy for Rosamond. And from the position of analogy, Dorothea turns quickly to the plural first person, "we." This turn toward shared subjectivity marks the complete accession of comparative understanding, as Dorothea begins to narrate for them both. Fellow feeling secures Dorothea's pedagogy; by narrating for both of them, Dorothea extends her own sympathetic understanding to Rosamond's, tacitly asserting the commonality that mutual comparison struggles to perceive. Dorothea continues: "Even if we loved some one else better than—than those we were married to, it would be no use. . . . I mean, marriage drinks up all our power of giving or getting any blessedness in that sort of love. . . . it is so hard, it may seem like death to part with [this feeling]—and we are weak—I am weak—" (797).

The passage meditates upon the challenge of accurate sympathetic comparison. Dorothea insists on a larger analogy, between, on the one hand, her own marriage and reputed affair with Will and, on the other, Rosamond's marriage and her own potential affair: a pluralized love triangle. Insofar as Dorothea is speaking *for* Rosamond, she works to *formalize* their relation. Insofar as she collaborates *with* Rosamond in their relation, in a harmonics of understanding, it is shared, parallel, serial work. But it seems that the plural "we" has pushed Dorothea's analogy to its breaking point. In asserting her power to speak for both, Dorothea may have overstepped her authority: the "we" can collapse two individuals into a single subject for only so long before threatening precisely that mistaken identification which formed the bases for their flawed marriages to begin with. (Readers who dislike George Eliot's narrators misread them in precisely this way.) This collapse of authority is what Dorothea notes as she revises herself—"we are weak—I am weak." It marks a tension between what is believed and what may actually be true that, in Aaron Matz's account, looks satirical disillusion in the face.[86] Dorothea can push the sympathetic analogy only so far herself—she can no more force Rosamond's understanding than she could Casaubon's. There is a dilemma of correspondence between her representation and the truth of their experiences—her point grasps for counterpoint.

The sympathetic relationship is saved from collapsing into Dorothea's singular experience by a completely unanticipated turn of events: *Rosamond* begins to sympathize with *Dorothea*, and "taken hold of by an emotion stronger than her own," she assumes Dorothea's role. Rosamond narrates: "'You are thinking what is not true'" she tells Dorothea, "'When you came in yesterday, it was not as you thought'" (798). Most immediately, this means that Dorothea's analogy is wrong, that her fear that "we" was only "I" is well founded. Rosamond continues by describing the misunderstanding and reveals that Will remains as much as ever Dorothea's lover. By doing so, Rosamond dismantles the analogy shared by the two women in order to rebuild it, salvaging Will from his hazardous role as extramarital lover and refurbishing him as Dorothea's future husband. Hence, just as Dorothea has interceded, on the basis of analogous experiences, by redescribing Rosamond's marriage with Lydgate, Rosamond intercedes, through a surprised perception of common grounds, by redescribing Dorothea's future marriage to Will. In effect, Rosamond harmonizes Dorothea's sympathetic analogy and stabilizes their serial relation. And in a perhaps more striking move, Rosamond manages, for the moment, to produce a new response in the reader. Her action secures a network of sympathetic engagement in which we are deeply implicated and which she has the power to destroy. To our surprise, it turns out that Rosamond plays the central role in this last portion of the novel, and in review, it is clear that the novel has worked hard to alienate her from our sympathy—to establish her as the definitive social other—in order to prepare this final climactic moment of insight.[87]

It worked. For most of the novel, Rosamond is a toxic character. (And she largely remains so, as shown in the posthumous report that Lydgate called her his basil, because "basil was a plant which had flourished wonderfully on a murdered man's brains" [835].) When Eliot's publisher, John Blackwood, finished book six of eight he wrote, "You have painted that heartless . . . obstinate devil with a vengeance."[88] For these reasons the potential projected by this final moment of sympathetic understanding does not bear full fruit; no matter how much sympathy Dorothea lavishes upon Rosamond, she is locked in a relatively unhappy marriage. But it is equally important to stress that *Rosamond's* moment of sympathy catalyzes Dorothea's marriage to Will. As Sharon Marcus notes, in an important reading of this scene, without Rosamond's intervention, "Dorothea would never know that Will loves her and would not admit her love to him." For Marcus, the scene provides a key example of the complex and supportive relation between the "plot of female amity" and the marriage plot.[89] More generally, the scene marks, for the reader, how female friendship can model a startling instance of wider harmonic identification. By these means, we, as readers, are surprised (as we are surprised by Aunt Glegg's charity in *The Mill on the Floss*) by a sudden discovery of fellow feeling. The moment of understanding here is temporary, but it is real; beyond fellow feeling, we share a moment of fellow knowledge with someone carefully and extensively rendered as our other.[90]

Rosamond's intervention dramatizes a potential moment of failure or, more properly, a dialectical movement between the experience of similarity and a crisis of dissimilarity that drives us toward better understanding. By these means, analogy is iterated in cycles of extension and revision, reflecting what I recognize as an empirically vested comparative procedure. These scenes make clear the dependence of comparative inquiry on the coordination of analogy with its failure. The essential differentia of Eliot's use of analogy is illustrated here—she leans heavily on analogy's error as an avenue to common understanding. In *disanalogy*, the comparison established between two sets of relations is leveraged to prove dissimilarity. In such negative analogies, what is *not* common trumps what *is*. Comparative inquiry uses disanalogies to produce a negative hermeneutic that stands in critical relation to constructive analogies.

This critical relation is central to the formal tension of *Middlemarch*: the climactic engagement between Rosamond and Dorothea constitutes the most important link between the split plots of "Middlemarch" and "Dorothea Brooke." It is striking that so many readers of *Middlemarch* remember the novel as a romance between Dorothea and Lydgate. Such a possibility is sketched in the novel only to underline a more basic point: Eliot is more thoroughly committed to a comparison drawn *between* histories (as against the unitary possibilities of romantic history) than we are prepared to realize. This formal dualism replaces comedic closure with one of the richest extended parallelisms in British literature—a complement to the realist novel's effort, explored recently by

Hilary Schor, to open up the marriage plot to larger possibilities.[91] Underwriting this binary form is a commitment to the power of comparative historicism to generate insight, though not resolution. A model for broad social understanding is supplied where the marriage plot will not suffice. All of Eliot's later novels capitalized on some version of this analogy of relationships, with their tantalizing possibilities and pointed juxtapositions, misunderstandings, and reconciliations. Their relationships frustrate and actively critique the possibilities of direct identification, forcing moments of frisson, in which character and reader alike take a larger comparative stance toward understanding.

Form and the Entangled Word

Your pier-glass [or mirror] made to be rubbed by a housemaid, will be minutely and multitudinously scratched in all directions; but place now against it a lighted candle as a centre of illumination, and lo! the scratches will seem to arrange themselves in a fine series of concentric circles round that little sun. It is demonstrable that the scratches are going everywhere impartially and it is only your candle which produces the flattering illusion of a concentric arrangement, its light falling with an exclusive optical selection. These things are a parable.

George Eliot, *Middlemarch* (1871–2)

Moments of sympathetic analogy in Eliot's fiction work to destabilize the distinction between participant and observer, between character and reader, between object and subject, between fiction and realism. As we have seen, Eliot's novels activate comparative historicism to engage the historically embedded other as an "equivalent center of self," and this is only possible insofar as they test the limits of language's ability to get at history. More particularly, her novels recognize the formal features of analogy (like metaphor, allegory, and parable) as potentially distortional abstractions—the threat of "sympathy ready-made." This helps explain why, as in the preceding epigraph, Eliot cares more for scientific "parables" than procedures. Such parables demonstrate Eliot's continued interest in how specific linguistic forms—whether parables, metaphors, or symbols—shape our understanding of the world and document historical change. For this reason, *Middlemarch* underlines the connection between the search for a "primary tissue" (which Lydgate dreams will explain the development of organic tissues and organs) and the "key to all mythologies" (that Casaubon imagines will finally prove the paternal relation between Christian doctrine and other religions). Lydgate, like Casaubon, is misled by a formal model that overwrites the complexity of his studies.

As the narrator puts it, "we all of us, grave or light, get our thoughts entangled in metaphors, and act fatally on the strength of them" (85). Eliot is concerned not with metaphor in itself but rather with the tacit assumption that singular perspectives and ready-made formal models will suit all new occasions. Conceived as a struggle for understanding, science provides a dramatic illustration of the struggle to interpret patterns in the world. And science itself, *Middlemarch* argues, would benefit from a better understanding of the social world it operates within. After all, it's only when Lydgate consults with Dorothea that he begins "to see things again in their larger, quieter masses, and to believe that [he] too can be seen and judged in the wholeness of [his] character" (756).

We are led into error when we overlook the historical implication of forms, whether scientific or social. Such thinking regresses toward Platonic idealism, imaging that forms have a status independent of their manifestation in time through concrete examples. I have already alluded to Eliot's concise but profound "Notes on Form," and here I want to return to its analysis and the observation that "poetic form was not begotten by thinking it out or framing it as a shell which should hold emotional expression, any more than the shell of an animal arises before the living creature; . . . [both] grow & are limited by the simple rhythmic conditions of . . . life." Though Eliot's reference is to "poetic" form, her critique of Lydgate indicates that this observation applies equally to scientific modeling, in its dependence on repeated observation and adjustment. In fact, this gives an alternative way of reading the pier-glass analogy in the preceding epigraph, as a dynamic rather than static model of illusion. We are surprised by the sudden appearance ("lo!") of the "concentric circles," when a single light source is applied, and even more, by their seeming movement and reorganization, and their subsequent disappearance when diffused light returns. The insight is generated by the comparison between the different forms taken by experience over a succession of moments. The pier-glass is a parable, in other words, not merely of "flattering illusion" but of the way different forms of experience can produce startling new insights when juxtaposed, culminating in a movement of *disillusion* that gathers new insight into the world.

Eliot's "Notes on Form" sum up her career-long exploration of how forms adjust to new content. David Kurnick has emphasized Eliot's sense of the dynamic interplay between form and content; the point of her many analogies to the growth of shells was that form emerges in the lived relations of social and biological systems.[92] The bivalve metaphor was particularly effective because it established a productive analogy between the emergent formal properties of mindless biological systems and the contingent change of social forms over time. Taking up Catherine Gallagher's contention that two forms of organicism underpin nineteenth-century thought, a Romantic organicism rooted in the integrity of the biological body, and a political organicism organized in terms of political economy and the history of the social body, we might say that Eliot's shell metaphor navigates between both. For Eliot, the mollusk shell articulates the relation between the seeming integrity of synchronic form and its historical incoherence as the product of contending corporeal and environmental forces. As Eliot put this in her Ilfracombe journal, the analogy between the mollusk shell and houses on a cliff could alternatively make human habitations seem natural outgrowths of the land and reveal mankind as a "parasitic animal—an epizoon making his abode in the skin of the planetary organism."[93] In Eliot's view, time and new situations put pressure on forms that can never perfectly accommodate them; there is a necessary misalignment in time between form and content that reflects the process of growth and change.

The misalignment of form and content also creates possibility: it gives form a capacity to address new contents and adapt to them, and so affords the mollusk shell—when recognized as a model that has important distinctions from human society—to disclose something about humanity's (potentially catastrophic) impact on nature. Eliot's discussions of organic form, in the longer context of her concern for how social forms evolve in human history, expose her deep concern for form's *epistemology*, her interest in how forms allow us to know things about the world and about the past. As I argue in the prelude, analogy demands we rethink form's relation to the formation of new knowledge. In the very long view, arguments about form have traditionally been concerned with form's *ontology*, or *being*—whether forms are "universals" that exist outside of the physical world and so transcend history, or whether forms are immanent features of the world around us, and so determined by social, material, and historical conditions.[94] Does form have its own (perhaps superior) kind of being, or is it subject to the kind of being we find all around us (people, trees, electrical currents), and hence without a being of its own? We might summarize this as the question of whether form is a structural "whole" that predetermines its contents, or whether form is simply a felt sense of wholeness, our attempt to posit a comforting totality within the succession of changing events.

Recently, the literary scholar Caroline Levine has worked to address this disagreement, arguing that there are many kinds of forms, including "wholes" which impose a sense of totality, and "rhythms" which remind us of succession and the ephemerality of patterns in time. In Levine's reading, all such forms act in a complicated way to constrain the material world—"imposing order"—even as they are subject to constraints "imposed by materiality itself."[95] This formulation addresses the relationship Eliot recognizes in her mollusk analogy, whereby the "living creature" and its "shell . . . *grow* & are *limited* by the simple rhythmic conditions of . . . life" (emphasis added). For Levine, this problem is best grasped as a relation of mutual "affordance"—a term she derives from histories of technology (6). On this view, both specific forms and specific materials have intrinsic possibilities; just as a hammer affords certain kinds of work by virtue of its design, copper affords certain kinds of malleability and conductivity by virtue of its properties as a distinct kind of matter. Forms and types of matter "constrain" by virtue of their nature—in other words, by virtue of their specific kinds of being, one structural, the other material. For this reason, a claim for form's ontology is implicit when Levine argues that each form can be specified by the peculiar and inherent *constraints* it places on the world: even if forms are "portable" and can be applied to new situations, any new application is always "latent"—predetermined by the constraints that form affords.

This view recognizes the profound entanglement of the mollusk and its shell and the implication of negative constraint, but does not take full measure of the generative possibilities produced by this interaction. For Eliot, though forms are "limited," it is equally important that they "grow." I take this to mean that forms have possibilities that are *not* latent. If we view forms in terms of constraint, they only have the capacity to pick out predetermined patterns. So, in *formal analogy*, one system of relations is used to model the patterns of another. Analogy is a powerful analytic of form itself, insofar as it can be used to expose instances of any recognized form (this is why the language of analogy threads throughout Levine's study). At the same time, analogies (especially what I have termed *harmonic analogies*) have the capacity to disclose *new* forms. My point (elaborated in the prelude as *entangled reference*) is that forms have epistemological possibilities, as well as ontological "affordances," and this means that forms can disclose other forms and other ways of knowing without predetermining their shape.

The question of form's epistemology is central to this study because my key contention is that literary forms—and, particularly, the analogies that constituted comparative historicism—furnish new ways of understanding social and natural history. Forms are not only structures but also methods that allow us to derive new knowledge from the world, and among Eliot's contemporaries, the nineteenth-century thinker who perhaps understood this best was George Henry Lewes. Lewes was both Eliot's life partner and the most intimate interlocutor for her thinking. They were closely involved with each other's work, and after his death, Eliot redrafted and published the last two portions of his major philosophical study, *Problems of Life and Mind* (1875–9). That work, which takes up the analogy between physical and mental science, is fundamentally concerned with the comparative method, both in its capacity to tell us about the mind and in its ability to derive knowledge from the world in general. This was because, to Lewes's mind, the comparative method, as organized through analogies, was the best way to deal with the basic uncertainty of knowledge, especially knowledge of mental processes. As he puts it: "In every question, from that presented by the growth of a blade of grass to that presented in the evolution of the social organism, from the chemical union of two gases to the formation of ideal types, there must necessarily be certain transcendental elements, not determinable by us, *unexplored remainders* after the most exhaustive exploration."[96]

Lewes insists that these "unexplored remainders" are an implication of abstraction in general, and he argues that the problematic misprision between formal abstraction and "remainders" can only be addressed through comparatism, which allows one to focus on the patterns between phenomena in place of perfectly characterizing any individual instance: "We may compare one vital phenomenon with another, or with its conditions, as we compare one sphere with another, or any one function of unknown quantity with another; and the comparison may yield exact results, although we remain eternally ignorant of the excluded elements" (1:44). His primary example is π, the ratio between the circumference and radius of the circle; while it is not possible to specify its ratio explicitly, in natural numbers, it works perfectly well to express a common "remainder" between proportions. Lewes is hewing very closely here to the mathematical basis of analogy as a ratio of ratios; as I have noted both in the introduction and in other venues, it is precisely as a form of mathematical comparison that analogy first took hold of the Greek imagination and assumed its early place as a transformative epistemology for pattern analysis.[97] The operative point is that, once it was recognized through an analogy drawn between two circles and their respective radii, π could be put to powerful work.[98]

This analysis places Lewes within a larger contemporary shift toward what Christopher Herbert has termed "Victorian relativity," a framework for understanding in which "the true subject matter of social science is not so-called concrete facts considered as entities unto themselves but, rather, 'the relations of structures and reciprocities of functions.'"[99] This relativity, directed in *Problems of Life and Mind* toward the relationship between mental life and physical science, runs into difficulties when we attempt to compare humanity with non-human life. Lewes explains that he began his research by drawing trans-species connections between comparative physiology and comparative psychology. But he was initially misled by the assumption that these comparative sciences were equally suited to address the relation between humans and other life. As noted in chapter 3, comparative physiology, which studies physical specimens, made profound advances in the early nineteenth century. However, insofar as we have limited access to the mental life of other organisms, Lewes argues that comparative psychology relies heavily on a misguided "anthropomorphism," filling in the "remainder" by attributing human intents and desires to nonhuman actors.[100] As an example, he draws a comparison between what happens when we eat, when a dog devours its prey, and when a polyp captures and digests other microscopic creatures. From a physical, metabolic perspective, these events bear a strong analogy. And psychologically, we might feel justified were we to infer that the dog, like us, was hungry and meant to eat. But what would it mean to say that the polyp felt hunger or intended to catch its meal? Lewes diagnoses the problem succinctly: "From a certain objective [i.e., material] similarity we have inferred that the two cases were similar *throughout*" (1:126). As we have seen, Eliot is equally attentive to such difference—marked in the distinction between the bivalve shell and human home. The "remainder" cannot be ignored.

The implication would seem to be that comparatism is more useful the more similar its objects, but Eliot reads this example in precisely the opposite way within *Middlemarch*. In chapter 4 of *Middlemarch*, in an effort to characterize Mrs. Cadwallader's busy interference in Dorothea's marriage, the narrator observes:

> Even with a microscope directed on a water-drop we find ourselves making interpretations which turn out to be rather coarse; for whereas under a weak lens you may seem to see a creature exhibiting an active voracity into which other smaller creatures actively play as if they were so many animated tax-pennies, a stronger lens reveals to you certain tiniest hairlets which make vortices for these victims while the swallower waits passively at his receipt of custom. In this way, metaphorically speaking, a strong lens applied to Mrs. Cadwallader's match-making will show a play of minute causes producing what may be called thought and speech vortices to bring her the sort of food she needed. (59–60)

Here we have an example of what Ian Duncan terms the "strangeness of scientific language in George Eliot's fiction," which, he notes, "oscillat[es] between literal and figurative registers."[101] More than one contemporary complained that the figurative oscillation of the scientific "parables" in her novels gave them vertigo.[102] The analogy here moves in complex ways. When we first see the polyp under the microscope, we see a drama of activity, in which creatures throw themselves willingly into the active mouth of the polyp. On closer look, we find a subtler tension between passive victims and inert swallower and the busy intermediation of "tiny hairlets" that make vortices between them. As is often the case, Eliot's point is formal: the "thought and speech vortices" which condition human interaction are social forms that have a complex relation to their contemporary conditions and history, and so they operated beyond the exclusive control of any actor, including Cadwallader. In making this point, the narrator insists that neither Cadwallader nor the people she visits act exclusively to interfere in Dorothea's marriage; instead, their social positions and their collective social mores condition and even encourage their business. If it is an awkward analogy (and the phrase "thought and speech vortices" is certainly awkward), this is because it foregrounds the complex and not entirely happy adjustment between forms and the conditions they address and, more generally, between scientific models and social situations they can never perfectly grasp. This is what gives such analogies their power to grow and take surprising turns; this is also what lends these analogies the ability to oscillate, as Duncan notes, between a "figurative" sense that requires only that we recognize a formal similarity and a "literal" sense, where we recognize elements of a true relationship that disclose something about the world. As we move from seeing the analogy as an imaginative figure—a literary description ontologically distinct from the object it describes—to a new epistemic perception of both nature and society, the analogy itself swings from a formal to harmonic relation. What seems to be an analogy that specifies some relation between victim and prey, or between natural and social systems, turns out instead to flatten these distinctions, making a general case for the distribution of power that characterizes the interplay of individual actors and the "vortices" of common forms.

We don't have to fully absolve Cadwallader in order to take Eliot's larger, insistently social point: that power and intent are more distributed than we conventionally assume. Eliot takes Lewes's insight—that polyps do not have as much intent as we might suspect—and works it through from the opposite end, as an insight (derived from the remoteness of microbial life) into the limits of our intention as humans. There are truly oppressive implications for the novel here, conceived in more conventional terms as an engine of subject formation and self-definition. To afford the social so much power, one line of criticism contends, leaves the subject with few options but resignation to their determined fate. If, as we have seen, individual forms are deeply entangled in their conditions and their objects, this is even more true of social forms, and our ability to make sense of the role they play in our interaction.

Conclusion: Against Origins

It is interesting to contemplate an entangled bank, clothed with many plants of many kinds, with birds singing on the bushes, with various insects flitting about, and with worms crawling through the damp earth, and to reflect that these elaborately constructed forms, so different from each other, and dependent on each other in so complex a manner, have all been produced by laws acting around us.

Charles Darwin, *On the Origin of Species* (1859)

I want to focus on the dynamic interaction of Eliot's social forms, whirling in space as "vortices," "entangled" in our encounter with the world. *Middlemarch's* polyp analogy insists on an ecological reading of the social world, and it does so by focusing our attention on the forms that mediate social interaction, insisting that forms bear a complex, uneven, and evolving relation to contemporary conditions and history. I have been arguing that Eliot drew this insight from her long study of the German higher critics, and their attention to the complex transformations of figurative language over time. But I also want to emphasize how closely her understanding aligns with the ecological vision offered by Charles Darwin in *On the Origin of Species* (1859). As Darwin puts it in the foregoing epigraph, an evocative reflection offered at the close of his most famous work: "It is interesting to . . . reflect that these elaborately constructed forms, so different from each other, and dependent on each other in so complex a manner, have all been produced by laws acting around us."[103] In light of the previous discussion, we can recognize the importance of "constructed forms" here; they both provide the illusory sense that the scene is an aesthetic and amicable whole and ultimately provide the entry point for Darwin's reflections on the constant adjustment between any natural "form," other competing forms, and their respective conditions. Darwin agreed that forms which, on first glance, seem to be "th[ought] out" or "fram[ed]" (as Eliot put it of the mollusk), in fact "grow & are limited by the simple rhythmic conditions" of life. Later in his career, in works like *The Descent of Man* (1871), Darwin took up the implicit and perhaps more troubling insight he alludes to here: if such laws are always "acting around us" they may continue to shape us, or at least, shed light on the origins of social forms and elements of human society. "Thus, from the war of nature," Darwin elaborates cautiously, "the most exalted object . . . the production of higher animals, directly follows."[104]

As we will see in chapter 5, Darwin shared with Eliot a deep insight into the power of comparative analysis to address the complex interrelation between human society and nature, between their respective forms and their conditions. He was powerfully attuned to the contribution of figurative or "literary" language to the scientific imagination. If Lewes saw anthropomorphism as a deeply flawed aspect of our attempt to understand nature, Darwin recuperated anthropomorphism as an analytic rooted in the kinship of all life. Darwin, like Eliot, recognized that we are always "entangled" in figurative language but also that comparative observations within and over time give us a corrective sense for how words themselves are entangled in historical experience, forming a part of living history's encounter with the world.

Eliot, however, was skeptical of Darwin's study of "origins," and (looking forward to the discussion of Darwin that follows) I want to close this chapter by considering why. Ultimately, Eliot's critique of Darwin reflected a basic disagreement over the nature of comparative historicism and its ability to tell us about the past. Among nineteenth-century novelists, Eliot is now famous as a writer deeply concerned with the implications of contemporary science and the empirical method. I myself, as a onetime student of biology, was first drawn to Eliot (and, through her, to nineteenth-century fiction) by her expansive scientific metaphors, which demonstrate her close knowledge of contemporary discoveries. In despite of the general consensus that Eliot's fiction is deeply engaged with contemporary science, her private comments suggest her distrust of contemporary scientific theories. Much has been made of the remark, recorded in her diary, about *On the Origin of Species* (1859): the book "makes an epoch, as the expression of [Darwin's] thorough adhesion, over long years of study, to the Doctrine of Development."[105] For Eliot, this was no complement. Her fiction abounds with characters, from Silas Marner, to Tertius Lydgate, to Isaac Casaubon, to Ezra Mordecai Cohen, who are crippled by a "thorough adhesion, over long years of study" to a single theory or creed. Eliot interpreted Darwin (as did her longtime partner and lover, George Henry Lewes) as another adherent to the speculative theory of transformation first sketched out by Erasmus Darwin and Jean-Baptiste Lamarck. In Eliot's view, such "adhesion" demonstrated an inability to fully weigh the uncertainty of experience and the impossibility of fully knowing the past—whether human or natural.

If Eliot complained that the *Origin of Species* was "sadly wanting in facts," we unfortunately do not know what she made of Darwin's more outlandish theories, like pangenesis. In the *Variation of Animals and Plants under Domestication* (1868), Darwin postulated that sexual reproduction was mediated by pollen-like "gemmules," particles of heritable information that collect from all the tissues of the parent and ensure inheritance of the various "characters" that govern growth and development of the body. This pangenesis was an eccentric idea, but it was Darwin's best attempt, absent a modern understanding of genetics, to shore up the theory of natural selection and explain how inheritance could work.

His theory was widely criticized as entirely speculative, more suited to the wide-eyed theorizations of Darwin's grandfather than to the eminent naturalist who, only four years earlier, had received the Royal Society's Copley medal, its most prestigious award. Alfred Tennyson evidently found pangenesis amusing. In response, he wrote a short and waggish poem:

> Curse you, you wandering gemmule,
> And nail you fast in Hell!
> You gave me gout & bandy legs
> You beast, you wanted a cell!
> Gout, & gravel, & evil days—
>   (Theology speaks)
>   (Shaking her head)
> But there is one who knows your ways![106]

The verse fragment marks Tennyson's ongoing and intimate engagement with contemporary scientific developments. The poet, who both suffered from gout and once feared he would inherit his father's mental illness, was necessarily interested in Darwin's discussion of the relation between pangenesis and "inherited malformations," which included, in Darwin's view, both gout and insanity.[107] The tonal oddity of Tennyson's poem marks the discomfort of the speaker. "Theology" seems to both endorse Darwin's theory and—feeling superannuated in the age of Darwin—bewail her latter-day maladies. I'm tempted to see also, in the confusion between scientific and Christian thought, a wink at the way that Darwin himself repurposes theological language, insofar as his theory reformulates the origin of life (genesis) as the origin of genealogy (pangenesis). However one reads these lines—whether it is Darwin who "knows your ways," or God, or both—they indicate Tennyson's serious interest in Darwin's theories and his willingness to entertain their (sometimes comical) implications.

Eliot, by contrast, was less interested in the nuances of Darwin's theories than in how they reflected a common (and, in her view, flawed) way of reading present patterns as evidence of common descent. As K. K. Collins notes, Eliot was studying various Darwinian approaches to the study of social history in the 1870s. In scattered notes from that period, Eliot elaborates the line of thinking captured in her "Notes on Form in Art": if (as she previously argued) forms "grow & are limited by the simple rhythmic conditions of . . . life," this implies that similar conditions will tend to grow similar forms. Yet contemporary anthropologists ignore the implications. Eliot records her frustration: "I wish it were borne in mind & practically applied in the investigation & discussion of origins. For then we should not see that irrational exclusiveness of theory which insists on interpreting all resemblances between the mythologies & legends of different peoples as result of tradition or identity of descent" (389–90).

As Collins observes, this is a critique of Darwinian thinking in general, and it is rooted in Lewes's own critical analysis of Darwin's approach.[108] In a multipart review of Darwin's *Variation* (including both the theory of natural selection and the theory of pangenesis), Lewes explores his own thinking about the nature of biological change. In particular, Lewes attacks Darwin's confidence that "organic resemblance" implies "kinship"—rejecting Darwin's belief that the history of biological forms can be read off from the patterns of relationship between existing species.[109] As discussed in chapter 3, Richard Owen (a friend to Lewes) addressed this problem by discriminating between "homology" (a contemporary pattern rooted in a true community of structure) and "analogy" (a superficial similarity based in a common function). In light of *On the Origin of Species* and Darwin's argument for the relation between species and their histories (as I noted in chapter 3), Owen later reinterpreted this distinction as the difference between historical relations and responses to present conditions: homologies demonstrated true kinship (e.g., vertebrates), whereas analogies showed similar adaptive strategies (e.g., flying).

In his review of the *Variation*, Lewes makes the perceptive point that, given the essential uncertainty of evolutionary history, especially when it comes to the origin of life, it is exceedingly hard to distinguish between analogy and homology: "Homologies may be thought more decisively to point to kinship, and very often they do so beyond a doubt; but we shall see how impossible it is to draw a line between homologies and analogies in many cases."[110] The problem, as Lewes recognizes, is that comparative historicism can cut either way. Citing Goethe, Lewes notes that the comparatist can choose to be either "synthetic" or "analytic" (in Darwin's terms, "lumpers" or "splitters"): shared patterns can be formalized as a common history, or they can point to serial or parallel histories of how species respond to similar environments.[111] As Lewes puts it, "Are the resemblances observable among organic forms due to remote kinship, and their diversities to the divergencies caused by adaptation to new conditions? or are the resemblances due to similarities and the diversities to dissimilarities in the *history* of organic beings?"[112]

Comparative historicism differs from traditional history because it is about histories, about origins and ends rather than this origin or that end. Lewes, like Eliot, insists that comparative historicism studies the pattern *between histories* and that this undercuts its capacity to elucidate a unitary past. Eliot's further point is that, if anatomists sometimes confuse functional similarity for kinship, analogy for homology, anthropologists do so, too. In analyzing Darwinian thought, both Eliot and Lewes foreground the increasing visibility of comparative historicism as a broadly shared method that is characterized by uncertainty—uncertainty about the past, but also uncertainty about the relation between different possible accounts of the past. For this reason, comparatism also furnished a way to talk about *historicisms*.

Eliot consistently recognized comparison as a way to critically examine different historical visions. Such multiplicity of view helped guard against the "irrational exclusiveness of theory" that she ascribed to both Darwin and contemporary anthropology. As her narrator puts this in *Felix Holt* (1866), "None of our theories are quite large enough for all the disclosures of time" (86). The quintessentially historiographic move in Eliot's analysis of the past was to compare historicisms, placing them in history as analogous and analogously limited systems for understanding. The historiographic turn means that even comparative historicism itself could become an object of her concern. *Middlemarch*, in fact, begins with a reflection on the contrast between the "epic life" of Saint Theresa of Avila and the "common" perception of "later-born Theresas" who had "no coherent social faith and order." Through comparison, the editor insists, we recognize how even profound contrasts, like that between the epic histories of saints and quotidian histories of "tangled circumstance," obscure the deeper continuities of the "history of man."

Unlike Darwin, Eliot understood comparative historicism as a particularizing, negative hermeneutic; a critical check to confident generalization, whether about people or history. Darwin used comparatism as a critical method, too, but he was more hopeful of its productive capacities—its ability to generate new insights that were generally valid, rather than specific to particular moments of encounter. In chapter 5, I will explore how Darwin deployed comparative historicism, in all of its harmonic, formal, and critical complexity, to articulate a new understanding of natural relations, and the patterns that organized species development. *On the Origin of Species*, every bit as much as one of Eliot's novels, is driven by a careful exploration of the problems of abstraction, formalism, and the particular commitments of figurative language. Whereas, for Darwin, analogy proved invaluable for redescribing the external world of natural relationships, for Eliot, analogy was a distinctive personal epistemology, a way to map the internal development of the individual, and an anticipation of modernism's "turn inward." In Eliot's novels, the power of sympathetic perception is proved not through disquisition and analysis but through moments of contact and the comparison of self with other—moments of surprised fellow feeling. Eliot's talents compound such moments of stunning error and painful recognition, and they accrue in experiences that collapse representation's distance. I am persuaded by these encounters, impressed by what Eliot describes as "an idea wrought back to the directness of sense, like the solidity of objects": within Eliot's novels, there lies "an equivalent centre of self, whence the lights and shadows must always fall with a certain difference."[113]

CHAPTER FIVE

The *Origin* of Charles Darwin's *Orchids*

A long castle in the air, is as hard work (abstracting it being done in open air, with exercise &c and with no organs of sense being required) as the closest train of geological thought.—the capability of such trains of thought makes a discoverer, & therefore (independent of improving powers of invention) such castles in the air are highly advantageous, before real train of inventive thoughts are brought into play & then perhaps the sooner castles in the air are banished the better.

Charles Darwin, Notebook M (1838)

This book begins by imagining the moment that Charles Darwin sat down and privately acknowledged the legacy of his grandfather, scrawling "Zoonomia" at the top of his first evolution notebook. It is a charged moment—truly world changing, considering all that would follow. But while Darwin clearly recognized he was heading down a new path, he didn't know where it would lead. His notebooks gave space to gather scattered thoughts, collect notes, and speculate; to daydream and, as he often put it, build "castles in the air." This phrase suggests a critical idiom that emphasizes the quixotic instability of imaginative architecture, but Darwin reads it differently. In this chapter's epigraph (from a later point in his evolution notebooks), Darwin muses about the meaningful place imagination holds in his work. It balances a sense of the significance that fancy holds for scientific discovery (for Darwin it is what "makes a discoverer") alongside a sense of the corollary need to check this fictive play against the "real train of inventive thoughts" drawn from experience.[1] Such checks banish false castles ("the sooner . . . the better") but implicitly sustain those valuable discoveries that have purchase on the ground.

In chapter 4, I noted the importance of disillusion in George Eliot's fiction and the characteristic way that imaginative constructions are continually refuted and then revised in the face of new experience. Like Eliot's fictions, Darwin's aerial castles hinge on a strategy of falsification that anticipates Karl Popper's theory of scientific method. In such examples, Darwin recognizes the productive relationship between imaginative labor and the extension of knowledge. In print, Darwin assiduously claimed to follow strict Baconian induction, but the notebooks show that he treasured imaginative theorization and fanciful language for their ability to escape from the immediate consequences of direct experience, fashioning possibilities that experience could later test. The epigraph asks us to pay attention to imagination's "train." Like "inventive thoughts" drawn from life, there is a structure, a sequence to the procedures that draft a strong hypothesis. The task of this chapter is to explore what gives that "train" its form. At the outset, it is tempting to assume that Darwin had something like logical entailment in mind, in a movement (to adopt Popper's terminology) from abstract universal statements to the particular assertions that would be necessary consequences. But rather than framing out a universal thesis in his writings, Darwin usually asks us to imagine particular stories that can plausibly explain what we now see.

An example: early in Darwin's most famous work, *On the Origin of Species by Means of Natural Selection* (1859), Darwin "beg[s] permission to give one or two imaginary illustrations."[2] He asks us to imagine two different populations, one a pack of wolves, the other a bank of flowers, loose communities struggling to survive at some hazy point in the past. In the wolf pack some cubs are better adapted to certain kinds of prey, others, for those lean times when prey is scarce. Over time, Darwin asks us to imagine, such differences and the resulting competition might found distinct communities, even distinct species. The flowers present a "more complex case"—one that highlights cooperation over competition. Suppose the flowers secrete nectar at the "inner base of the petals. In this case insects in seeking the nectar would get dusted with pollen, and would certainly often transport the pollen from one flower to the stigma of another flower" (91–2). This is of immediate benefit to both; the insects who "seek" the nectar get nourishment, while "unintentionally" pollinating flowers; the flowers get a cross-fertilization that is known to "produce very vigorous seedlings." Over time, and with variation, the anatomy of some flowers might change to encourage only certain bees to visit them, and certain bees might grow curved proboscises better tuned to the peculiar geometry of the flowers. This would help the flowers secure a dedicated courier for their pollen, and it would help the adapted bee, allowing it to "obtain its food more quickly, and so have a better chance of living and leaving descendants." He summarizes: "Thus I can understand how a flower and a bee might slowly become, either simultaneously or one after the other, modified and adapted in the most perfect manner to each other, by the continued preservation of individuals presenting mutual and slightly favourable deviations of structure" (94–5). Imagined stories furnish the reader with the ability to envision natural selection, sketching it out as a possible feature of the world.

There is a *harmonic analogy* between the "imaginary illustrations" of the wolves and the flowers, but it is uneven.[3] At this point in the *Origin*, the reader does not realize that this common pattern is the product of a shared history; Darwin saves until the closing pages his speculation that all life is descended from a single progenitor. These stories take their force instead from common patterns of struggle, variation, and success. At the same time, and even though natural selection produces these common patterns, the results widely diverge. This seems to be Darwin's point in choosing these two examples. Wolves compete, with themselves and with their prey. Flowers and pollinating insects cooperate. The lupine story presents one differentiated narrative; but to grasp the interaction between flower and bee, we must keep two narratives in mind and explore their interplay.

These two examples point up the importance of *comparative historicism* to Darwin's science, conceived as the study of how different actors and distinct narratives interact in history. The second example is more complex because, though it's easy to imagine the wolves struggling to capture their prey, the source of floral change is harder to pin down. Floral evolution presents the coordinated and increasingly elaborate development of *both* bee and flower. In drawing these two narratives together, Darwin institutes a problem of priority. Does this kind of adaptation happen "simultaneously or one after the other"? Who changes first—the plant or the insect? In his discussion, Darwin discriminates between what the bee "intends" and what it accomplishes "unintentionally." But he carefully avoids the complementary question—what does the flower mean to do? The passive construction of Darwin's conclusion demonstrates how Darwin's prose diffuses this basic problem of agency. The flowers and bees "become . . . modified and adapted"; there is "continued preservation." *Who* chooses, *who* "preserves"? The flower or the bee?

Darwin's larger answer is neither—it is "nature" or "natural selection" that acts upon these organisms. But this answer only defers the problem. His description of natural selection as something that acts upon creatures, something that functions like an agent yet operates without intent, created problems for contemporaries who recognized the language of William Paley's divine watchmaker and the evidence of design. For some, like Asa Gray, this intentionality suggested that evolution and divine teleology were consonant; for others, like T. H. Huxley, it was problematic shorthand for a strictly mindless material process. In the third edition of the *Origin*, Darwin argues that such "metaphorical expressions" are common in science and are "almost necessary for brevity"; if he seems to be "personifying" nature, he only means "the aggregate action and product of many natural laws" (85).

Modern readers generally overlook the importance of the language of natural intent in Darwin's work, but this intentional language has left a lasting mark on evolutionary theory. The zoologist John O. Reiss has argued that intentional teleology thrives in evolutionary accounts, and he locates this teleology in the mistake, traceable to Darwin, of assuming human intent is "analogous" to the representation of intent in nature.[4] Similarly, the philosopher Jerry Fodor has argued that historical accounts of natural selection are inconsistent because they confuse mental intent with a description of what actually happened in a natural event. While we might say that nature "selected for" some specific trait at a moment in the past, it is always plausible that some *other* trait proved useful, and that the feature we are trying to account for was accidentally brought along for the ride.[5]

When I was pursuing a career as a molecular biologist, I saw this problem firsthand. At the time, I was collaborating on an elaborate experiment in artificial evolution.[6] We were creating mutated versions of an enzyme that originally cut proteins that had a specific pattern (let's call it A-B-C-D), and we selected for enzymes that cut an entirely new sequence (A-E-F-D). It turned out that the successful mutants, while they were better at cutting the new sequence, often gained the ability in unexpected ways. Virtually all the new mutants still cut the old sequence as well as the new. And most were now able to cut lots of different sequences; instead of choosing one sequence over another, they were cutting many (A-*-*-C). Some of the mutated enzymes may have been so effective at catalyzing other enzymatic reactions that they actually killed their host cells before we could identify and test them. We thought we were selecting for one thing, yet in each case, it turned out we were selecting for many different things, only some of which could be identified.

For Reiss and Fodor, this is the sign of a false analogy between human intent and a mindless natural process. But even if this analogy is false, it is tremendously productive. By treating selection *as if* it were an intentional process, Darwin shows us how to construct plausible fictions about the past and increasingly sophisticated evolutionary descriptions of how nature took its modern form. By asking what flowers and insects mean to do, we get a better sense of what they actually accomplish. Darwin lodges intent in the heart of natural enquiry, as a necessary technique for imagining how natural selection operates. The question of intent allows us to enter into the different narratives that compose history, testing their possibilities as explanations for past events. In previous chapters,

I have emphasized the importance of imaginative projection for comparative historicism. The dialogue that it opens between present and past is constituted through analogy—described by the novelist Edward Bulwer Lytton as the "analogical hypothesis," but more generally known through such terms as Johann Winckelmann's "imitation," Walter Scott's "living memory," and George Eliot's "sympathy." Comparative historicism went beyond Charles Lyell's assumption of continuity (in which present conditions can account for past change) to the more complex assertion that both the similarities and the differences between present and past societies can be understood as a complex analogy between what we do now and what we would have done then. As we will see, Darwin's method also turns on a mode of sympathetic identification. To understand natural history on this view is to locate an intentional self within it—one modeled on our own experience.

The imagination is necessary for this activity because, as Darwin notes, castles in the air require "no organs of sense"; they are framed in the mind, not in the world. The peculiarly descriptive, narrative, and even novelistic features that many have recognized in *On the Origin of Species* are intimately related to the energetic textuality of Darwin's theory. While historians Steven Shapin and Simon Schaffer have emphasized the importance of practical, paratextual knowledge in the development and communication of scientific labor,[7] the *Origin* confronts its reader in a peculiarly nonpractical way. Unlike, say, the fashioning and operation of elaborate experimental technologies, it required no "*Fingerspitzengefühl*," no technical feel for the science.[8] This is one reason that Darwin's science fits awkwardly in the framework of scientific history laid out by Lorraine Daston and Peter Galison.[9] While they perceive the nineteenth century as a high point in the cultivation of objectivity as an ethic organized around exemplary objects and the discipline of the scientific eye, the *Origin of Species*, with its peculiar insistence on imaginative stories that, by their nature, cannot be seen, seeks instead to cultivate a specific sense for how evolutionary narratives can be told. It cultivates, in the mind's eye, a feel for the contours of an effective evolutionary narrative—the "train" of natural selection.

In place of presenting extensive experiments or exemplary objects for inspection, the *Origin* offers a wide-ranging survey of reported observations and imaginative reconstructions. Evolutionary science, in Darwin's time and in our own, has exceptional features that distinguish it from the traditional physical sciences. It is peculiarly "historical" and "descriptive" in character, relying upon a reconstruction of past events rather than being "hypothesis-driven" or "experimental."[10] This is generally true even though, as we will find later, hypothesis and experiment played an important role. "Ambivalences and multivalences," the necessity of seeing things in different ways, are central to the *Origin* and its science. These more literary features of Darwin's theory are not merely functional but intrinsic; as literary historian Gillian Beer puts it, they can't be "skimmed off."[11] If we now recognize Darwin as a vibrant writer, this is in part because of the successful efforts of Beer, George Levine, Robert Richards, and many others to make us look again to Darwin's words.[12] As Levine has put it: "Alive with metaphor, twists, and hesitations, the prose is saturated with aesthetic, intellectual, and ethical energy, and with the sorts of ambivalences and multivalences characteristic of literature."[13] Insofar as I am also making a case for the centrality of Darwin's writing to both the formation and the presentation of his theories, I emphasize Darwin's power as a writer and the writerly force of his science. As we will see, imaginative language was crucial to the impact of the *Origin*, by most accounts, the most influential scientific treatise of the nineteenth century.

At the same time, the *Origin* did not present a full-fledged research program. As Darwin described the *Origin* project in a letter to John Murray, "It is the result of more than 20 years work; but as here given, is only a popular abstract of a larger work on the same subject, without references to authorities & without long catalogues of facts on which my conclusions are based."[14] Reduced to this core argument, the *Origin* unveils history as an extensive web of analogies between past and present, between domestic and natural species—connections made intelligible by natural selection and its narratives of variation, selection, and change. No writer examined in the present study, except perhaps his grandfather Erasmus, spent as much time thinking about the importance and complexity of analogy as Charles Darwin. The *Origin* launches with an extended series of comparisons that organize and recast previous natural observations as part of a new pattern, before condensing this pattern into complex and contingent histories of change.[15] In doing so, it retraces and reforms the critical procedures central to historical fiction. If Beer has convinced us that "Metaphor, Analogy, and Narrative" (the title of one chapter from *Darwin's Plots*) are important in the *Origin*, I want to show precisely how the three fit together, knitted in sequences that draw comparative analysis into coherent narratives evaluated by the central agent of his works—natural selection.

As an "abstract," Darwin never intended the *Origin* as a plan of practical scientific action and investigation. That effort was saved for a later work—his never-completed "big book on species"—but it was in fact first fully realized in the remarkable monograph that succeeded the *Origin*, Darwin's evocative and meticulous study of orchids, *On the Various Contrivances by which British and Foreign Orchids are Fertilised by Insects* (1862). Whereas the *Origin* provided a series of imaginative sketches that suggested how natural selection might have operated, while parrying central objections to that theory, the *Orchids* showed how to organize a research program around the hypothesis that natural selection was real. If the *Origin* sketched out the extensive dimensions of Darwin's castle in the air, the *Orchids* showed what that castle might look like if built on solid empirical grounds.

The first section of this chapter is devoted to exploring Darwin's evolving sense of analogy as a central tool for thinking through both the implications of comparative historicism and the more particular features of natural selection in the evolution notebooks. In the section that follows, I track Darwin's discussion of analogy into the *Origin*, exploring how analogy is displaced, within that work, into his largest, most speculative statements. The third movement is to study the novels that Darwin used to formulate a theory of narrative that could sustain his running survey of comparisons. What place can "imaginative illustrations" find in scientific practice? What gives imaginative theory a convincing pattern, or "train"? What convinces a reader that there are shared patterns in evolutionary history? Ultimately I will argue that it is a sense of shared motive, an ability to invest the reader in the outcome of his stories, that organizes Darwin's fiction. Central to that strategy is the interpolation of intent within nature. And so, this chapter finally turns to the *Orchids* to explore how intent (and, especially, the intentions of plants) helped to shape the protocols of Darwin's evolutionary science. In my view, the study of plant life gave Darwin, like his grandfather, his most powerful argument for the common origin and the shared sensibility of all life.

The *Origin* and the *Orchids* divide two features of historical fiction between them: on the one hand, comparative history is an analytic organized by analogy, on the other, sympathetic and interpersonal analogies allow reader and writer to bridge the historical gap, allowing the agents of historical fiction to cross significant social and natural divides. If the *Origin* studiously avoids asking what flowers mean to do, such questions are the central focus of the *Orchids*, tangled up in the "contrivances" of curious plants that engage the researcher directly both in their extraordinary aesthetic display and in stunning, sometimes explosive behavior. In addressing these marvelous flowers, Darwin worked surprisingly hard to construct a new model of natural intention that was specifically engineered to be amenable to both materialist and natural theological accounts. I see this as part of his complex relation to contemporary belief within and without scientific circles, as he worked both to secure the place of natural selection in the contemporary investigation of nature (still very much influenced by the natural theology of William Paley, William Buckland, and Charles Bell) and to make natural selection a bigger tent for natural investigation. If terms like *intent* and *teleology* have become casual terms of abuse—to a degree inconsistent (I suspect) with our own investments as readers and teachers—perhaps a more Darwinian sense of how we might use them can enliven our sense of their critical power.

The Analogy Notebooks

Analogy is *sure guide* & my theory explains why it is sure guide.

Charles Darwin, Notebook E (1838)

Darwin's key strategy in *The Origin* is described by a broad shift from extended comparisons in the opening chapters to their explicit displacement into unitary analogy at its close. It is generally recognized that William Paley's *Natural Theology* (1802), a masterful study of intelligent design in nature, was an important influence on Darwin's thinking. Though the term *analogy* does not appear until the third chapter of that work, Paley's study turns on the insistent analogy, established over the first two chapters, between the evident design of clockwork and the complexity and organization of living forms. This clearly provided a model for how Darwin establishes the basic analogy between domestic and natural selection in the *Origin*.

But Darwin's notebooks show that his readings of Paley (as well as Erasmus Darwin) were just the beginning; he spent over twenty years honing his use of analogy in its many forms and in dialogue with a variety of other works. In the "A" and "B" notebooks Darwin initially stages analogy as a genre of broader speculation—a position with close kinship to the theories of analogy advanced by a range of philosophers, from David Hume to his contemporary John Stuart Mill. As Darwin drills into the particular strategies of analogy used by systematists, he becomes increasingly intrigued by its more technical uses. Midway through the "B" notebook he turns to the entomologist William Sharp Macleay's work and is excited to find a way to think about deviations in phylogenetic structure—departures from common type—in response to environmental conditions: "The relations of Analogy of Macleay &c., appears to me the same, as the irregularities in the degradation of structure of Lamarck, which he says depends upon external influences.—For instance he says wings of bat are from external influence.—Hence name of analogy, the structures of the two animals bearing relation to a third body, or common end of structure."[16] Macleay had a circular theory of species development—something like Venn diagrams—that argued for a radiation and subsequent overlap of distinct genera in which similar environmental niches produce analogous physical forms. Earlier in the notebook, in an attempt to understand what would cause individuals of a species to diverge enough to establish other distinct species, Darwin had discussed "affinities from three elements"—differential adaptation to more aqueous, terrestrial, or aerial environments.[17] Macleay's analogies gave Darwin a way to think about how species are acted upon by extrinsic rather than internal factors—the common relation to a "third body."

Darwin saw Macleay's analogy as an important improvement over Lamarck's earlier observations (discussed in chapter 3) regarding the differentiation of the ideal succession of forms. For Lamarck, this deviation—famously theorized as the inheritance of acquired characteristics—was a pervasive problem that caused departure from systematic order. It was distinct from the larger progressive force that drove the development of the great chain of species. Macleay provides a way of locating Lamarckian differentiation in the context of a class of structural analogies identified by anatomists like Saint-Hilaire and Owen— those similarities of structural modification toward an apparent "common end" evident, for example, in the commonalities between bat and bird wings as organs of flight. Darwin, reading through Macleay, sees these analogies of structure as a key solution to the problem of speciation, describing how organisms of the same species would develop *differently* in different environments. Hence, "analogy," in the specific biological sense of what would later be termed "convergent evolution," simultaneously marked similarity between very distinct groups of organisms and identified a potential engine of dissimilarity between closely-related individuals. Darwin emphasizes this when he later observes in the "C" notebook that "the relationship of analogy is a *divellent* power & tends to make forms remote antagonist powers.—Every animal in cold country has some analogy in hot gaudy colours so all changes may be considered in this light" (emphasis added).[18] In contemporary chemistry, "divellent" compounds like alcohol and other volatiles vaporize and disperse easily.[19]

As analogy evolves in the pages of Darwin's notebooks, it becomes increasingly active. For Macleay as well as Lamarck, Georges Cuvier, and Geoffroy Saint-Hilaire, analogy was a structural description—a paradigmatic relationship between different genera. For Darwin, it becomes a "power" that "make[s]," a way to understand the action that produces "all changes." Ultimately, this power will be ascribed to natural selection itself—and here analogy, in a kind of regency, is vested with some of its chief capacities. At certain moments analogy seems to outstrip the capabilities of natural selection entirely. Here, it provides one of Darwin's earliest ways of theorizing *sexual* selection: "Whether species may not be made by a little more[,] vigour being given to the chance offspring who have any slight peculiarity of structure[,] hence seals take victorious seals, hence deer victorious deer, hence males armed & pugnacious all order[s]; . . . in any class those points which are different from each other, & resemble some other class, analogy."[20] Darwin sees sexual difference in deer and chickens as a further example of "dissemblances [of] analogy," and cites the philosopher John Abercrombie's distinction between resemblances and analogy: "When there is a close agreement between two events or classes of events, it constitutes resemblance; where there are [also] points of difference, it is analogy."[21] The key point is that differences, once established, become a fulcrum for further differentiation and, ultimately, speciation.

At this point in the notebooks, analogy serves at least three purposes: a broad way to extend theory to a larger scale; a more focused way to think about convergent evolution between genera; and a way to imagine differential adaptation within genera, including sexual differentiation. As the notebooks progress, Darwin begins to mix these senses almost fancifully, as when Darwin thinks about using an analogy between species and social structures: "A species," muses Darwin, "must be compared to family entirely separated from any degree" so that, while "tailor[s]" from different families "would be analogous to each other &c &c," they actually belong directly and exclusively only to the actual families from which they descend.[22] Even as he continues to argue against those who seek to exclude analogy from scientific practice (Notebook E insists that "analogy is *sure guide* & my theory explains why it is sure guide"),[23] he seems to acknowledge that from a practical perspective the term is overburdened. In later notebooks, as his thinking takes firmer shape, analogy recedes again into a nominal form of extended speculation. Darwin retains its other senses too, but works to find more specialized language for describing structural correspondence, divergence, and sexual difference, as well as identifying important causal factors: geographic isolation, geological time, the overproduction of offspring, the spontaneous variation of young. Darwin reworks the paradigmatic network of comparative description into a sharper narrative retrospect.

By explaining how systems brought into comparative view originated from a common point and then differentiated along unique paths, Darwin is able to gather a wide-ranging network of observations into a filiated collection of stories. The most important strategy for displacing analogy into more specific language is the term *natural selection* itself, a neologism that solidifies the analogy between the domestication of species and the selective forces of nature, adding the crucial component of change over time. Yet all of these closely related ideas emerge from an initial constellation, temporarily held together by "analogy."

By the close of the notebooks, the complexity and range of registers within which "analogy" operates make it too unwieldy a term to use reliably (a problem I sympathize with). As John Stuart Mill put it in his *System of Logic*, "There is no word . . . which is used more loosely, or in a greater variety of senses, than Analogy."[24] Darwin's evolving relationship to analogy recapitulates the larger nineteenth-century shift detailed in previous chapters, by which philologists, anatomists, and historians reconfigured analogy as a comparative mode of historical analysis, even as they exchanged its name for more specialized vocabularies (comparative grammar and Franz Bopp's theory of "family affinity"; comparative anatomy and Richard Owen's theories of homology; comparative history and Thomas Babington Macaulay's emphasis on "conjunction and intermixture").[25] One of the key dimensions of Darwin's thinking—the one that the most recent generation of scholars has been most focused on recuperating—was his emphasis on indeterminism, uncertainty, and complexity. Natural selection provides a vision of natural history that is characterized both by its profuse multiplicity of forms and by the uncertainty of their destination. As Elizabeth Grosz puts it, "The movement of evolution is in principle unpredictable, which is to say that it is historical."[26]

Darwin's case for natural selection gives one of the most profound nineteenth-century attempts to wrestle with the implications of chance. The present book is designed to show that many of the most important formal principles of his writings—a focus on plots over plot, multiplicity over unity, and conjectural history—are derived from comparative historicism and organized to register precisely the contingency and uncertainty of past events. In recognizing Darwin's insistently comparative studies of the past, I underline the formal innovations historical fiction furnished to natural history.

On the Origin of Species and the Curation of Analogy

Analogy would lead me one step further, namely, to the belief that all animals and plants have descended from some one prototype. But analogy may be a deceitful guide.

Charles Darwin, *On the Origin of Species* (1859)

The foregoing discussion helps to explain the attenuated place of analogy in the *Origin*, with many of its features displaced into comparisons and the figure of natural selection itself. Comparison and analogy are often nearly interchangeable in mid-nineteenth-century comparative anatomy. As chapter 3 notes, Owen responded to this problem when he proposed that the various forms of analogy theorized by Geoffroy Saint-Hilaire should be reorganized as distinct classes of "homology" as well as analogy. As of 1859 Owen's distinction between homology and analogy was not yet universal; Darwin ignores it in the first edition of the *Origin*. This creates an important flexibility, allowing Darwin to draw analogy and comparison together as interrelated modes that discern the patterns of natural selection.

Though the first few chapters of the *Origin* are filled with complicated and suggestive comparisons, Darwin generally reserves the term analogy for explicit speculations that extrapolate beyond specific observations. As we saw in chapter 1, Erasmus Darwin was similarly fascinated by the relation between the "looser" analogies furnished by the imagination and the "stricter" analogies of science. If, in the movement from Charles Darwin's notebooks to the *Origin*, analogy evolves from being a "sure guide" (in the notebooks) to a "deceitful guide" (in the preceding epigraph), this obscures the profound role that analogies retain within the vast network of comparisons that sustain the argument for natural selection.

It is in a movement from comparison to "analogy" that Darwin creates space for one of the most controversial speculations in the work, in a longer passage that includes this epigraph:

> Analogy would lead me one step further, namely, to the belief that all animals and plants have descended from some one prototype. But analogy may be a deceitful guide. Nevertheless all living beings have much in common, in their chemical composition, their germinal vesicles, their cellular structure, and their laws of growth and reproduction. We see this even in so trifling a circumstance as that the same poison often similarly affects plants and animals; or that the poison secreted by the gallfly produces monstrous growths on the wild rose or oak-tree. Therefore I should infer from analogy that probably all organic beings which have ever lived on this earth have descended from some one primordial form, into which life was first breathed. (455)

The status of analogy in this passage is largely ambiguous: What, exactly, *is* common in these germinal vesicles, or in the laws of growth and reproduction? What is analogous about the monstrous growths of the gallfly, and how does that provide evidence for "one primordial form"? We may speculate—noting, for instance, that these examples make a case for the common nature of plant and animal life—but the meaning is unclear, the analogies notional, allusive. Strictly speaking, the grammar of an explicit analogy (A is to B as C is to D) requires at minimum two clauses—two subjects and two predicates. Instead of such syntactic parallels, this passage uses denotative plurals.

Analogy serves a curatorial function in the move from the notebooks to the *Origin*, both in helping to analyze which natural patterns, which possible stories are significant and in focusing the reader's attention on the larger speculative import of his theory and away from the implicit analogies of his teeming network of comparisons. As I have been arguing throughout this book, to do away with the term "analogy" in comparative histories nevertheless means to retain its syntactic and semantic forms, even as they are applied to new problems and operate within new framing assumptions. Retained as an explicit critical term, analogy remains important as a way to organize the wealth of observations Darwin brings to bear on the larger questions of speciation.

For one, analogy pairs observations drawn from across a vast network of printed books and correspondents. These curatorial analogies highlight the importance of the bibliographic imagination to Darwin's science. Shapin and Schaffer have suggested the importance of "virtual witnessing" to scientific experiment, making the case that a triumvirate of "instrumental," "social," and "textual" technologies establish the conditions within which the reader of a scientific paper can set aside the distinction between imagining and replicating an experiment. Virtual witnessing allows a scientific reader to accept the experiment, as it is described, as if they were in the room. The *Origin* does not work that way. Instrumental technology is reduced to the basic instruments of modern European life—the printing industry, a reliable and standardized post, the maritime technologies that allow observations and occasionally specimens to circumnavigate. The social technology deployed by Darwin, secluded in his house at Down, is almost exclusively the product of its textual forms. Even the site of witnessing, the point of contact between experiment and observer, is skewed within the coordinates of analogy, suspended somewhere between the disparate locations where the individual observations are made. I like to think of this as the peculiar form of *textual witnessing*—elucidating a pattern that is effectively generated by its textual encounter. If we are used to thinking of textual technology as a medium for the interaction of individuals and experimental objects—the network within which they act—in Darwin's science, the texts constitute actors in their own right.

The other major function for analogy in the *Origin*, beyond the kind of broad speculations we've been exploring, is in highlighting the cognitive movement by which systematists organize natural kinds, "judging by analogy" where the boundaries lie between genus, species, and variety (56). The most important of these paradigmatic analogies is the discrimination between domestic species and their natural counterparts. Darwin spent two decades developing the analogy between domestic and natural varieties that appears at the front of his earliest sketches of natural selection (1842 and 1844). This analogy also cut to the heart of the distinction between Darwin's theory and Alfred Russel Wallace's own speculation about the origin of species. In his Ternate Essay (the famous essay that was presented with one of Darwin's early sketches to the Royal Society in 1858), Wallace bases his approach on the error made by naturalists in asserting an analogy between domestic and natural species:

> This argument [for the fixity of species] rests entirely on the assumption, that varieties occurring in a state of nature are in all respects analogous to or even identical with those of domestic animals, and are governed by the same laws as regards their permanence or further variation. But it is the object of the present paper to show that this assumption is altogether false. . . . [N]o inferences as to varieties in a state of nature can be deduced from the observation of those occurring among domestic animals.[27]

Where Wallace argued for radical, systematic difference, Darwin saw a dynamic analogy—a relation conceived in terms of similar processes. Darwin's most important task was to reconfigure a relationship conceived in terms of formal disjunction as a harmonic relationship organized by processes in time. The *Origin* foregrounds this purpose consistently, from the first sentence of chapter one ("When we look to the individuals of the same variety or sub-variety of our older cultivated plants and animals, one of the first points which strikes us, is, that they generally differ much more from each other, than do the individuals of any one species or variety in a state of nature" [7]) to the mirrored titles of its initial chapters: "Variation under Domestication," and "Variation under Nature."

One way of getting at the difference between such different applications of analogy is to make a generic distinction (set out in my introduction and explored throughout the previous chapters) between the formal and harmonic dimensions of analogy. This important distinction has been overlooked by theorists within the history of science and cognitive philosophy, but it is a powerful way to understand Darwin's strategy. *Formal analogy* uses one-half of the comparison as a model for the other, a relationship described as "mapping" or "modeling" in linguistic and cognitive literature. *Harmonic analogy*, by contrast, casts correspondences into an indeterminate encounter, so that neither half of the comparison takes precedence as model for the other. To pair "Variation under Domestication" with "Variation under Nature" is to open a dialogue between these two subjects, a dialogue that will turn up new patterns and can't be predetermined. Harmonic analogy captures an important countermovement in what comparison can achieve. Insofar as the analytic of comparison draws from the previous discourses of analogy, it pushes toward an evenhanded balance between the objects of juxtaposition. The peculiar power of analogy, a power that Darwin uses masterfully in the *Origin*, accrues in the movement between these two relations; from the hierarchical implications of formal analogy, to equable serial engagement, and back again.

An example will help demonstrate analogy's changefulness, particularly as it filiates with other comparisons. Darwin observes: "I find in the domestic duck that the bones of the wing weigh less and the bones of the leg more, in proportion to the whole skeleton, than do the same bones in the wild-duck" (11). The initial *disanalogy* in domestic proportions is generated through comparison to the analogous proportions in wild ducks (the copula "x:" marks disanalogy):

domestic wings : weight x: domestic legs : weight

when compared to

wild wings : weight :: wild legs : weight.

Here, then, is the consensus argument foregrounded by Wallace, derived from systematic analysis, and conceived as a formal distinction between domestic and natural species. But Darwin continues: "And I presume that this change may be safely attributed to the domestic duck flying much less, and walking more, than its wild parent." By throwing in the "third body" of environmental habit, the disjunction folds over into parallel analogies that share a common dynamic. Brought into harmony, the serial comparison already implies a conclusion that Darwin will advance in later chapters, suggesting the ultimate analogy:

domestic ducks : [domestic selection] :: wild ducks : [natural selection].

This ultimate analogy between domestic and natural selection has a strongly formal character, insofar as husbandry helps Darwin to model what natural selection looks like. But Darwin is able to get there only after demonstrating, through a series of repeated comparisons like those just presented, that domestic and wild species in fact have much in common. This reflexive pattern is shown in the assiduous way he argues for the infinite malleability of domestic species. Against contemporary belief (shared by Wallace) in the "limits of variation" that established boundaries around what could be done to change domestic animals like sheep and dogs, Darwin's comparison to natural selection convinced him that domestic breeding could ultimately produce new species. The harmonic analogy between domestic life and nature suggested that Wallace and others were wrong about species in captivity. The current of the initial chapters of the *Origin*, then, runs from deformation-disanalogy to comparison-harmonic analogy, to modeling-formal analogy. This kind of analogical drift features throughout *Origin*, but it is especially apparent in the earlier chapters.

This powers the repeated moments of catachresis in the *Origin*, passages that disclose new meanings in old words. A network of shifting correspondence breaks down the distinctions between ostensibly stable terms: species, race, variety, and subvariety; the natural and the domestic; genus and species. As noted in chapter 1, Carl Linnaeus had cemented the conceit that schemes of classification, from genera through families, might be arbitrary, but species were fixed. Darwin's theory of natural selection rewired the stable hierarchies of classification—species, genus, family, order, class—into a historicizing gradient of terms that index increasingly remote ancestral relations. Darwin's theory exchanges ontological distinction—the great chain of being—for the continuity of transformation in time.

As I note in the prelude, this collapse of ontological distinction into continuity has been used by Manuel DeLanda as the chief example of a "flat ontology."[28] At the close of this chapter, I will explore the epistemological implications of this flat ontology; here, I want to underline Darwin's sense that this flattening is deeply entangled with the figurative properties of scientific language. Darwin doesn't coin new words. Instead, he works to change the reader's understanding of old terms—either outright (as in the case of "species," "variety," "variation," or even "nature") or via new combinations, particularly "natural selection," and "sexual selection." Because the interpretation of meaning is contextual, it is difficult to grasp a word used in a new way—to dislodge the reader from what we might term, following Hans Robert Jauss, the horizon of semantic expectation. Simply redefining a term does not suffice.[29] The scale of the *Origin*'s task and the innovative interpretation that drives it required redundancy, that is, repeated uses and instances of translation from older norms—moments that reinforce new interpretations of old phenomena.

In the first edition of the *Origin*, Darwin looks ahead to a day when naturalists generally accept his theory and the semantic reorganization it requires:

"The terms used by naturalists of affinity, relationship, community of type, paternity, morphology, adaptive characters, rudimentary and aborted organs, &c., will cease to be metaphorical, and will have a plain signification. When we no longer look at an organic being as a savage looks at a ship, as at something wholly beyond his comprehension; when we regard every production of nature as one which has had a history . . . how far more interesting, I speak from experience, will the study of natural history become!" (485–6). This extraordinary passage links the ontology of natural selection (the perception of the historical nature of "organic being") to shifts in how language is used and understood. The transition from "metaphorical" into "plain signification" mirrors the transition from hierarchical to flat ontology. Moreover, in noting the metaphorical dimension of terms like "affinity" and "community of type," Darwin not only underlines the relation between literary language and scientific practice but indicates the substantive contribution of figurative language to our understanding of the world. Darwin's insight here enlarges an observation made by his grandfather in *The Botanic Garden* (1789–91). If Erasmus Darwin argued that the "imagery of poetry" presents "looser analogies" that "lead" to the "stricter ones" of science (*CW* 1, v), Charles Darwin sees this as a possibility of figurative language in general. The profound implication is that, in using this language figuratively, naturalists were reaching toward a perception of the historical nature of natural order that they could not yet fully apprehend.

A key strategy for the *Origin* is to use a network of "looser" and "stricter" analogies, and the network of continuities and discontinuities these analogies mapped, to organize figurative language into a substantive scientific understanding. I've found it challenging as well as illuminating to work out the structure of Darwin's analogies, and their difficulty elevates the central problem of stability. What features of an analogy are important? Which comparisons or copula should we take as primary? Are important features tacit? In some cases, Darwin activates implicit analogies at a later point in the text. He draws upon this virtual network when he sums up his variety-species argument by asserting that "the species of large genera present a strong analogy with varieties" (113). Other virtual connections remain tacit—implicit in Darwin's network of observation and description. Later in the work, analogy sometimes serves as a shorthand for absent examples—a casualty of analogy's curatorial function: "Many analogous facts could be given" (399). The extraordinary network of comparisons that Darwin stages and interweaves in the *Origin* are always changing, reworked, revised, rewired as circumstances and Darwin's evolving argument take shape. This movement can be characterized as the ebb and flow of harmonic engagement and formal organization—the movement between new observation and sharper formulation.

It is a founding concern of literary studies, particularly after structuralism, that formalizations are sticky things. A strong model makes it hard to see what other ways of knowing might be out there. The movement between *harmonic* and *formal* pattern articulates analogy's peculiar capacity to write itself in and out of stable forms. My prelude explains why I have found Bruno Latour's description of the circulating reference a useful way to characterize this dynamic.[30] On this view signification is a chain, structured through form-content relations that move down toward more particular observations, up toward broader generalizations. This nested series of relations helps get at the movement between observation and formalization throughout Darwin's work and gets to the heart of his claim that "no one could be a good observer unless he was an active theoriser."[31] Darwin shared with George Eliot a deep sense for the interplay between form and interaction in time. In chapter 4, we saw how Eliot, in emphasizing how forms "grow & are limited by the simple rhythmic conditions of . . . life," sought to underline the dynamic interplay between previous experience and fresh encounter, between the given and the new. Forms do not only afford certain kinds of structure, as Caroline Levine has argued, but also make new structures and new insights possible.[32] Scientific models, like all formalizations, can be revised with respect to new observations, but more important is the way that a strong model makes new observations visible. The peculiarity of analogy, from this perspective, is that it provides a link between distinct chains of signification, a form of *entangled reference* that makes it possible to radically reshape apprehension of the world from the ground up.[33] To draw from a preceding example, the interior proportions of wing and leg mass in the domestic duck hardly signify until someone brings them into contact with their wild counterparts (and vice versa). Even as he provides a transformative account of how biological form adjusts to the "rhythmic conditions" of life, through the continuous adjustment between structure and function, between adaptation and environ, Darwin also demonstrates how literary forms apprehend that dynamic in time. In this way, Darwin serves as one of the nineteenth century's most important theorists of the epistemological possibilities of literary form.

The most important thing that analogy does in the *Origin* is disappear—it is absorbed into new language, new modes of understanding, new models for engaging natural history. The chief example of such formalization is in "natural selection" itself. We have already seen how natural selection draws together key elements of Darwin's evolving use of analogy to theorize speciation. Here it is worth taking some time to see what this looks like in practice. Darwin uses the phrase in the initial chapters as a placeholder—designating a theory that will later be elaborated. It's not until the fourth chapter that Darwin has drawn together enough examples, particularly of the analogy between domestic and natural species, that he finally sets out a formal statement of his argument. Drawing from his observations in the Galápagos, Darwin invokes an island in which species are isolated from competitive immigration: "In such case, every slight modification, which in the course of ages chanced to arise, and which in any way favoured the individuals of any of the species, by better adapting them to their altered conditions, would tend to be preserved; and natural selection would thus have free scope for the work of improvement" (82). Structurally the sentence mimics the syntax of analogy—two independent clauses linked by a semicolon. But narrative movement takes the place of correspondent pattern. Each phrase of the sentence is loaded with previous discussions: (1) the "such case" adverbial link to the island scenario; (2) the noun phrase "slight modification"; (3) the adjectival description of the modification "which had chanced to arise"; (4) the linked description "which in any way had favoured the individuals" (5); the adverbial clarification of that description, "by better adapting them to their altered conditions (6); and finally, the base predicate "would tend to be preserved." All of these propositions, save the first, are packed into "natural selection." And of the six propositions absorbed by natural selection, all (except perhaps 3) are rooted in previous analogies.

This definition of natural selection collects a massive network of comparative observations, gradations of similarity and difference across time and within natural systems, into a unitary figure. As an alembic, this sentence condenses and consolidates various systems of comparison in order to distill their interaction. This consolidation is the mechanism by which, as Beer puts it, "gradually and retrospectively . . . the force of the argument emerge[s] from the profusion of example" (42).[34] By evoking the paradigmatic situation of selection, "natural selection" coordinates the distinct lives of different living beings as a parallel network of analogous narratives. This coordinates chronologies of descent and paradigmatic relations of competition within a single instance of narrative decision, of choice. The multiple reaches a singular moment of contact. Later, I'll take up the implication of this sense of intention for Darwin's imagination of nature, in particular, how it recognizes forms of sentience and activity that are not strictly human. Here, my point is that "selection" metaphorically unites two diagnostic perspectives upon evolution (perspectives detailed at length in the first three chapters of the *Origin*): the recognition that later generations are related to but different from past generations, and the visible variation that exists between organisms at the same time. To put this differently, "natural selection" formalizes an analogy to domestic selection, translating two different kinds of series—diachronic series across time and synchronic series across specific moments—into narratives of success and loss, in a movement from the organizing principle of lists to the narrative principle of stories.[35] "Natural selection" is the central narrative agent of the *Origin*, moving across the uneven texture of natural history in order to articulate the contingency of events as a system of resonant, if differentiated pattern.

Darwin and the Novels

Walter Scott, Miss Austen, and Mrs. Gaskell, were read and re-read till they could be read no more.

Francis Darwin, *The Life and Letters of Charles Darwin* (1887)

As we have seen, Darwin's engagement with analogy was extensive, and he fashioned his theory through an extensive dialogue with contemporary theorists like Macleay, Saint-Hilaire, and Owen. Yet these naturalists, with their emphasis on physical structure and systematic classification, did not provide strong examples for how to fashion such analogies into extended narratives, much less how to weave those narratives into a comprehensive vision of the past. Moreover, in emphasizing the relations *between* narratives over a singular story of continuous transformation (like that favored by his grandfather), Darwin's natural history drew from the comparative historicism of contemporary fiction after Walter Scott. As emphasized throughout this book, historical fiction featured analogies drawn between stories over unitary narratives or systems. In his major study, *Darwin and the Novelists*, George Levine has extensively discussed the effect of Darwinian narratives on contemporary Victorian fiction, and I see this as evidence of Darwin's investment, alongside his contemporaries, in the narrative strategies of comparative historicism.[36] In the later chapters of the *Origin*, Darwin is perhaps closest to the novelists, as he seeks patiently to clarify the coherence of the demystified natural world he envisions, providing an anthology of particular narratives (some more, some less convincing to contemporary readers), that bind his theory more closely to biological phenomena. Like Scott's narratives, the central features of Darwin's stories are moments of translation, between conflicting vocabularies, and between different ways of understanding the natural world. (Perhaps the most famous example is his imaginative account for how the "perfection" of the eye, identified by William

Paley with the design of the telescope, might be produced through a series of "transitional" improvements.)

Though it is now common to insist that Darwin's narratives are a central feature of his theory of natural selection, largely due to foundational studies by Beer and Levine, it remains hard to see where Darwin drew these stories from. One clue is to follow the influence of a contemporary theorist of narrative (also a close friend of Darwin's brother), Harriet Martineau. Darwin came to know Martineau during his time in London after returning from the *Beagle* voyage. He went on to read her historical novel *The Hour and the Man* (1841), as well as some of her travel literature, but he likely first encountered Martineau's work during his famous circumnavigation. Her *Illustrations of Political Economy* (1832–4), an extraordinary series of twenty-four novellas illustrating the various economic theories of Adam Smith, Thomas Malthus, and David Ricardo, was the only work of fiction besides Samuel Richardson's *History of Sir Charles Grandison* (1753) in the ship's library.[37]

Though it is explicitly organized as an attempt to translate the language of political economy into daily experience, Martineau's *Illustrations* also provide a sophisticated analysis of the relation between narrative, pedagogy, and living experience. In the first issue, Martineau argues that stories have the capacity to draw together the "utmost pains-taking" in scientific fact with the "strongest attachment to the subject by the reader." The "form of narrative," Martineau argues, is best for teaching "moral science" because it demonstrates by example how abstract principles impact individual lives—including the reader's. Narratives excite interest in both fact and theory by connecting them to the reader's own concerns and anxieties. They give the reader a powerful medium in which to "observe and compare and reflect and take to heart."[38]

Darwin's own "imaginary illustrations"—like the example of the wolves and the flowers with which we began—take precisely this form. They ask us to *observe*, in our mind's eye, a situation in the past, *compare* that situation to others, *reflect* on what it means more generally, and, yet, *take to heart* the struggles, both the successes and the failures, of its agents. This is the basic nexus of comparative historicism, as drawn together in the historical novel.[39] Darwin's relationship to literary history remains a challenging problem for scholars. His agonized sense that his aesthetic faculty had withered later in life, the collapse of his interest in poetry and drama, and the rarity with which he references literature in his many scientific writings, have prompted scholars recently to question whether studies of Darwin and literature (as opposed to studies of Darwin *as* literature) make sense. If the major studies have made us alive to the profound vision, energy, and enchantment of Darwin's writing, as well as their broader impact, they have struggled to convince us that Darwin drew these capacities from literary resources rather than from his extensive reading of other richly imaginative scientific writings, like those of Alexander von Humboldt, Charles Lyell, and William Buckland. Gowan Dawson's suggestion that "Darwin's scientific thought might in fact have been shaped much less by his literary reading than that of contemporaneous naturalists" serves to highlight the larger critical question of Darwin's place as a major object for studies of science and literature.

In part, the challenge comes from taking Darwin's statements about poetry as representative of his literary engagement in general. For one, Darwin's early investment in poetry is rooted in an older (likely inherited) understanding of poetry's relation to scientific description and observation. As he aged, and found it impossible to integrate poetry into his scientific life, Darwin remained an avid reader of fiction, especially once he'd settled with Emma at Down House. The Darwins devoured novels together. Reading, as in many Victorian homes, was a shared family event. They often mixed in biography, history, or one of many travel narratives, but novels made up the majority of what they read for entertainment. Generally, they were read aloud, either in the drawing room or upstairs, where he and Emma read to each other on the sofa, and they were read at several points of the day; during the midmorning break from work, after lunch, before dinner, before retiring for bed.[40]

The true extent of his novel reading is hard to estimate fully, though it is the single literary genre that he certainly read throughout his life.[41] Darwin kept two notebooks to track his reading from 1837 to 1860, and we can use them to draw larger conclusions about his reading habits. If we collect all the titles designated as "read" in the reading lists and categorize them by subject, it gives an intriguing map of Darwin's interests in the roughly twenty-year period before publishing the *Origin*. Most important are works of hard science and travel narratives, as well as works on horticulture and animal husbandry, along with literature, including both poetry and fiction (figure 8a). We can further break down the literary subjects to see how, as Darwin would later record in his *Autobiography*, his interest in poetry waned, as compared to novels and especially historical fiction.

But this picture fails to capture the sustained attention to the novel described by his family members, particularly his son Francis. The reading notebooks give a distorted picture of the extent and emphasis of Darwin's literary reading. As Francis Darwin described it, "Walter Scott, Miss Austen, and Mrs. Gaskell were read and re-read till they could be read no more."[42] Between 1838 and 1860,

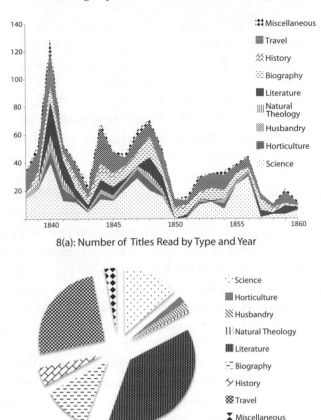

8(a): Number of Titles Read by Type and Year

8(b): Time Spent Reading per Day

Figure 8. Charles Darwin's reading habits.

however, Darwin recorded reading only five novels by these authors, each only once. We know, for instance, from Emma Darwin's diary that they finished reading Scott's *Guy Mannering* aloud on February 17, 1856, yet the work is never mentioned in his notes.

It is more accurate to understand the reading notebooks as documenting two differentiated reading practices. On the one hand, there are the technical works that Darwin kept assiduous track of as he worked, noting them for future consultation, writing short abstracts and summaries as he read them, checking them off in his journal when completed, while distributing the abstracts to files organized by working subject. Literary works, on the other hand, are rarely noted unless they were previously unfamiliar, in which case the notebooks could jog his memory. With only a few exceptions (including Lockhart's *Life of Scott*), Darwin almost never records a second reading of the same work. Above all, the reading notebooks were an aid to his scientific research—a way to keep track of the titles he needed to get to, as well as a catalog of those he'd already covered and indexed. And so, even though Darwin had almost certainly read most of Walter Scott's novels by the time they moved to Down, and his works were continually "read and re-read" in the house, he's largely absent from Darwin's reading lists.

One way to understand the reading lists, then, is that they document in the negative how Scott's novels, and novel reading in general, were background conditions of Darwin's literary environment: they went unremarked because to him they were unexceptional, a key element of the domestic atmosphere rather than objects of critical attention. Even so, novels were central features of their imaginative life at Down. Francis Darwin provides the most detailed description of Darwin's reading habits in the biographical essay prefixed to *The Life and Letters of Charles Darwin* (1887). His father reserved only about an hour a day for reading scientific and specialized technical works. In contrast, Darwin spent somewhere between three to four and a half hours a day on recreational literature, which included biography, travel narratives, histories, and, above all, works of fiction. All accounts suggest that novels were the primary fare of these recreational readings.

We can use this description to readjust the picture given in Darwin's reading notebooks. In order to imagine how Darwin actually divvied up his time in these years, I've used the relative mixture of genres from his comprehensive reading list and adjusted it on the basis of Francis Darwin's account of how his father spent his time in figure 8b. (According to Francis, he spent about three quarters of his time reading recreational literature, with the remainder split between technical reading and newspapers; and his recreational reading was split evenly between fiction reading and other forms of literature.) Even if Francis Darwin's estimates are a little off, and if we allow that different kinds of reading work at different speeds and with different degrees of attention, the resulting picture gives a qualitative estimate of how much interest Darwin gave each genre as measured by time.

This chart describes Darwin's textual environment. Novels probably made up nearly a third of what Darwin read in an average year. Perhaps a third of those were works of historical fiction. Combined with biographies and works of history, about a quarter of Darwin's reading was devoted to works about past people and societies. By contrast, only an eighth of Darwin's reading time was given over to the monographs and articles on geology, botany, and zoology that were foundational to his scientific work. In fact, according to Francis Darwin's account, his father spent nearly the same amount of time reading novels that he spent in his study, performing the taxing work of dissection, experimentation, and observation that constituted his scientific labor. Even if I am a little envious, I do not mean to scant Darwin's enormous productivity. If anything, it's a reminder of the extraordinary economy of his professional labor. Elizabeth Darwin describes his "horror of losing time," the extraordinary "zeal in what he was doing at the moment, [that] made him careful not to be obliged unnecessarily to read anything a second time."[43] If novels were read and read again, scientific studies were rarely so treated. Everything in Darwin's study was a tool. The bookshelf that lined the southeast wall, within reach of his working table and swivel stool, had a row of drawers filled with specimens and technical instruments. Above were shelves piled with books to be read and books read but not indexed. As Elizabeth later observed, "His library was not ornamental, but was striking from being so evidently a working collection of books."

Here, I pose a question central to Darwin studies, namely, how might we interpret the novels Darwin read as part of his "working collection of books"? Beer has consistently persuaded that "none of Darwin's reading seems to have been in vain. It was all useable, and used, though relatively little of it was undertaken in a utilitarian spirit."[44] Darwin's extraordinary, lifelong attachment to novels encourages us to test Darwin and his writings against his literary engagements. Poetry has usually been given pride of place in thinking about the genres that shaped Darwin's literary influences. The point is not, strictly speaking, that Darwin understood *On the Origin of Species* as a novel but, rather, that Darwin's extensive investment in the novel, particularly in historical fiction, inculcated a comparative understanding of the past that emphasized the complexity and indeterminacy of previous events. As I argued in chapters 2 and 4, to talk of the nineteenth-century novel after Scott means talking about historical fiction at one remove, insofar as the realist novel, particularly as it was fashioned by a writer like George Eliot, depended upon the extension of comparative strategies that were initiated by historical novels.

That said, Darwin recognized the novel as a key genre for transformative experience. Adelene Buckland has deepened our understanding of the rich interplay between natural science and historical fiction in the early nineteenth century, especially after Scott took the office of president of the Royal Society of Edinburgh in 1820.[45] Geologists like Thomas Dick Lauder and Charles Lapworth read Scott as they tromped around Scotland looking for geological evidence, retracing the paths of his principle characters. Lauder even wrote his own historically derived "Scotch novels," complete with rich descriptions of the physical features of the Scottish Highlands.

Darwin's description of his own "astonished" experience in the Highlands as a "kind of geological novel" relies on his sense, derived from Scott, that the novel offers a series of narrative episodes that change how we understand the world through complexity and contradiction. His extensive reading of historical fiction, by Scott as well as Martineau, Edward Bulwer Lytton, and Henry Whitelock Torrens, drove this lesson home. One passage in the *Origin* seems to suggest that at times Darwin had historical fiction in mind in that work. Addressing the problem of the imperfect geological record, and the "missing links" between distinct fossil series, Darwin explains: "I look at the natural geological record, as a history of the world imperfectly kept, and written in a changing dialect; of this history we possess the last volume alone, relating only to two or three countries. Of this volume, only here and there a short chapter has been preserved; and of each page, only here and there a few lines" (310–1). The passage is usually read as an example of Darwin's bibliographic imagination—his ability to think creatively through the analogy between printed histories of human events and the geological histories he was working to secure. This seems clear, but it's particularly striking how the features of this parable derive from that old device of the historical novel, the recovered manuscript.

Though Scott's use of the found manuscript as a framing device was not, strictly speaking, innovative, its use in *Waverley* and his other novels made the trope hugely influential for historical fiction.[46] Wilhelm Meinhold's *Mary Schweidler, the Amber Witch* (1844; a novel Darwin read soon after publication) may have suggested this comparison to the geological record. As Meinhold puts it, "The manuscript, which was bound in vellum, was not only defective both at the beginning and at the end, but several leaves had even been torn out here and there in the middle."[47] In its place, Meinhold promises, after much research into the idioms and historical customs depicted within, to "restore those leaves which have been torn out of the middle, imitating, as accurately as I was able, the language and manner of the old biographer, in order that the difference between the original narrative and my own interpolations might not be too evident."[48] Even if Darwin imagines the historical novel as a model for geological interpolation, this does not mean he sees it as a genre of scientific writing. But it *does* mean he appreciates how (as noted in chapter 2) Scott's novels reconstruct a past that is strictly unavailable in the historical record. If Darwin did not see the *Origin* as a kind a novel, he still owed to historical fiction (by way of comparative history) a way to produce coherent narratives from scattered historical records and to aggregate differentiated histories as a network of patterns. By these means, comparative historicism pluralized Darwin's plots.

Orchids in Action

The effort to get the *Origin* completed and to press exhausted Darwin. While he asserts in closing that "there is grandeur in this view of life," the economy of its summary plan did not present many advantages for seeing it. Immediately after sending off the last proofs the family packed up and retired to Wells Terrace to recuperate. It didn't help. Darwin was trying to rest and gather energy for his next big push. Finally, he would finish his big book on species, tentatively titled "Natural Selection." When they returned to Down, Darwin launched into the new project doggedly.

The work progressed slowly over the course of 1860 with short breaks for side projects, revisions of the *Origin*, and, especially, a series of ongoing studies about the pollination of flowers by bees and moths. Darwin had been happy to find, after moving to Down in 1840, a collection of orchids in the nearby Cudham Valley, which the family referred to as "Orchis bank."[49] This launched a twenty-year period of sporadic interest, which advanced as Darwin drew on his growing network of botanist correspondents for specimens. After finishing the *Origin*, he returned to this work for relief. And as he came to look at orchids more closely, fascination grew into obsession: "You cannot conceive how orchids have delighted me," he wrote to Hooker.[50] "The Orchids are more play than real work."[51] "Orchids have interested me more than almost anything in my life," he confides to another correspondent.[52] "This subject is a passion with me."[53] By 1861 Darwin was completely absorbed with his work on orchids and was thinking about publishing a long study for the *Transactions of the Linnaean Society*. As the treatise grew and he felt compelled to commission a series of illustrative woodcuts, he changed his mind and wrote to John Murray, the publisher of the *Origin*, to see if he was interested in a book. Murray, who had seen the *Origin* through its third printing in less than two years, jumped at the opportunity. A year later, it appeared as *On the Various Contrivances by which British and Foreign Orchids are Fertilised by Insects* (1862).

Darwin used the study of orchids to demonstrate how natural selection would work in the field—articulating a powerful new research paradigm and changing how naturalists would study nature. As he put it to Murray on September 24, the *Fertilisation of Orchids* would "show I have worked hard at details, & . . . *perhaps* serve to illustrate how natural History might be worked under the belief of the modification of Species."[54] The *Orchids* records one of Darwin's most sustained efforts at experiment. In contrast to the curatorial strategy of the *Origin*, Darwin's own careful studies drive the argument. But the question of intent (by which I mean the teleological question that asks what specific adaptations are *meant* to do) would be central to this new science. Throughout the *Fertilisation of Orchids*, Darwin asks what specific "contrivances" are *contrived for*. What purpose do they serve? How do these adaptations interact with their environment, and what does this tell us about the ability of orchids to interpret the world around them and react? Rather than reading orchids into our world, Darwin argued, we must learn how they read and engage their own ecologies. The unsettling result is to imagine a world of nonhuman intent and distributed sentience that is far closer to humanity than previously imagined. As Lynn Voskuil observes, nineteenth-century British orchid culture "developed sophisticated forms of ecological awareness that suggest new models of agency and human responsibility."[55] In deploying this language of intent, Darwin effectively flattened the natural world, diffusing intentionality as an emergent property shared throughout nature. Darwin understood such purposefulness as the distributed result of natural selection's incessant adaptation rather than evidence of a divine plan and the *scala naturae*. As I will argue, this is part of Darwin's larger effort to expand the paradigm of natural selection so that it might incorporate the strategies of natural theology without endorsing Christian apologetics. Natural theology offered a powerful model for understanding the relation between natural order and our capacity to interpret it. Shorn by Darwin of the conceit of a higher order, this naturalized theology furnished a powerful means for reading ecology in its relation to human experience.[56]

Orchids—difficult to grow outside their natural habitat, wildly various, fertilized by complex, often unknown processes—captured the attention of naturalists as well as the wider public. It was famously expensive to import orchids from the far-flung locations where unusual specimens grew, in remote locations drawn from across the imperial and mercantile world. They testified to imperial power and wealth but also to a failure of horticulture. It was extraordinarily difficult to propagate orchids domestically. (In a private letter to Hooker, Darwin speculated that decaying plant matter was needed, but this was not confirmed until the end of the century.)[57] Specimens had to be shipped, alive, from their natural habitat. Enormous numbers died in transit. One contemporary enthusiast recalled a shipment of several hundred specimens in which only three survived, a loss of a thousand pounds of merchandise.[58] By 1860, the Victorian fever for orchid collecting was epidemic. From the mid-1830s, orchids had grown into an object of an intense green exoticism, both because of the extraordinary expense of collecting and maintaining them in the proper conditions within a hothouse and because their outrageous forms, so radically different from the handful of orchids that could be found in the English countryside, seemed to testify to the remote jungles and mountain valleys from which they were gathered.[59] An early reviewer of Darwin's orchid book described this orchid fever:

> Not very far from where we write, a public sale of Orchids is a not uncommon event. . . . If you are not an orchidaceous maniac, when the hammer descends and the purchase-money is fixed for some rare exotic, you should recollect that this is a rare Orchid—that even the Conservatory at Chatsworth has not many specimens— that in the next number of the—*Magazine* the cultivator's name will appear with honours—and that, of all the cardinal virtues, botanically considered, at present the display of a new or uncommon Orchid is the most cardinal and the most commendable.[60]

Though he would later complain of the reviewer's "very kind pity and contempt," Darwin sought to capitalize on this broad interest in "orchidomania."[61] Hoping to pique Murray's interest, Darwin explained that "the chief object is to show the perfection of the many contrivances in Orchids."[62]

I want to put particular weight on those "contrivances." Darwin's long fascination with orchids may have started with first reading of his grandfather's *Loves of the Plants* (1789), during his student years at Edinburgh. Orchids fascinated, in part, because their aesthetic and active engagement with the world and with the scientist confused the stable categories of scientific subject and natural object. As Theresa Kelley observes, in a study of Romantic botany that emphasizes the relation between the Darwins, "orchids embody the slippage between the scientific, the literary, and the aesthetic."[63] Darwin's use of "contrivance," particularly as it prompts curiosity, gets at the peculiarly expressive nature of orchid flowers, which, in Erasmus Darwin's account, provided evidence of communication between animals and plants. Erasmus described their extraordinary capacity for apparent mimicry in an update for the second edition of the *Loves*: "There is a curious contrivance attending the *Orphrys*, commonly called the *Bee-orchis* and the *Fly-orchis*, with some kinds of the *Delphinium*, called *Bee-larkspurs*, to preserve their honey; in these the nectary and petals resemble in form and colour the insects, which plunder them: and thus it may be supposed, they often escape these hourly robbers, by having the appearance of being pre-occupied."[64] As I argue in chapter 1, this concern for the anthropomorphic behavior of both animals and plants, presented as a poetic conceit in the first edition of the *Loves*, explored Erasmus Darwin's belief in the common origin of all life. To understand the "loves" of the plants, he argued, we must understand how we share features of their sexuality. We have already seen how *On the Origin of Species* responds to this sense of deep entanglement, speculating on the relation between plant and animal life, and insisting on the continuities between all living things. Elizabeth Grosz has most extensively explored Darwin's extraordinary capacity in later work to "reconceptualiz[e] the relations between the natural and the social, between the biological and the cultural, outside the dichotomous structure in which these terms are currently enmeshed."[65] To read the *Orchids* in dialogue with a work like *The Loves of the Plants* is to recognize how Charles Darwin worked to reshape these dichotomous visions of the natural, and their analogies to the social, at a much earlier stage of his career. Turning our attention to the almost human activity of these strange flowers, their peculiar sensitivity, vitality, and movement, Darwin creates space for the flowers to intervene within his account of nature.

The flowers had marvelous things to say. Darwin learned that some orchids use a pollinating structure, shaped like a lollipop, that holds the pollen away from the body of its fertilizing insect when it first adheres, and so avoids self-pollination as the insect brushed past the stigmatic surface of the same flower (figure 10). Then, as the insect flies to the next flower, the affixed stalk bends down into a joustlike position perfectly positioned to strike the stigmatic surface of a new flower, fertilizing it. Darwin found that several other genera of orchids, which had previously been assumed to lack nectar, in fact hid the nectar within secondary chambers, so that only the right kind of insect would be able to access it.[66] He speculated that the "bee orchid" (the same described by Erasmus), whose rostellum looks precisely like that insect, and which was known to be self-fertilizing, nevertheless must have its unusual form because it was adapted to attract bees (65).

Darwin spends considerable time with the *Catasetum* (or "seed thrower"), which was known to launch its pollinating body through the air after a foreign object brushes against one of its long, triggerlike sepals. He was able to model the physical motion of the projectile using a strip of whalebone, a split quill, and a pencil. Through his interactions with the flower, Darwin demonstrated that the triggering mechanism was not mechanical, instead mediated by some sort of chemical cell-to-cell signaling. As he puts it, "How then does nature act? She has endowed these plants with, what must be called for want of a better term, sensitiveness" (212). Erasmus Darwin had long argued that plants had embodied modes of affect analogous to animal sensations, including desire—this underwrote the central conceit of *The Loves of the Plants*. As I note in chapter 1, it was only after the later publication of his *Zoonomia* (1794), in which Erasmus explicitly recognizes this claim, that critics turned against his botanical epic. In the *Fertilisation of Orchids*, Charles is evidently pleased to find there was truth to his grandfather's sensational theory—he designates the *Catasetidae* "the most remarkable of all Orchids" (130).

Orchids, in their extraordinary activity, in their sensitive interaction with their environment, in the interspecies relationships they cultivated, afforded Darwin a powerful model for sociability of ecologies and of life in general. They also helped him to think about the place of his writing in the textual ecology of the print market. Like the orchid, Darwin's book would have to recruit interest—especially the support of experts and hobbyists who could furnish advice and materials for study—and use that support to succeed with the reading public. When Murray arranged publication of the *Origin*, he'd asked if Darwin might propose a design for the cover. Darwin demurred: "I cannot think of any fit ornament or Device for outside: the subjects treated are so many & so general, that no one illustration would serve." Two years later, Darwin reached out to Murray to see if he'd be willing to include a design for the *Fertilisation of Orchids*. Darwin oversaw the illustration and had the illustrator deliver it to Murray. On the plum cloth binding of the first edition appears a gilt illustration of *Cycnoches* (figure 9). The orchid depicted is only mentioned in passing within the text, as he speculates (correctly) that it's a second example of a trisexual genus like *Catasetum*, and yet it seems that Darwin found, in its complex nature, a "fit ornament" to summarize his new study.[67] Darwin supervised the production of all his illustrations scrupulously—inviting illustrator G. B. Sowerby to Down for a week, superintending the work, and paying the ten-pound fee out of pocket. The woodcuts within the pages of Darwin's *Fertilisation of Orchids* are assiduous diagrammatic representations of the orchids they depict, employing conventions, as Jonathan Smith points out, from several distinct genres of botanical illustration.[68] The cover, by contrast, seems to reference an earlier tradition of illuminated illustration that values artistic presentation over truth-to-nature. In part, this draws on previous illustrations of the *Cycnoches* or "swan-necked" orchid, which was a particular subject of fascination for Victorian readers; James Bateman's marvelous *Orchidaceae of Mexico & Guatemala* (1837) devotes extensive attention to its semblance to the bird, including woodcuts

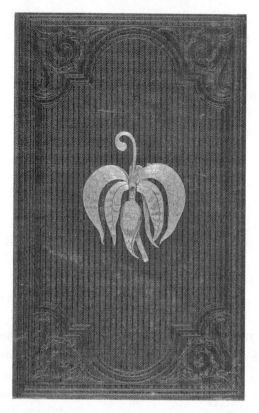

Figure 9. "Cycnoches." Gilt. By George Brettingham Sowerby II. From Charles Darwin, *Fertilisation of Orchids* (1862), 121997, cover. Courtesy of The Huntington Library, San Marino, California

in which its swanlike features are strikingly exaggerated.[69] The gilded image of the stylized *Cycnoches*, like Bateman's woodcuts, emphasizes aesthetic display over the objective eye.

This aesthetic engagement is supplemented by a presentation that places the reader in the prospective position of the insect that fertilizes it (an orientation that radically contorts the flower's vertical presentation in nature, extensively documented in earlier botanical illustrations). The welcoming orientation of the orchid's exotic display draws the reader in to the major thesis of Darwin's work: these extraordinary and beautiful physical adaptations are complements by the orchids to their insect of choice. We are attracted to orchids because orchids mean to attract. If contemporary Victorians were obsessed with orchids, they were, in some sense, only responding to the orchids' brilliant evolutionary strategies.[70] The cover illustration discloses how Darwin recruits the orchids into his science, studying the nuance of their physical position and attraction, their "marvelous" attention to the location, approach, and position of their pollinator, as strategies to engage his reader.

The orchids helped tease out the implications of Darwin's flat ontology. To recognize the continuities between orchid, animal, and human is to recognize their capacity to engage us, and so to communicate their nature and help write Darwinian science. In such moments, the *Fertilisation of Orchids* puts the reader in contact with the flowers themselves. This flattening of distinction gains substantial power through the mundane technologies and wide range of sources Darwin used to consult the orchids, leveling distinctions between elite science and enthusiastic hobby, between social communication and natural interaction. Darwin regularly suspends his analysis to give fulsome thanks to merchants like James Veitch, as well as the hobbyists, gardeners, and botanists who lent him specimens and the resources of their expensive hothouses (a cottage industry in itself). It would be hard to overstate the importance of this material aid to Darwin's work. On the basis of contemporary accounts in which foreign orchids fetched prices anywhere from one to one hundred pounds at market, the dozens of specimens that Darwin received for examination and dissection represented a donation of materials that dwarfed his own financial investment in the work and perhaps equaled the entire cost of the first print run.[71] The book testifies to the openness of Darwin's science, speaking not only to the deep collaboration of mid-nineteenth century naturalism but also to Darwin's consistent efforts to work that collaboration across already well-entrenched divisions between naturalist and layman, between merchant, enthusiast, and expert. James Secord ascribes Darwin's eclecticism to "the very essence of [Darwin's] theory of transmutation [which] demanded that he transgress the boundaries of his own scientific community."[72] In flattening natural distinction, Darwin's science also flattened the social world.

A striking feature of Darwin's study is the contrast between the extraordinary expense and exoticism of the plants and the almost absurdly prosaic instruments he used to study them. Even as the preciousness of orchids elevated them into an almost ethereal object of green envy in the Victorian period, the mundane implements that Darwin used to divine their secrets put them in contact with terra firma. The *Orchids* records one of Darwin's most sustained efforts at experiment; its arguments supported largely by his own investigations, in contrast to the curatorial strategy of the *Origin*. Little was known about the mechanisms of fertilization in the wild. Readers of the *Orchids* marveled

Fig. II.

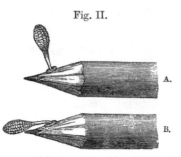

A. Pollen-mass of O. mascula, when first attached.
B.   do.   do.   after the act of depression.

Figure 10. "Pollinia of Orchis mascula." Woodcut. By George Brettingham Sowerby II. From Charles Darwin, *Fertilisation of Orchids* (1862), 604448, p. 15. Courtesy of The Huntington Library, San Marino, California

at Darwin's revelation that, in twenty years of observation, he had seen an insect visiting an orchid only on two separate occasions (34). This provided extraordinarily slim observational data—not enough to finally put to rest the alternative hypotheses that orchids were largely self-fertilized or fertilized by wind-born pollen. Darwin's experimental strategy, elegant in its simplicity, was yet groundbreaking. He would play the part of the insect, mimicking its behavior in order to fertilize the orchids himself. Casting himself as a pollinator, deducing through imitation the structure and behavior of the pollinating insect, Darwin then imagines the evolutionary history that has brought the flower and its pollinator to their present form. Slowly, he figures out what kind of pollinator he needs to be for the flower to engage him, simulating the parts of the flower and the proboscises of suitor insects using a variety of pencils, pens, and needles, strands of hair, and bits of bone and wire. It's science in a junk drawer.

The illustration of the *Orchis mascula* pollen body, the first illustration to appear in the *Orchids*, drives the point home (figure 10).[73] As a register of scale for the viewer, the pencils dwarf their new appendages. They are carefully whittled and sharp; the lead points glitter dully. Physically and visually unnecessary, these leads have no present purpose, and so I speculate whether the pencils were used in the execution of any of the other three dozen drawings used in the book, or perhaps, in the writing, or later editing, or even later proofing of the work itself. As a register of labor, they point to the close imbrication of experiment, composition, and drawing that produced the volume now in the reader's hand. Like the presence of Hallam within Tennyson's *In Memoriam* (discussed in chapter 3), the impulse of the orchid is registered through Darwin's pencil, in the shaded illustrations and beguiling text that sketch out its engagement with the world. Through such representations, ecology shades into the collaborations of scientific discovery and the network of book production. These pencils seem to hint how Darwin's new science will work: a ready-to-hand science that shares authority with its objects, registering the hand of nature in the composition of his books.

Flat Theology and Reading for Intent

The language of *contrivance* is at the center of Darwin's effort to see orchids in action; he uses the term with peculiar frequency in his study of orchids and largely abandons it in later works. Contrivance marries intent and innovation; it marks what we now term adaptations and emphasizes a semblance of intent in the physical and behavioral strategies that help organisms thrive. He adopted the term from his grandfather: in *The Loves of the Plants*, Erasmus Darwin described the adaptation of the "Bee-orchis" as a "curious contrivance."[74] This was not the only available vocabulary; in his *Das entdeckte Geheimniss der Natur im Bau und in der Befruchtung der Blumen* (1793)—which Charles Darwin carefully studied and annotated—C. K. Sprengel had described such adaptations as a "*bestimmte Art*" or "certain/given manner." Charles prefers his grandfather's term. "Contrivance" is especially prominent in the sixth chapter of the *Origin*, where Darwin tackles the apparent design of complicated organs, as foregrounded in works of natural theology by William Paley and Charles Bell. That same section of the *Origin* contains Darwin's only previous use of Erasmus's "*curious* contrivance," with reference to orchid pollination (193).

The *intentive* vocabulary of "contrivance" provides a key point of translation between the language of natural theology and the new imaginative regime of natural selection.[75] It engages the reader's sympathetic imagination in the work of constructing persuasive evolutionary accounts. Darwin underlined this shorthand in the full title of the orchid study: "On the *Various Contrivances* by which British and Foreign Orchids are Fertilised by Insects" (emphasis added). By these means, Darwin elevated the problem of design by foregrounding his grandfather's purposive language. This was intentional: "Like a Bridge-water Treatise," Darwin explained in proposing the orchid study to Murray, "the chief object is to show the perfection of the many contrivances in Orchids."[76]

The Bridgewater Treatises were a famous series of eight natural theological tracts, sponsored by the Earl of Bridgewater, in which influential naturalists and theologians sought to demonstrate the evidence of God's hand in nature.

Darwin well knew that, in a work he described to Murray as something like a ninth "Bridgewater," contrivance implied intent.

He'd made a careful study of the Bridgewater treatises and later described William Paley's *Natural Theology* as a formative influence. According to his reading notebooks, he read at least seventeen works of natural theology over a twelve-year period from 1839 onward. For these reasons, the *Orchids* must be recognized as a critical object for the ongoing discussion of Darwin's engagement with natural theology. Scholars of Darwin have struggled to untangle his extensive engagement with what has been characterized as the "rhetoric" of natural theology.[77] In Darwin's own time criticism was directed at the term *natural selection* itself, as a kind of anthropomorphism that ascribed powers to nature that echoed the capacities of an intelligent creator. The first edition of the *Origin* encourages this line of thinking, particularly Darwin's insistence that nature's powers are nearly omnipotent alongside humanity's: natural selection, he argued, is "a power incessantly ready for action, and is as immeasurably superior to man's feeble efforts, as the works of Nature are to those of Art" (61). As John Leifchild put it, with some justice, "I do not see how the conclusion can be evaded, that if Natural Selection is merely a metaphorical expression, the whole of what is attributed to it is attributed to a metaphor, which must be a veil for some unexpressed reality."[78]

As we have already seen, Darwin recognized the capacity of "metaphorical" language to apprehend features of the natural world that were incompletely grasped by science. The anthropomorphic language of intent casts this dynamic in sharp relief, forcing us to ask, in effect, whether the ascription of human motivations to natural systems distorts those systems or captures something foundational to their operation, asking whether the language of natural theology—even absent belief in a creator—has something to teach us about the world. Leifchild's point is that natural selection, as a metaphor for a process that chooses the fittest individuals in nature, indexes an intentional process. And in a larger perspective that frames Darwin's writings within the coordinates of secularization and disenchantment, commentators have tended to handle such moments in Darwin's works as a facile, if unfortunate, deployment of theological language. On this reading, Darwin coopts theological arguments into a materialist agenda.[79] Few argue that the language is anything more than convenient.[80] And this argument finds support in the *Origin*, where explicit alliances between natural selection and intentional language occur only in passing.

But in the *Orchids* the connections between intent and adaptation are part of the basic rhythm of "how natural History may be worked under the belief of the

modification of Species." We can see an example of this in Darwin's speculations about the foot-long nectary of the *Angræcum sesquipedale* of Madagascar:
"What can be the use, it may be asked, of a nectary of such disproportional
length? We shall, I think, see that the fertilisation of the plant depends on this
length and on nectar being contained only within the lower and attenuated ex-
tremity. It is, however, surprising that any insect should be able to reach the
nectar: our English sphinxes have probosces as long as their bodies: but in Madagascar there must be moths with probosces capable of extension to a length of
between ten and eleven inches!" Darwin's prediction of a yet unknown species
of Moth—not confirmed until 1903—is a famous chapter in the wider adoption
of natural selection as a model for empirical research (197–8). I am more interested in Darwin's approach. It begins with a question (What can be the use?)
that is rooted in the conceit that such structures have uses, the kinds of built-in
purposes that are central to the argument from design.

In his *Critique of Pure Reason* (1781), Immanuel Kant gave extensive attention
to the place of such "purposive causality" in science, arguing that it provides a
"regulative principle" necessary for the study of organic bodies. In such cases,
organs and the organism itself seem locked in a relation of mutual necessity, so
that each is the means of the other as end. This presupposition authorizes the
assumption "that everything in an animal has its utility and good aim," but this
is not "constitutive"—the conceit of purpose does not tell us anything about the
actual structure of nature.[81] But, as we will see, Darwin's turn from the singular organic form toward ecological histories—toward the relation *between* beings and *over time*—undermines the distinct status of human reason. This means
that "purpose" itself is as an immanent and so "constitutive" feature of organisms under the common regime of natural selection. To frame the "what for"
question in terms of purpose, in Darwin's analysis, is to call forward problems
of mechanism and action that must be answered through imaginative narratives: stories, derived from comparative analysis and extensive analogy, that apply what we have seen done to what we could do and move from English sphinxes
to Madagascar moths. This extension launches the experiment. After bristles
and needles fail, and on the assumption that longer probosces need thicker support, "I took a cylinder, one-tenth of an inch in diameter, and pushed it down
through the cleft in the rostellum. . . . By this means alone I succeeded in each
case in withdrawing the pollinia; and it cannot, I think, be doubted that a large
moth must thus act" (199–200).

By dilating purpose to include both moth and flower, Darwin's solution
raises an important new problem. The purpose of the long nectary is to select

for the unique long-nosed moth, yet circularly the purpose of that long proboscis is to help the moth with the uniquely long nectary of the *Angræcum*. In the Kantian formula, each means is the other's end. In the *Origin*, the object of selection is usually response to some external condition, a "third body." Here there is no third body, simply a reciprocal, analogous relation between insect and flower. It's easy to imagine how a flower would be modified to better suit a given condition, so also for the moth—but why would both be reciprocally modified in a fashion that forces them into an exclusive relationship? The *Angræcum* provides Darwin with the first sharply defined case of "coevolution," and he gets there through the imaginative analogy of the *Angræcum* orchid, its moth, and allied variants. Darwin's solution to the puzzle bears recording at length:

> As certain moths of Madagascar became larger through natural selection in rela-
> tion to their general conditions of life, either in the larval or mature state, or as the
> proboscis alone was lengthened to obtain honey from the Angræcum and other
> deep tubular flowers, those individual plants of the Angræcum which had the
> longest nectaries (and the nectary varies much in length in some Orchids), and
> which, consequently, compelled the moths to insert their probosces up to the very
> base, would be fertilised. These plants would yield most seed, and the seedlings
> would generally inherit longer nectaries; and so it would be in successive genera-
> tions of the plant and moth. Thus it would appear that there has been a race in
> gaining length between the nectary of the Angræcum and the proboscis of cer-
> tain moths. (202–3)

Darwin's explanation is incomplete (as he admits: "We can thus *partially* understand how the astonishing length of the nectary may have been acquired" [202, emphasis added]). What causes the initial differentiation between the shorter variants of the *Angræcum*, with their modest moths, and these new lines of monsters? And once you have distinct long varieties of each, what is it that drives the "race"? (The implicit answer is that, by compelling moths to bring their "base" into contact with the stigmatic surface, the flower ensures the greatest possible transfer of pollen.) But in spite of these gaps in Darwin's history of the unusual flower, the account satisfies through the convincing turn to narrative. Darwin doesn't make the process seem inevitable, but he succeeds in making it imaginable as a speculative history in which flowers "compel." In effect, the turn toward comparative historicism—the analysis of the relation between different flowers and moths over time—makes it possible to understand how organisms read each other's history. Though natural selection does not feature explicitly, its narrative agency, worked out at length in the *Origin,* is the powerful engine of this success.

To me, emphasis upon competition distorts an otherwise far more coopera- tive view of natural selection than offered in the *Origin of Species.* Darwin's work on *Orchids* added an important chapter to natural selection by exploring in finest detail the mutually beneficial adjustments of flower and insect that pro- mote collective success. The preceding passage illustrates Darwin's intense in- vestment in sharply defined narrative values. Though we generally characterize Darwinian nature in terms of brutal competition, Darwin was more interested in those collaborative engagements through which organisms successfully in- teract: a three-legged race in place of the "war of all against all." Darwin some- times frames this in terms of precarious interdependence: "If such great moths were to become extinct in Madagascar, assuredly the Angræcum would become extinct. On the other hand, as the nectar, at least in the lower part of the nec- tary, is stored safe from depredation by other insects, the extinction of the An- græcum would probably be a serious loss to these moths" (201–2). But more often it is "wonderful" how the orchid contrives, "marvelous" how moths, in serving, serve themselves. Always, the language of intent is retained as the ava- tar of differentiation; the anthropomorphism of striving insects is retained as a narrative stand-in for natural selection.

To this point, I have been loosely working Jerome Bruner's distinction between paradigm and narrative.[82] For Bruner, they are characteristic of two distinct modes of explanation, one systematic and scientific, the other diachronic and discursive. Bruner's contrast helps isolate what is so peculiar about natural se- lection. Poised between paradigmatic and narrative systems of understanding, natural selection translates between a comparative view of systemic relation and narratives of success and loss, between analysis and valuation.[83] This inter- play of perspective is a major function of comparative historicism, insofar as it aligns stories—sometimes competitive, sometimes collaborative—in order to explore larger patterns.

Reviewers of the *Orchids* who were sympathetic to the argument from design celebrated Darwin's intentional language: "The whole series of the Bridgewater Treatises will not afford so striking a set of arguments in favour of natural the- ology as those which he has here displayed," marveled the *London Review.*[84] The *British Quarterly* found "it is full of the marvels of Divine handiwork . . . [that] prove fascinating beyond all comparison."[85] Joseph Campbell, the Duke of Argyll, provided one of the most influential perspectives in the *Edinburgh Review,* rolling Darwin's work into an omnibus review of works on the "Supernatural." Campbell is an attentive reader of Darwin's language: "He exhausts every form of words and of illustration by which intention or mental purpose can be described. 'Contrivance,'—'curious contrivance'—'beautiful contrivance'—these are expressions which recur over and over again."[86] Campbell's review makes the case that the language of intent is ingrained in "natural selection," which, in turn, entertains a view of the "natural" that includes powers which would otherwise be termed "supernatural": a " 'nature' in the largest sense. . . . We must understand it as including every agency which we see entering, or can conceive from analogy as capable of entering, into the causation of the world. First and foremost among these is the agency of our own mind and will" (380). Darwin's *Orchids* makes it clear that, as Campbell puts it, "Purpose and intention . . . meet us at every turn, and are more or less readily recognised by our own intelligence as corresponding to conceptions familiar to our own minds" (395). For Campbell, this is sufficient grounds for believing in the "Author"; for Darwin, it reflects the extraordinary analogy between natural selection and the consideration of nature *as if* designed, a correspondence between Darwinian evolution and the intentive imaginary.

At the heart of this disagreement is a dispute over what kind of analogy intent presents. For Campbell, as for other advocates of design, intent demonstrates formal correspondence, identifying synecdochal parts of the order of being. We recognize intent because it was built into us, the family resemblance between natural and human order that is the common legacy of the greater power from which they are derived. For Darwin, however, the analogy is harmonic, the correspondence between human intent and its epiphenomenal register characterized by their distribution in a world that does not require higher organizing principles. When the orchid, in a moment of mutual recognition, responds to Darwin's pen, it is because, in this basic sense, it recognizes his integration into its world.

Note how far the ground of analogy has shifted from the *Origin*. The term "analogy" never appears in *Orchids*; its residue collects in scattered references to the curatorial role it had served in *Origin*: as ever, "Analogous facts . . . could be given."[87] Primarily, analogy's analytic force is focused into a series of intersubjective engagements: Darwin and insect, Darwin and flower, flower and pollinator. All demonstrate how "natural history may be worked" across the analogy between domestic and natural action. We must read ourselves into nature and activate sympathy as a critical lens. George Eliot would approve.

Much like one of Eliot's fictions, Darwin's *Orchids* does not coopt theology through the ambiguity of intent; rather, it expands this ambiguity as a powerful meeting ground between materialist and theological perspectives on nature. Darwin's relation to religion was truly complex; in his autobiographical note he records his wavering between the deism of his grandfather, father, and brothers and the more reserved (newly minted) agnosticism of Thomas Henry Huxley. For Huxley, agnosticism was a skeptical hedge that disarmed the charge of thorough atheism, but for Darwin it represented an important intermediate position, a position where a religious understanding might retain value. His love for his wife, the pain it gave him to feel at odds with her devotion, is well known, as is his sincere respect for William Buckland, Charles Lyell, and the other eminent British naturalists of the previous generation, who, almost without exception, believed that natural inquiry reinforced religious belief. By drawing natural selection into a bigger tent, as an engine of intent that countenanced both skeptical and illuminated perspectives on natural order, Darwin preserved a shared space for enthusiasm and marvel, marking out a pluralist domain of scientific enquiry.

Darwin was excited to see one review of the *Fertilisation of Orchids* above all. Asa Gray, the American botanist, had written a series of articles in 1860 that argued for the compatibility of natural selection and natural theology. Gray was the ideal reader for *Orchids*. When the American naturalist proposed a review, Darwin wrote immediately to Murray, encouraging the loan of three woodcuts for illustration. Gray's two-part summary did not disappoint: "The present book is almost wholly a record of observed facts, of curious interest, irrespective of all theories of origination, and perhaps as readily harmonized with old views as with new—with direct as well as indirect creation."[88] In the sequel, he continues:

> We cannot close without an expression of gratitude to Darwin for having brought back teleological considerations into botany. . . . Mr. Darwin,—who does not pretend to be a botanist—has given new eyes to botanists, and inaugurated a new era in the science. Hereafter teleology must go hand in hand with morphology, functions must be studied as well as forms, and useful ends presumed, whether ascertained or not, in every permanent modification of every organ. In all this we *faithfully* believe that both natural science and natural theology will richly gain, and equally gain, whether we view each varied form as original, or whether we come to conclude, with Mr. Darwin, that they are derived.[89]

Gray's emphasis on teleology is important to contemporary debates; there had been an ongoing argument, most recently between Owen and Huxley, about the place of teleology in the reconstruction of extinct species. Huxley believed reconstruction should be strictly empirical; Owen held that functional correlation—the assumption that specific features of the skeleton suited specific ends—was necessary and demanded a teleological perspective.[90]

I think Gray is right to see Darwin as neither a scold for intelligent design nor the pundit who would insist we accept that the watchmaker is blind.[91] To be clear, Darwin consistently makes the case, in *Orchids* as elsewhere, that natural selection makes divine intervention unnecessary. George Levine has made the most persuasive argument that Darwin's brand of naturalism is thoroughly secular while reproducing the enchanted sensibility of natural theology.[92] I think Darwin's commitment to intent gives the modes of natural theology an even larger role. Natural selection preserves a way of talking about teleology while bracketing the problem of supernatural agents, a *flat theology* that, like agnosticism, abjures the question of a higher power. Reading Darwin this way watches him create space within scientific thought for this softer seculariza-tion, characterized by Charles Taylor in terms of pluralism and sustained by epistemological modesty and the sense that any individual view is one among many available positions.[93] I once studied under a research biologist who was also an initial creationist, and so believed that life was started by divine act and then developed through evolution. At the time I found his view disturbingly out of step with his commitment to natural selection. Now I am not so sure. What might Darwin's reception have looked like had *Orchids* been published first, if the numbers of Murray's collected Darwin ran *Voyage of the Beagle—Orchids—Origin?* Asa Gray speculates that "it would have been a treasury of new illustra-tions for the natural theologians, and its author, perhaps, rather canonized than anathematized, even by many of those whom his treatise on the origin of spe-cies so seriously alarmed."[94] If the *Orchids* failed to succeed with the public as Darwin and Murray hoped (its gradual sales did not turn a profit until the 1870s), it may be that, as Gray indicates, Darwin had already alienated a good portion of its potential readership. We can only imagine what the "Big Species Book" might have looked like, had Darwin not been forced to reduce it to an "abstract," stripped of the wonderful contrivances that serve as common ground in *Orchids*. It would have certainly proved more challenging for intermediaries like Huxley to absorb Darwin's theories into the cause of a harder-nosed scien-tific naturalism.[95] It's my sense that this was an important object of Darwin's continuing study of natural theology during the twenty-year effort to work out his theory of natural selection. In addition to searching for the right language for disclosing that theory, he was sussing out the right vantage from which to imagine a wide, wonderful, pluralist nature.

I think Gray's approach to Darwin's teleology speaks generally to what we might term *intentive reading*. Darwin's studies of the origin of species and the fertilization of orchids lay out how an indeterminate or "soft" teleology, characterized by the strategic use of intent, provides a powerful historicism that does not fall prey to deterministic thinking.[96] Histories of natural selection have conserved features, with specific comparative patterns, including paradigmatic and narrative dimensions, and these formal properties make it possible to predict as well as to look back. The Madagascar moth is a case in point. I locate this power in natural selection's use of analogy to disclose new features in the world. This knowledge is always tentative. At the heart of Darwin's histories, as in those of Scott or Eliot, is the importance of how things fell out, but also the important fact that they did not have to happen exactly the way they did. Another actor may have sufficed, conditions could have changed. Above all, meaningful effects are the sum of contingent causes.

If this is true for adaptation, I think it is truer of intentive accounts of writers and their works. In this chapter I've frequently conflated Darwin's intentions with those of his works. I do this in order to draw out a common critical practice by which we emphasize the activities of a text instead of its author as a way to avoid the problem of authorial intent, the error of assuming we can know what an author meant. Though the problem of intent took focus with the New Critical prohibition against the "intentional fallacy," it gained even higher visibility, and received a huge jolt of fresh energy, with the publication of Roland Barthes's famous essay on "The Death of the Author," and the subsequent publication of Michel Foucault's meditation on authorship as a "function of discourse." In Barthes's famous account, when we ascribe meaning to the author's intentions we limit the possibilities of a text, reducing its complexity and interest. As he puts it, "We know now that a text is not a line of words releasing a single 'theological' meaning (the 'message' of the Author-God) but a multi-dimensional space in which a variety of writings, none of them original, blend and clash."[97] Yet the present work, which (in chapter 2) asks how the dead continue to live through books, and (in this chapter) how flowers help to write them, seeks a wider dispensation.[98] By emphasizing the deeply social nature of authorship, its filiation with the collaborative nature of scientific inquiry, and the continuity between social and natural life, I hope to recognize the value of intent, the intents of authors as well as those of other human and nonhuman agents, in understanding how books work. If authors are not strictly "original," they are nevertheless primary nodes within that larger network. Intentive reading helps explain why. Intent and teleology are powerful heuristic fictions, in the sense that they are analogies between what we would do and what others may have done, between our motivations and motivations out there. The more accurate the analogy, the more persuasive it will be. Far from a simply formal analogy, intent attunes us to the world and opens the possibility of our harmonic interaction with it (whether we find that world between the pages of a book or beyond its jacket).

Intent brings to a head the concern, running throughout this book, for the problematic relation between humanity and nature, between social forms and the natural patterns they sometimes attempt to apprehend, between ontology (the nature of being) and epistemology (our ability to understand that being). I have been arguing, alongside Manuel De Landa, that Darwin's vision of nature flattens the ontology of life, placing humanity alongside other beings and so reading us into the history, as well as the ecology, of nature. As David Amigoni puts this, "the possibility that nature was always already 'cultured' (in being shaped, modified, supplemented) became a powerful yet troubling source of analogy for Charles Darwin."[99] For some, this marks the troubling anthropocentrism of Darwin's science; from the beginning, the theory of natural selection seemed to project human motivations onto nature, and so distort our apprehension of the natural world. But as Jane Bennett has recently argued, anthropomorphism does not necessarily entail anthropocentrism; on the contrary, anthropomorphic figures can help us to recognize aspects of the natural world that resonate with human experience.[100]

Along similar lines, I have been arguing that Darwin's language of intent flattens natural theology into nature. But the far more consequential implication is that, by reading anthropomorphic figures into the patterns of nature, and so recognizing the continuity between them, we flatten the *epistemology* of life. My reading here is inspired by recent work by a range of literary scholars, anthropologists, and philosophers of science, including Donna Haraway, Richard Doyle, Matthew Hall, Eduardo Kohn, and Jeffrey Nealon, that argues the continuity between our modes of experience and the natural world, including plant life.[101] Anthropomorphism, personification, and the figurative or poetic strategies by which we impute meaning in general disclose features of an intentive world that, absent the conceit of a creator, hold us within itself. If we tend to focus on Darwin's work in evolutionary biology, rather than his deep absorption in botany (including six monographs published in the decades following the *Fertilisation of Orchids*), one reason is that it is hard to grapple with the un-settling implications of Darwin's unnatural history of plant life: the story of how plants move, feel, and interact in ways we both recognize and share. Dar-win gravitated to uncanny plants—vegetation that moved, flowers that ate ani-mals, even plants that could hear bassoons—because, like his grandfather, he recognized plants as a limit case that proved the general intelligibility of life. By including us within that flatter world, natural selection also flattens our engage-ment with it—securing our ability to interpret our entanglement within nature. This is, I think, my most important point. Evolutionary research, in Darwin's plant studies, ultimately turns on a series of intersubjective, sympathetic engage-ments, demonstrations of how "natural selection might be worked" across the analogy between domestic and natural behavior. When we recognize our place in the world, we activate sympathy as a critical lens. By these means, our own experience furnishes, through analogy, insight into the nature of life itself. For a flat ontology, a flat epistemology. In a flatter world, a flatter view.

Conclusion: Epitaph for the Darwins

The readings I have developed here are wrapped around the central relationship between Erasmus and Charles Darwin, a cross-generational collaboration that one Darwin could not predict and the other generally refused to address. I don't mean to say that Charles Darwin intended his *Orchids* to complete *The Loves of the Plants* any more (though perhaps no less) than the *Origin* is designed to re-work *Zoonomia*. But Charles was certainly, thoughtfully, concerned with his grandfather's example as a naturalist and author. I want to imagine this as a collaboration across historical remove but outside the unidirectional parameters of influence, response, and resistance—something more like solidarity across time. I see the Darwins, as both naturalists and writers, as analogous figures—connected in ways both harmonic and formal. In recognizing this, I underline the subject of this book: the complex analogy between science and literature.

This interrelation is a problem that Charles addresses explicitly in portions of his manuscript biography of Erasmus, written as a preface to a short German essay on his grandfather by Ernst Krause. In passages like the following (ex-cised before publication by Henrietta Darwin, Charles's daughter), Darwin al-ludes to this complicated debt: "As we have been here considering how much or how little the same tastes and disposition prevail in the same family, I may be permitted to add that from my earliest days I had the strongest desire to collect objects of natural history; and this was certainly innate or spontaneous, being probably inherited from my grandfather. Some of my sons have also exhibited an apparently innate taste for science. Members of the family have been elected Fellows of the Royal Society for four successive generations."[102] Here the relation is framed in terms of familial inheritance, but Darwin intends more. Erasmus is no more responsible for inherited traits than his wives or his children. But a clear sense of pride intrudes, a sense that his grandfather started something of a family line, in the boast that family members have been in the Royal Society for "four successive generations."

Charles responds here to observations, offered by Krause, that trace the relationship between the Darwins: "We find the same indefatigable spirit of research, and almost the same biological tendency, as in his grandson; and we might, not without justice, assert that the latter has succeeded to an intellectual inheritance, and carried out a programme sketched forth and left behind by his grandfather" (132). Nearing the end of his life, and triangulated through Krause, Charles Darwin is finally able to create space to discuss, however carefully, his relationship to his grandfather. As Krause notes, it is an "intellectual inheritance" mediated by writing, by the "sketch[es]" that Erasmus "left behind." The German biographer asks us to consider how Erasmus's books are recast and reworked through contact with his grandson's writings, how the grandfather "divined" the grandson. It is a problem I've raised in the previous chapters, staged through scenes in which Scott, Tennyson, and Eliot rework their connection to friends and family lost to death and social divides. To imagine oneself as the analogue to someone else, to see figures paired within literary history, is to change our understanding of each, much as Darwin continued to adjust his picture of himself and his work in light of his grandfather's legacy. Readers of the Darwins have rarely drawn them so intimately together, and I suspect one reason is the eclipse of Erasmus in the triumphalist history of science to which Krause contributes in his essay. But at a more basic level, such encounters place teleology and historicism in a tense dialogue that prompts us to explore the present through its complex relation to the past.

In the new preface to the second edition of the *Life of Erasmus Darwin*, published five years after Charles's death, is printed an epitaph recently set in the cathedral at Lichfield:

ERASMUS DARWIN, M.D., F.R.S.

Physician, Philosopher, and Poet,

Author of the "Zoonomia," "Botanic Garden," and other works.

A skilful observer of Nature,

Vivid in imagination, indefatigable in research,

> Original and far-sighted in his views.
> His speculations were mainly directed to problems
> Which were afterwards more successfully solved by his
> Grandson
> CHARLES DARWIN
> An inheritor of many of his characteristics.

Written by Charles's cousin, Francis Galton, the inscription canonizes the family line. Erasmus Darwin, once vilified by the canon of that same cathedral for a family motto that ran *E conchis omnia*, or "everything from mollusks," would have appreciated the irony. In elevating Charles, in bold type, as the other subject of the epitaph, they are graven alongside each other in the afterlife of human history and discovery. Together, they continue to help us, because (as Eliot once put it) in their death, they are not divided.

Climate Science and the "No-Analog Future"

As we sail into the future, we need to forecast what lies ahead. . . . However, novel climates represent uncharted portions of climate space, where we have no observational data to parameterize and validate ecological forecasts. They are the climatic equivalent of uncharted regions of the world, to which early European cartographers supposedly applied the label, "Here there be dragons."

John W. Williams and Stephen T. Jackson, "Novel Climates" (2007)

Analogies are ubiquitous in our intellectual life. They are a key feature of how we address new experience, how we work to understand others, how we interpret the past and set it in relation to the present. But what happens when analogies fail, when no suitable model, no identifiable pattern of resonance, is available?

This is the problem that climate scientists are now trying to address. Recent evidence shows that warming has exceeded previous estimates and that if carbon emissions continue to increase at the current rate, global warming is likely to reach two degrees Celsius by 2050, and more than four degrees by the close of this century.[1] The rest of us are slowly coming to realize that this would be a catastrophic event for both human and much nonhuman life. As I write, representatives from virtually every nation on earth are gathering in Paris for the COP21 Global Climate Talks—the eleventh meeting of the signatories to the (so-far largely ineffectual) Kyoto Protocol. We can only hope that these representatives of the world, and those from the United States in particular, accomplish more than they did almost twenty years ago.

In the ten years I've worked on this book, it has become increasingly clear that global warming is the defining challenge of the twenty-first century and, potentially, of this millennium. And I have worried that these ten years of labor have done nothing to address that problem. Worse, I've realized that, in the process of living my life, and in my travel to the institutions, archives, and conferences that have sustained my research, I've in fact made the problem worse.

The nineteenth century (in some accounts) marks the beginning of the Anthropocene and witnessed the first widespread discussions of anthropogenic climate change.[2] We still have a lot to learn from it. Yet, if my book conceives of the nineteenth century as the "Age of Analogy," climate scientists are increasingly theorizing the twenty-first-century as an age of *no* analogy. This is why, as Williams and Jackson worry in the epigraph, climate scientists might be tempted to look to the future, and the increasingly outrageous predictions of their models, and say, "Here there be dragons." As Williams and Jackson put it, in their foundational study, "How do you study an ecosystem no ecologist has ever seen?"[3] In other words, how do you look into the "no-analog future"?[4]

But within climate circles, and for those who study the history of ecologies, our present dilemma has prompted growing interest in studying the history of "no-analog ecologies" and "no-analog climates." The solution Williams and Jackson suggested, and which ecologists and climatologists continue to pursue, is to look at "no-analog" situations in the past, developing models for how these extremely unusual climates and ecologies develop, and eventually to extend those models into the future. Much as George Eliot viewed disanalogy as a key moment in the formulation of new knowledge, Williams and Jackson argue that, to address climate change, we must draw an analogy between the no-analog future and the no-analog past.

This is comparative historicism writ large and with the highest possible stakes. This book is not about climate change, but it *is* about the role analogy plays in understanding the past and developing models that interpret our present condition and address our future experience. It may be foolish, but I can't help hoping that, in the careful study of how analogies operate and, particularly, in examining how they disclose new features in the world, this study might supplement the wide belief that analogies are simply models with a finer-grained understanding of how analogies disclose new patterns (and perhaps) new solutions. In studying our own place within this larger world, and the modes of being we share with other forms of life, we may find new ways to adapt and cope. As Williams and Jackson put it, "Past no-analog climates differ from those we will encounter in the future, but they can be used to test [our predictions]." Perhaps these climates, through the resonance and interaction organized by their common features and their disanalogies, will someday discover new ways for all of us to live.

Introduction · Analogy under a Different Form

1. C. Darwin, *The Autobiography of Charles Darwin*, 49.

2. Carlyle, "The Memoirs of the Life of Sir Walter Scott, Baronet," 337.

3. The most influential Anglo-American account of nineteenth-century historicism is presented in *The Idea of History*. Collingwood famously sums up source criticism as "scissors-and-paste" history, but a more nuanced and recent evaluation of nineteenth-century empiricism and its impact on modern historical thinking can be found in Tosh, *The Pursuit of History*. Michael McKeon suggests that this distinction can be understood as the productive interplay between "external explanation" and "internal interpretation." McKeon, "Theory and Practice in Historical Method," 52.

4. Phillips, *Society and Sentiment*, 320.

5. My argument here echoes Jonathan Sachs's contention that the Romantic period saw a transition from an "exemplary" reading of history "where historical examples are removed from specific contexts and made to teach lessons for all times" to historicism. Sachs, *Romantic Antiquity*, 39.

6. The standardization of time has an uncertain relation to the comparative turn I describe here. I would note, however, that at least in Zemka's account, standard time does not take firm hold until later, in the mid-nineteenth century. And I'm tempted to speculate that the wide influence of thinking comparatively about other times and other places would heighten the perceived need for standardization. B. Anderson, *Imagined Communities*; Zemka, *Time and the Moment*.

7. For a summary of mid-twentieth-century comparative history, and its reaction to Marc Bloch's work, see Sewell, "Marc Bloch and the Logic of Comparative History." For an analysis of comparative anthropology, see Wylie, "The Reaction against Analogy."

8. Daston, "Historical Epistemology."

9. Culler, *The Victorian Mirror of History*.

10. Grady, *Shakespeare, Machiavelli, and Montaigne*, 2.

11. Thomas Henry Huxley's *Science and Culture* (1881), *On Science and Art in Relation to Education* (1883), and *Evolution and Ethics* (1893); Matthew Arnold's "Literature and Science" (1882).

12. A. Huxley, *Literature and Science*; Hyman, *The Tangled Bank*; Barrett and Gruber, *Darwin on Man*.

13. Snow, *The Two Cultures and the Scientific Revolution*. Gowan Dawson and Bernard Lightman have recently shed fresh light on the problematic label "science and literature"; they prefer "science *as* literature" but nevertheless reluctantly retain the "and" in the title of their series. Dawson and Lightman, *Victorian Science and Literature*, 1:x–xi.

14. Beer, *Darwin's Plots*; Levine, *Darwin and the Novelists*; Levine, *Darwin Loves You*; Levine, *Darwin the Writer*; Shuttleworth, *George Eliot and Nineteenth-Century Science*; Shuttleworth, *The Mind of the Child*. More recent work by David Amigoni, Gowan Dawson, Bernard Lightman, James Secord, and Jonathan Smith has complicated this picture by emphasizing the complex social and economic conditions for interaction between literary writing, print publication, and scientific practice. Amigoni, *Cults, Colonies, and Evolution*; Dawson, *Darwin, Literature and Victorian Respectability*; Lightman, *Victorian Science in*

Context and *Victorian Popularizers of Science*; Secord, *Victorian Sensation*; J. Smith, *Charles Darwin and Victorian Visual Culture*.

15. Said, *Culture and Imperialism*, chap. 3. This argument adapts Gowan Dawson's, which criticizes the "one culture" model for assuming that "such discursive interchanges [between science and literature] are, even when unintended, invariably creative and positive for both science and literature, providing fresh ways for writers and scientists to imagine the world as well as new modes of organizing their insights." While I, too, emphasize the "creative and positive" dimensions of this interchange, I recognize it as part of a collective effort to address the wide perception that both the social and natural worlds were growing more complex, rather than coherent. Interrogation of the complex and uneven analogy between science and literature can give us a better handle on how this worked. Dawson, *Darwin, Literature and Victorian Respectability*, 193.

16. In a similar vein, Anne DeWitt has recently explored the inability of the "one culture" model to address the differentiation of science and its popular reception in the nineteenth century. DeWitt, *Moral Authority, Men of Science, and the Victorian Novel*.

17. Buckland, *Novel Science*; J. Smith, *Charles Darwin and Victorian Visual Culture*; Dawson, *Darwin, Literature and Victorian Respectability*; Klancher, *Transfiguring the Arts and Sciences*.

18. Even when authors are working against each other, they can be said to collaborate, insofar as the tension they define is work they produce together. Only authors or works that have nothing to say to each other can be said not to collaborate at all.

19. I intend the term to place Daston and Galison's work on collective empiricism, and the work of Adrian Johns on the material scientific text, within actor-network theory and in dialogue with models of collaborative authorship. Daston and Galison, *Objectivity*; Johns, *The Nature of the Book*. See, for instance, Nicholas and Ruder, "In Search of the Collective Author."

20. Latour, *Science in Action*; Darnton, "What Is the History of Books?"

21. Shapin and Schaffer, *Leviathan and the Air-Pump*; Benedict, *Curiosity*; Daston and Park, *Wonders and the Order of Nature*; Daston and Galison, *Objectivity*.

22. Latour, *Pandora's Hope*, chap. 2.

23. In part, this is because natural selection works on subtle distinctions that are often not expressed in a visible phenotype: "Man can act only on external and visible characters: nature cares nothing for appearances, except in so far as they may be useful to any being." Darwin, *On the Origin of Species by Means of Natural Selection* (1859), 84. All further references are given parenthetically by page number.

24. George H. Sabine summarized the traditional distinction between descriptive and normative science in "Descriptive and Normative Sciences."

25. For an evaluation of the contrast between "historical" and "experimental" science, see Cleland, "Historical Science, Experimental Science, and the Scientific Method." For a tendentious account of the "descriptive" versus "hypothesis driven" contrast, see Casadevall and Fang, "Descriptive Science."

26. Grimaldi and Engel, "Why Descriptive Science Still Matters."

27. Fodor, "Special Sciences."

28. See Jim Endersby's discussion of the implications of Darwin's use of "lumpers" and "splitters" in his correspondence. Endersby, "Lumpers and Splitters." *The Age of Analogy* understands lumping and splitting as one possible vocabulary for the role of

what I term formal and harmonic analogy in evolutionary science, and one of the many strategies Darwin pursued in his integration of comparative history into evolutionary narratives.

29. Foucault, *The Order of Things*, xii–xiii. I note that, even though Foucault describes his method as "comparative" and studies the system of "analogies" that are articulated by given systems of order, he does not connect this to his extensive study of Renaissance semiotics and the analogies of things. This is in part, I believe, because he did not mark the way such earlier articulations of analogy are reformulated in the modern comparative method, nor did he explicitly note their intrinsic connection in the modern configurations of comparative historicism. As an "archaeology of structuralism," *The Order of Things* suffers from a failure to recognize structuralism itself as both a response and complement to comparative methodologies. See my discussion of Ferdinand de Saussure and comparatism in the prelude.

30. Chakrabarty, *Provincializing Europe*, 7–23, 249. We might recognize Chakrabarty's more recent call to recognize global warming as a "planetary conjuncture" as an effort, in line with Edward Said's "contrapuntalism," to activate comparative historicism as a politics—a way to imagine communities of interest across difference with the present, as well as solidarities across time. Chakrabarty, "The Climate of History"; Said, *Culture and Imperialism*.

31. Tosh, *The Pursuit of History*, 6–7. Scott's emphasis on the uneven and persistent multiplicity of history draws a strong contrast to Georg Wilhelm Friedrich Hegel's historical syntheses; and this distinct conception of history as an open-ended process, I argue, is rooted in Scott's use of harmonic analogy in place of Hegel's dialectic. See my discussion in the prelude, note 15.

32. It's fair to say that such comparisons implied similarity as well as substantive contrast, much as analogy presumed difference. John Donne's eighth elegy, generally known as "The Comparison" (1633), turns on precisely the excavation of this implicit possibility. But this potential was not fully realized and generalized until the comparative and historical turn I identify here. So the "Table of Degrees or of Comparisons" that Francis Bacon deploys as a stage of induction in his *New Organon* (1620) is designed to elucidate distinctions in different instances of a given phenomenon.

33. Manning, *Poetics of Character*, xi–xiii, 13–5.

34. Dickens, *A Tale of Two Cities*, 5. See the discussion in Griffiths, "The Comparative History of 'A Tale of Two Cities.'"

35. I use *metahistory*, a term borrowed from Hayden White, to refer to historiography as the "philosophy of history" that emerged in the nineteenth century; White, *Metahistory*. Lorraine Daston and Peter Galison characterize a longer movement in the practice of science as "collective empiricism," while studies like Jim Endersby's *Imperial Nature* demonstrate the comparative turn these efforts took in the nineteenth century. Daston and Galison, *Objectivity*; Endersby, *Imperial Nature*. On the demise of rhetoric, see Bender and Wellbery, *The Ends of Rhetoric*; Guillory, "Enlightening Mediation."

36. Braudy, *The Plot of Time*; White, *Metahistory*.

37. Woloch, *The One vs. the Many*, 18.

38. Schor, *Curious Subjects*.

39. James, "Daniel Deronda: A Conversation," 684.

40. Buzard, *Disorienting Fiction*. The capacity of the novel to internalize other genres is a critical commonplace, but it is given a formative discussion by Mikhail Bakhtin in *The Dialogic Imagination*.

41. Hutcheon, *A Poetics of Postmodernism*, 109.

42. C. Taylor, *A Secular Age*.

43. For an extended discussion of how the idea of culture developed in relation to contemporary natural science and evolutionary thinking, see Amigoni, *Cults, Colonies, and Evolution*.

44. In later works he identified culture's "very slowly acquired habit of mind" with "a very wide experience from comparative observation in many directions." Arnold, *Literature & Dogma*, 282. I also have in mind the vast comparative anthropologies organized in works like E. B. Tylor's *Primitive Culture*, a book that suggests a dynamic relation between primitive and modern cultures and, implicitly, within them.

45. Said, *Orientalism*.

46. Stocking, *Victorian Anthropology*.

47. Du Bois, "Races," 157.

48. Eliot, *Essays of George Eliot*, 435; Eliot, *Poems, Essays, and Leaves from a Note-Book*, 192–4.

49. Eliot, *Middlemarch*, 293.

50. Eliot, *Poems, Essays, and Leaves from a Note-Book*, 192.

51. Said, *Culture and Imperialism*.

52. Shih derives the concept of relational comparison from Édouard Glissant's *Poetics of Relation*. Shih, "Comparison as Relation," 436; Cole, *The Birth of Theory*, 34–41.

53. Chakrabarty, "Radical Histories and the Question of Enlightenment Rationalism," 275.

54. Melas, *All the Difference in the World*.

55. "Contemporary Literature." In this way I read comparatism, along with Haun Saussy, as "defined by the search for its proper objects," because the question of what those objects will look like and what histories they have to tell is central to the search for a sensitive, open engagement with the past and its politics. Saussy, "Exquisite Cadavers Stitched from Fresh Nightmares," 12.

56. Where Liu contrasts New Historicism to the histories of Michel Foucault, seeing analogy as a catalyst for the mutation of contiguity into the "quasi-magical 'sympathy'" of text and context, I see a sibling rivalry. Liu, "The Power of Formalism." See also the discussion of Robert Darnton's new historicism in the introduction to Chandler, Davidson, and Harootunian, *Questions of Evidence*.

57. See Greenblatt, *Renaissance Self-Fashioning*; Gallagher, *Nobody's Story*. See also their reflections on new historicism in Gallagher and Greenblatt, *Practicing New Historicism*.

58. Burguière, *The Annales School*. See also Furet, *In the Workshop of History*, 41–2.

59. See Flynn, *Sartre, Foucault, and Historical Reason*, 2:5–14. Foucault, *The Archaeology of Knowledge; and, the Discourse on Language*, 160. Foucault saw analogy as an important mode of thinking in the early modern period but emphasized its basic contrast to comparatism, and so he was unable to acknowledge its role in his own methodology.

60. Owen's personal seal described such natural patterns as "The One in the Manifold." Owen, *The Life of Richard Owen*, 1:388.

61. Gallagher, "George Eliot," 63.

62. Theresa M. Kelley explores Charles's engagement with his grandfather's science at the close of her extended study of Romantic botany. Kelley, *Clandestine Marriage*, 261–5.

Prelude · Thinking through Analogy

1. Machiavelli, "Letter to Vettori," 3; Michelet, *On History*, 140, 143; Collingwood, *The Idea of History*, 293.

2. Simmel, *The Problems of the Philosophy of History*, 67.

3. Ricœur, *Time and Narrative*, 1:x.

4. Simmel, *The Problems of the Philosophy of History*, 24.

5. Felski and Friedman, introduction to *Comparison*, 1–2.

6. Classical rhetoricians, particularly Cicero and Quintilian, understood comparison as a rhetorical figure—that is, a way of articulating an effective argument. This tied comparison to composition rather than substantive thought: it was on the latter side of the classical distinction between *logos* and *lexis*, *res* and *verba*, articulated in rhetoric as a distinction between "invention" (the formulation of the substance of an argument) and "disposition" (the translation of this argument into language). Rather than helping decide what to argue, comparison helped reinforce the point. And despite its literal meaning—to "pair" or "make equal" (L. *comparatio*—the substantive of *comparare*), comparison primarily drew contrasts to other examples. Cicero discusses it as a feature of judicial rhetoric, as judges weighed the crimes of the accused against others, and defendants examined the intent of their actions by contrasting the outcome with the alternative. Matsen, Rollinson, and Sousa, *Readings from Classical Rhetoric*, 183. Quintilian elaborates on comparison's contrastive function extensively and emphasizes its value for the rhetoric of praise and blame, and its power to amplify the superiority or inferiority of the case through distinction. Quintilian, *Institutes of Oratory*, 1:105, 353.

Rhetoricians of the early modern period discuss comparison in a variety of contexts but emphasize its power to amplify difference. Though they often acknowledge comparison's connection to thought, its conception as a rhetorical figure prevented extensive discussion beyond specific strategies for the disposition of arguments. Dudley Fenner's *Artes of logike and rethorike* (1584) explores comparison as a feature of both. Considered briefly as a logical operation, it helps evaluate the quality of "compared arguments" of reason. But when analyzed as a rhetorical figure, he dwells on cases of contrast, adducing dozens of classical and biblical examples. Similarly, in the first edition of his *Art of English Poesie* (1577), Henry Peacham follows Quintilian in describing *comparatio* as a scheme of amplification, emphasizing contrast and the comparison of unlike things. In reviewing comparison for his second edition (1596), he still follows Quintilian's lead, while acknowledging the need for a wider understanding of comparison as a way of thinking about the operation of tropes generally. In this, Peacham is representative of theorists who understood comparison as a mode of juxtaposition associated with and often governing a wide range of tropes and figures of similarity—including similitude, simile, metaphor, parabole, and allegory, to name a few. His new introductory note observes that "Comparatio is a word of large and ample comprehension, and therefore it may stand as a generall head and principall of many figures, but namely those which do tend most especially to amplifie or diminish by forme of comparisen." Peacham, *The Garden of Eloquence*, 1577, q3v; Peacham, *The Garden of Eloquence*, 1593, 156.

Rhetoricians used comparison to bolster a given argument, rather than find new features. Contrast is sharp, similarity more subtle. Early modern writers appreciated the power of comparison to advance trenchant contrasts and gave it an increasingly important role in "epideictic" rhetoric, the rhetoric of praise and blame. By the eighteenth century, such distinctions elevate comparison into a generic denomination for tracts that contrast entities divided by doctrine, politics, nation, or economic interest. Examples abound. To list a few: *The Devil turn'd round-head; or, Pluto become a Brownist. Being a just comparison, how the Devil is become a round-head?* (1642); *A comparison between the true and false ministers in their calling* (1675); *A comparison of the penal laws of France against protestants, with those of England against papists* (1717); *An essay on the agreement betwixt ancient and modern physicians; or, a comparison between the practice of Hippocrates, Galen, Sydenham, and Boerhaave, in acute diseases* (1747); *The contrast; or, a comparison between our woollen, linen, cotton, and silk manufactures* (1782). As a genre, "comparison" is as an engine for nation formation, sectarian and tractarian debate, trade policy, and perhaps the quintessential form for the eighteenth-century debate between the ancients and the moderns. Its use as a strategy of contrast in this period was so pervasive that, in its trim adverbial mode, it became shorthand for relief against a tacit background of cases, as in this early example: "The vegetables being comparatively lighter than the ordinary terrestrial matter of the globe." Woodward, *An Essay towards a Natural History of the Earth*, 262.

7. Swift, *Gulliver's Travels*, 1:103.

8. See Griffiths, "The Intuitions of Analogy in Erasmus Darwin's Poetics." Analogy helped wrestle with the complexly allusive language of scripture and draw positive insights into the obscure order of the divine economy. The early church fathers engineered its intricate hermeneutic workings, and it is worth tinkering with them to evaluate what this complexity can tell us about its practical use. For early interpreters of the Christian accounts, a defining challenge was to select the coherent set of interpretations from among the diverse ways that these accounts could be understood. Paul provided an influential precedent in his epistle to the Romans (12:6), suggesting that prophecy should be evaluated by "analogy" (Gr. αναλογια) to the larger faith—in other words, our confidence in the insight should be weighed in terms of its correspondence to the received system of Christian belief. Saint Augustine adopted this "analogy of faith" as a way to address the problem of reconciling the evident distinctions between what was now understood to be the "Old" and "New" Testaments.

Whereas Paul worried over which testaments to accept, Augustine emphasized the difficulty of evaluating whether specific passages in the received testaments were to be understood literally or figuratively. In *On Christian Doctrine*, he argues that, insofar as scripture is catholic and all-inclusive, it gives a unified vision of the past, present, and future, and every interpretation should ultimately reinforce the guiding message of charity. In this fashion, Christian doctrine internalized the Pauline analogy of faith as a central principle of the Bible. By selecting the appropriate literal or figurative interpretation for each account, the analogy of faith unified the Bible as a coherent doctrinal argument. In *The Utility of Belief*, Augustine elaborates the analogical method with respect to both the Old and New Testaments, arguing that there are four distinct senses that a passage can have: historical (the literal sense of what is narrated), etiological (focusing on the cause of the events described), allegorical (when the account is to be taken figuratively), and analogical (the sense of the passage that connects it to the accounts of the other testament).

The Scholastics, especially Thomas Aquinas, continued to revise Augustine's fourfold system of interpretation and expanded the function of analogy (Wippel, "Metaphysics"). The effect, as Victor Harris has argued, was to elevate the place of analogy in mapping connections both within the Bible and between the Bible and the world. As Harris puts it, by the eighteenth century analogy became the immanent method of natural theology and displaced allegory and other interpretive methods as the definitive test of Christian doctrine: "The final stage in this process coincided with the resurgence of 'divine analogies' between spirit and nature. Allegories evolved from figure to argument, and survived only by translation into the more precise idiom of analogy." V. Harris, "Allegory to Analogy in the Interpretation of Scripture," 20.

9. Lovejoy classically described the *scala naturae* as a "great chain of being" articulating the relation between the hierarchies of natural and supernatural order, while Foucault explored how early modern medicine and magic sought the basic "sympathies" that constituted the powerful interrelation of all things. Lovejoy, *The Great Chain of Being*; Foucault, *The Order of Things*.

10. Brontë, *Shirley*, 228.

11. Eliot, *Middlemarch*, 20.

12. I. A. Richards, *The Philosophy of Rhetoric*. In cognitive and computer science, analogy is treated almost exclusively as a mapping relation. See Dedre Gentner's influential "structure mapping" theory, Keith J. Holyoak and Paul Thagard's neural network-based mappings, and Douglas Hofstadter, Robert French, and Melanie Mitchell's "stochastic" analogy making programs, "Tabletop" and "Copycat." Gentner, "Structure-Mapping"; Holyoak and Thagard, "Analogical Mapping by Constraint Satisfaction"; French and Hofstadter, "Tabletop"; Hofstadter and Mitchell, "The Copycat Project."

13. Bohr, "On the Constitution of Atoms and Molecules. Pt. 1," 12, 7. All further references are given parenthetically by page number.

14. Bohr does make a brief reference to Hantaro Nagaoka's "Saturnian theory" but does not explore it; it's only as the Rutherford-Bohr model is disseminated through broader organs like *Science* and *Popular Science* that the "solar system" model is provided as a standard account of the Rutherford-Bohr theory.

15. Though Andrew Cole has recently emphasized the Neoplatonic origins of the Hegelian dialectic, tracing it to a more basic interplay between "identity/difference, likeness/sameness," I want to underline the distinction here between harmonic analogy, in its connection to comparative historicism, and the Hegelian dialectic, which moves toward historical synthesis (Cole, *The Birth of Theory*, 35). As I argue in chapter 2, a writer like Walter Scott uses analogy to open up history to alternative perspectives and the multiplicity of plots—emphasizing the essential and continuing plurality of history. I see a strong contrast here to Hegel's use of the dialectic to subordinate the multiplicity of history to unitary progress, or, as Cole puts it, "procession and return." In fact, Cole's larger ambition might be described as the attempt to expand dialectic as an alternative vocabulary for thinking about analogy as the study of similarity across difference, and so advancing dialectic's claim as a fundamental structure of critical thought (a point argued recently by C. D. Blanton in "Theory by Analogy").

16. It should be noted that, though Hofstadter relies on a formal or "mapping" model of analogy, he emphasizes analogy's capacity to generate new models and new "emergent" insights. See *Fluid Concepts and Creative Analogies*.

17. For a discussion of the Thomist theory of analogy in *De Veritae*, see Ricœur, *The Rule of Metaphor*, 325–7. Another possibility would be the distinction Colin Jager draws, with respect to William Wordsworth's *Prelude*, between "extrinsic analogies," imposed on nature or the self from without, and "intrinsic analogies," generated by the subject of experience. Jager, *The Book of God*, 180–1.

18. In my understanding, harmonic analogy is a place where Giorgio Agamben's "whatever being" meets the "comparatisme quand même" that Emily Apter finds in Alain Badiou's discussions of singularity. See Agamben, *The Coming Community*; Apter, " 'Je ne crois pas Beaucoup à la littérature comparée': Universal Poetics and Postcolonial Comparatism."

19. Spencer, *The Principles of Psychology*, 141; Mill, *A System of Logic, Ratiocinative and Inductive*, 2:96.

20. "ἀναλογ-έω " Liddell and Scott, *A Greek-English Lexicon*.

21. Aristotle, *The Complete Works of Aristotle*, 1:122. All future references are given parenthetically by volume and page.

22. He frames the analogy between domains adjectivally here, as ἀναλογίαν—the commonality is "analogical." Aristotle, *Posterior Analytics*, 68.

23. Aristotle, *Complete Works*, 1:162. Aristotle focused on the parallel relations of harmonic analogy, but one can as easily imagine formal analogies that violate the hierarchy of knowledge by applying patterns from one genus to species of another, unrelated genus.

24. Aristotle manages to deal with analogy's problematic status by insisting that any insight gleaned by an analogy across divorced genera must then be verified by the categorical knowledge internal to each domain, restoring their integrity.

25. Though it is also a form of "speculative realism," Graham Harman's object-oriented ontology is rooted in Martin Heidegger's phenomenology rather than mathematics and so does not concern me here. Harman, *Tool-Being*. At the same time, I note his coauthorship, with Levi Bryant and Nick Srnicek, of the introduction to *The Speculative Turn*, which emphasizes their choice of title "as a counterpoint to the now tiresome 'Linguistic Turn'" which they summarize as "anti-realism." Bryant, Srnicek, and Harman, *The Speculative Turn*, 1, 4.

26. Nealon, *Post-Postmodernism*, 137–47.

27. Badiou, *Manifesto for Philosophy*, 35.

28. DeLanda, *Intensive Science and Virtual Philosophy*, 16.

29. C. Darwin, *On the Origin of Species by Means of Natural Selection* (1859), 478. All further references to this edition are given parenthetically by page number.

30. C. Darwin, *On the Origin of Species by Means of Natural Selection* (1861), 85.

31. Quentin Meillassoux backdates this insight in two important ways. First, he argues that the dilemma of poststructural philosophy, the sense in which it divides us from access to the world in itself, did not emerge through structural linguistics or (in Badiou's account) the way Martin Heidegger "deconstruct[e]d the theme of the subject" but, rather, in Immanuel Kant's original and hugely influential analysis of experience as a relation between objects (in the world) and the perceiving subject. Badiou, *Manifesto for*

Philosophy, 73. In Meillassoux's analysis, this forced philosophy to think of experience as a kind of adequation between subject and object. He terms this "correlationism," "according to which we only ever have access to the correlation between thinking and being, and never to either term considered apart from the other." Meillassoux, *After Finitude*, 5. According to this line of thinking, Meillassoux argues, we could have no knowledge of a time before the human subject existed, because in such cases a correlation between subject and object is impossible.

Meillassoux's other major backdating of Badiou's thought can be found in his insistence that we *do*, in fact, have access to an "ancestral realm" before humans, and we reach it by means of science and mathematical description. We know about the Big Bang because of the work of scientists to articulate it through mathematics, giving science purchase on an ontology without a human subject. Note that, by emphasizing the location of this insight with respect to human history (and with respect to the historically located observers who made models of the Big Bang possible), we occupy a perspective that seems far from ahistorical. In chapter 5, I will explain at some length how Darwin uses imaginative language, and especially analogy, to imagine natural history before humans and subsequently to read humans into nature, undermining their status as the exceptional subjects of experience.

32. Harman, "The Well-Wrought Broken Hammer," 187.

33. Strictly speaking, the only "formal" languages are formal logics, including logical descriptions of mathematics, but as I will soon argue, the history of mathematics can be conceived in terms of a continuing process of formalization.

34. Meillassoux, *The Number and the Siren*, 49.

35. Dutilh Novaes, "The Different Ways in Which Logic Is (Said to Be) Formal," 318.

36. Dutilh Novaes, *Formal Languages in Logic*, 97–101.

37. Ibid., 83.

38. Ovsienko and Tabachnikov, *Projective Differential Geometry Old and New*.

39. Poncelet, *Traité des propriétés projectives des figures*, xi. My translation.

40. Chapter 3 explores the language of homology in connection with comparative anatomy.

41. Gernsheim and Gernsheim, *The History of Photography*; Darrigol, *A History of Optics*; Snyder, *Eye of the Beholder*.

42. Silverman, *The Miracle of Analogy*, 14–7. For a long view of analogy as a visual phenomenon, see Stafford, *Visual Analogy*.

43. Silverman, *The Miracle of Analogy*, 11.

44. Saussure, *Course in General Linguistics*, 2–5. All further references are given parenthetically by page number.

45. See Saussure's extended discussion of "Analogy and Evolution" in his fifth chapter.

46. Harry Shaw has made a similar argument with respect to natural kinds, in which a specific word is assumed to mark a single "real" feature of the world: "Natural kinds . . . erode the notion that we can make a hard and fast distinction between language (or the signifieds) on one hand, and a 'world beyond language' (or the referent) on the other. Or to put it another way: they remind us that such a distinction convincingly arises only when we subscribe to the fiction of synchronicity, only when we bracket the inherently historical nature of our existence." Shaw, *Narrating Reality*, 71.

47. I'm thinking here of Derek Attridge's argument for the place of literary encounters with alterity. Attridge, *The Singularity of Literature*.

48. Latour, *Pandora's Hope*, 39, 55.

49. One way of thinking about this dynamic is to use analogy to ask how the relation between reading the literary object for "surface" and for "depth" might be generalized for the dynamic interaction of self and world. See Best and Marcus, "Surface Reading."

50. Brassier, "Concepts and Objects," 53.

51. For a discussion of Lewes's dual or double-aspect philosophy, see Ryan, *Thinking without Thinking in the Victorian Novel*, 91–3; Davis, *George Eliot and Nineteenth-Century Psychology*, 17–8.

52. Attridge, *The Singularity of Literature*.

Chapter 1 · Erasmus Darwin, Enlightenment History, and the Crisis of Analogy

Portions of this chapter are reprinted with permission from *SEL: Studies in English Literature, 1500–1900* 51, no. 3 (2011): 645–65.

1. Coleridge, *Letters*, 1:152.

2. For an outline of comparative historicism, see the introduction and prelude to this book. *Historicism* is generally used to denote a way of thinking *about* history rather than simply history itself and, in this way, is closely linked to *historiography*. For a discussion of its relation to nineteenth-century German historicism (*Historismus*), see chapter 2.

3. By *élan vital* I mark the affinity between Erasmus Darwin's vitalism and the "vital energy" postulated by Henri Bergson, and other critical vitalists of the early twentieth century, including Hans Driesch. See Bennett, *Vibrant Matter*, chap. 5.

4. D'Israeli, *Curiosities of Literature* (1823), 3:27. The epigraph is drawn from D'Israeli, *Curiosities of Literature* (1794), 2:58.

5. I don't dismiss the importance of these larger contexts: these include coordinated government assault on the radical press; the suspension of habeas corpus in 1794; and the prosecution of Darwin's publisher in 1798, as well as the successful institution of a reactionary press, spearheaded by George Canning's *Anti-Jacobin*. For more on the radical press and its reaction, see Gilmartin, *Writing against Revolution*.

6. Coleridge, *Biographia Literaria*, 255; Shelley, *Essays*, 57.

7. Fulford, "Darwin in the Air."

8. Manning, *Poetics of Character*, 13–6.

9. Abrams, *The Mirror and the Lamp*; Wasserman, *The Subtler Language*; de Man, "The Rhetoric of Temporality."

10. See Priestman, *Romantic Atheism*, 67–74.

11. See Hacking, "Styles of Scientific Reasoning."

12. Budge, "Introduction: Science and Soul in the Midlands Enlightenment." For a discussion of Darwin's influence on Coleridge and Wordsworth, see two additional articles from that special issue, Vickers, "Coleridge and the Idea of 'Psychological' Criticism"; and Budge, "Erasmus Darwin and the Poetics of William Wordsworth."

13. Secord, *Victorian Sensation*, 461; Amigoni and Elwick, *The Evolutionary Epic*, ix–xxi.

14. Tucker, *Epic*, 19–20; Amigoni and Elwick, *The Evolutionary Epic*, xiv.

15. Stuart Harris has argued that *The Economy of Vegetation*, *The Loves of the Plants*, and *The Temple of Nature* together constitute one coherent epic, and he has reprinted them as a single work, the "Cosmologia"—see Harris, *Erasmus Darwin's Enlightenment*

Epic; and the introduction to E. Darwin and Harris, *Cosmologia*. This overemphasizes coherence, papering over the publication history of the works and the uneven evolution of their genres (in the case of the *Cosmologia*, at the expense of printing Darwin's poems without the elaborate notes, footnotes, and interludes). The substantive differences between these three poems indicate Darwin's ongoing search for a genre adequate to his theory of evolution.

16. Tucker, *Epic*, 22. In his time, Erasmus Darwin was celebrated for reinventing didactic verse and epic in his *Botanic Garden* and (to a lesser degree) in his *Temple of Nature*. It should be noted that until 1789 Darwin had been almost exclusively a poet of elegies. This is not generally known because, with the exception of a short elegy written in college, all of Darwin's poetry had been published anonymously or through collaborations with others (this includes extensive collaborations—only some acknowledged—with Anna Seward). By turning away from elegy (always a kind of comparative historicism, even when it is organized by an attempt to escape its unsettling insights), Darwin initiated a struggle with the synchronous analysis of social forms and the narrative exploration of their development that would dominate the rest of his work.

17. Benjamin, "Theses on the Philosophy of History," 161; Zemka, *Time and the Moment*, 2.

18. Anon., "The Botanic Garden," February 1795, 77.

19. E. Darwin, "Letter to Joseph Johnson," May 23, 1784.

20. Cited in Bewel, "Erasmus Darwin's Cosmopolitan Nature," 19.

21. C. Darwin, "Letter to Joseph Hooker," August 1, 1857.

22. Coleridge, "Letter to Robert Southey," July 29, 1802.

23. The distinct but filiated networks of production within which Darwin operated—helping found both the Lunar Society of Birmingham and the Lichfield Botanical Society, and as a formal member of the London Royal Society and participant in the less formal Lichfield poetry circle—gave him an acute sense of disciplinarity in advance of the institutional consolidation of the sciences and humanities as distinct disciplines, and a strong grasp of how such distinctions were enforced in practice by social custom and the constraints of the print market. See Bewel, "Erasmus Darwin's Cosmopolitan Nature."

24. The *Economy* was actually published in 1792, though the title page reads 1791. See the discussion in Priestman, introduction to *The Collected Writings of Erasmus Darwin*, 1:iv.

25. Darwin moved the advertisement to the front of *The Economy of Vegetation* for the 1791 edition, as part of the two-volume collection called *The Botanic Garden*.

26. E. Darwin, *Collected Works*, 2, vi. All further references to this edition are given as *CW* with volume and page number for prose, or volume, canto, and line number for verse. Noel Jackson notes that the *camera obscura* also observes Darwin's debt to Lucretius and his *De Rerum Natura*, the formative model for didactic verse. Jackson, "Rhyme and Reason."

27. Silverman, *The Miracle of Analogy*, 14.

28. *CW* 1, v. Note that this "Advertisement" was moved from fronting the 1789 *Loves of the Plants* to the beginning of *The Economy of Vegetation* for the 1791 edition, the source for both volumes in the *Collected Writings*.

29. At the close of the work, a "Catalogue of the Poetic Exhibition" (really, an index) invoked contemporary artistic markets like the Boydell Shakespeare Gallery in order to remind the reader that the poem was a commercial and aesthetic (rather than scientific) work. For more on the Boydell galley, see Hawkings, "Reconstructing the

Boydell Shakespeare Gallery"; Burwick, "John Boydell's Shakespeare Gallery and the Stage"; Brylowe, "Romantic Arts & Letters."

30. Kelley, *Clandestine Marriage*; Allewaert, *Ariel's Ecology*; LaFleur, "Precipitous Sensations."

31. Anon., " 'The Botanic Garden,' " July 1789, 337.

32. Ibid., 5.

33. Anon., " 'The Botanic Garden,' " November 1789, 376, 375.

34. Linnaeus and Botanical Society at Lichfield, *The Families of Plants*, lxv.

35. For my more extended analysis of classification ontologies and information science at the British Museum, see Griffiths, "The Radical's Catalogue."

36. Milne emphasized how important dictionaries were to the distinction between "artificial" and "natural" strategies and made the case in favor of arbitrary systems of classification: "Artificial methods are to the natural, what the alphabetical dictionary is to the etymological." Milne and Linnaeus, *The Institutes of Botany*, 31.

37. Linnaeus, *The Families of Plants*, i. All further references to this edition are given parenthetically by page number.

38. D'Israeli, *Curiosities of Literature* (1794), 2:59.

39. Reprinted in C. Darwin and King-Hele, *The Life of Erasmus Darwin*, 32.

40. For my longer discussion of Darwin's theory of prosody, see Griffiths, "The Intuitions of Analogy in Erasmus Darwin's Poetics."

41. Seward, *Memoirs of the Life of Dr. Darwin*, 180–1.

42. Newton, *Sir Isaac Newton's "Daniel and the Apocalypse,"* 149.

43. Newport, *Apocalypse and Millennium*, 3–9.

44. Priestley characterizes "rational Christianity" in Priestley, *History of the Corruptions of Christianity*, 2:463.

45. Ibid., 1:xiv.

46. Ibid., 1:xi.

47. Butterfield, *The Whig Interpretation of History*, 11–2.

48. Kidd, "The Ideological Significance of the 'History of Scotland.' "

49. O'Brien, "Between Enlightenment and Stadial History," 53.

50. A. Smith, *An Inquiry into the Nature and Causes of the Wealth of Nations*, 1:30–5, 168–70, 415.

51. O'Brien, *Narratives of Enlightenment*, 133.

52. Ferguson, *An Essay on the History of Civil Society*, 13–4.

53. Hearne, *Ductor Historicus*, 1:540.

54. Priestley, *The History and Present State of Electricity with Original Experiments*, 2:13.

55. Mellor, *Blake's Human Form Divine*, 249. Blake's understanding of analogy, given largest exposition in *Jerusalem* (pls. 49 and 85), derives directly from a line of medieval ontology first articulated by Thomas Aquinas. Thomist metaphysics adapted the structural identification of analogy (A is to B as C is to D) to posit a similar correspondence between, on the one hand, the relationship between God and divine being and, on the other, humanity and terrestrial being. Humanity's relation to God is indirect, mediated by their similar relationship to their distinct modes of existence. Like the *scala naturae*, it posits a correspondence between those modes. At the same time, the Thomist analogy of being displaced that correspondence from the elements of the terrestrial and divine

worlds themselves onto corresponding patterns of disposition and interrelation: a great network, rather than great chain. Divine analogy serves the larger Thomist rapprochement between theology and physics by explaining how information derived from the structure of the physical world reflects the metaphysical economy.

56. Explored in Priestman, *Romantic Atheism*, 49–54. Fuseli's original sketch is displayed online at the British Museum: http://www.britishmuseum.org/research/collection _online/collection_object_details/collection_image_gallery.aspx?assetId=214792001 &objectId=749709&partId=1.

57. De Man, "The Rhetoric of Temporality," 193–5.

58. Wasserman, *The Subtler Language*. For another classic study of Romantic analogy, see Abrams, *The Mirror and the Lamp*.

59. Manning, *Poetics of Character*, 71–2.

60. Bacon, *The New Organon*, 146.

61. Snyder, *Reforming Philosophy*, 71–3.

62. McGann, *The Romantic Ideology*.

63. De Man, "The Rhetoric of Temporality"; Jager, *The Book of God*, 50–2.

64. Darwin's decision to hold publication of *The Economy* until he could get updated engravings of the famous Portland Vase (finally executed by Blake) capitalized on a groundswell of public interest in the famous cameo urn and in Josiah Wedgewood's celebrated replicas. In carefully detailing the "Eleusinian mysteries" he felt the vase expressed, Darwin explained the mysteries as a comprehensive narrative mythology—a total narrative of human society.

65. Various other editorial decisions—particularly new tributes to the author by contemporary poets, including Anna Seward's unacknowledged tribute to his "genius" at the opening of *The Economy of Vegetation*—support his claim by authorizing the work and elevating the poet as an important presiding figure in his own right. Though the poem was still printed anonymously, such hints meant that for a large subset of literary Britain, Darwin's official acknowledgment three years later was redundant.

66. Quoted in C. Darwin and King-Hele, *The Life of Erasmus Darwin*, 32.

67. This vision of an earth terraformed by the fertility of life (modeled, as I will shortly argue, on Darwin's analysis of vegetable seeds) aligns closely with Anne McClintock's classic analysis of patriarchal power and sexuality in empire and the "colonial contest." McClintock, *Imperial Leather*.

68. Reill, *Vitalizing Nature in the Enlightenment*; Packham, *Eighteenth-Century Vitalism*.

69. See Packham's discussion in *Eighteenth-Century Vitalism*, 6–7, 104–5.

70. This material vitalism should be distinguished from animism, the attribution of spirit or soul to nonhuman bodies and objects. Darwin's vitalism clearly sees this activity as an inherent property of physical systems rather than a supernatural force. This anticipates the way Gilles Deleuze and Félix Guattari would explore material vitalism as an imminent feature of matter and energy in "Treatise on Nomadology." See Bennett, *Vibrant Matter*, x; and also DeLanda, *Intensive Science and Virtual Philosophy*, 2–6. Alan Richardson and Packham see such moments as evidence for the unstable relation between vitalism and materialism in the period, yet it seems equally clear that Darwin intends a common dynamic that underwrites both. Richardson, *British Romanticism and the Science of the Mind*, 49–52; Packham, *Eighteenth-Century Vitalism*, 151–2.

71. Herschel, "Catalogue of a Second Thousand of New Nebulae and Clusters of Stars," 226. See also Martin Priestman's discussion in *Romantic Atheism*, 65–6, and Holmes, *The Age of Wonder*, 122–3.

72. Hutton, "Theory of the Earth." For more on Hutton's vitalism, see Furniss, "A Romantic Geology."

73. Mitchell, *Experimental Life*, 7–8.

74. Richards, *Romantic Conception of Life*, 300–1.

75. Stefani Engelstein has described how late eighteenth-century debates about the nature of complementarity turned on disputes over the essence of the organic unit. Kant, she argues, solved this problem by defining the organism as an "organized and self-organizing being," subordinating nature to the rational faculty of the discrete mind. Engelstein, "The Allure of Wholeness."

76. Leeuwenhoek, "Observationes D. Anthonii Lewenhoeck, de Natis è Semine Denitali Animaculis."

77. Anon., "Darwin's Zoonomia," 115, 117.

78. Mathias, *The Pursuits of Literature*, ll. 90–1.

79. Canning and Frere, "Loves of the Triangles," 132.

80. Wilberforce, "Darwin's 'On the Origin of Species,'" 255.

81. Duff, *Romanticism and the Uses of Genre*.

82. Tucker, *Epic*, 18.

83. Wordsworth, *The Prelude*, 623; Shelley, *Essays*, 12.

84. Erasmus Darwin comes closest to this new procedure in the extensive catalog of diseases provided in the second volume of *Zoonomia*, which sought to "reduce the facts belonging to Animal Life into classes, orders, genera, and species," so that "by comparing them with each other" he and his contemporaries could "unravel the theory of diseases" (*CW* 6, vii). This more acute comparatism abandoned conserved analogies in favor of specific pathologies of similarity and difference, and its utility impressed contemporaries.

Chapter 2 · *Crossing the Border with Walter Scott*

1. Studies of Scott's impact on the historical novel were given a decisive turn in György Lukács's *The Historical Novel*, the first extended modern analysis of his contribution to the genre. Ina Ferris's *The Achievement of Literary Authority* broke new ground in its analysis of Scott's impact on the literary marketplace and the contemporary imagination of the author, a study given considerable statistical extension in the analysis of publishing records provided by William St. Clair. Ian Duncan's *Modern Romance and the Transformations of the Novel* changed the terms of the genre discussion, reorienting studies of Scott toward the status of the Gothic, romance, and realism in his fiction. An increasing historiographic impulse came as James Chandler's *England in 1819* and Duncan's *Scott's Shadow* placed Scott's historicism in relation to new Romantic theories of history and their connection to the philosophy of the Scottish Enlightenment. Katie Trumpener's *Bardic Nationalism* is the most comprehensive reading of the generic antecedents to the historical novel itself, in both previous historical fiction and the national tale. Surveys by Murray Pittock and others, including a recent study by Ann Rigney, have explored the extraordinary international and global influence of Scott's works, while recent work by Adelene Buckland has explored Scott's influence on early nineteenth-century natural science, particularly geology. Lukács, *The Historical Novel*; Ferris, *The Achievement of Literary Author-*

ity; St. Clair, *The Reading Nation in the Romantic Period*; Duncan, *Modern Romance and Transformations of the Novel*; Chandler, *England in 1819*; Duncan, *Scott's Shadow*; Trumpener, *Bardic Nationalism*; Pittock, *The Reception of Sir Walter Scott in Europe*; Rigney, *The Afterlives of Walter Scott*; Buckland, *Novel Science*.

2. Carlyle, "The Memoirs of the Life of Sir Walter Scott, Baronet," 337.

3. This kind of social history, termed *histoire vue d'en bas* by Lucien Febvre and translated as "history from below" by E. P. Thompson, is often viewed as a twentieth-century focus, but it is clearly also an emphasis of Scott's novels (as we will see, a point emphasized by Thomas Babington Macaulay) and works that show his influence, for example, Jules Michelet's *Le peuple* (1846). Febvre, "Albert Mathiez: Un tempérament, une éducation"; Thompson, "History from Below."

4. Mill, "The Spirit of the Age," 51.

5. As Susan Manning puts this, later writers drew from Scott this strategy for "synchronically juxtaposed tales of doubled and fractured identities . . . invoking correspondential rather than determinative relationships between cause and effect." Manning, *Poetics of Character*, 30.

6. Scott, *Guy Mannering*, 244. All further references to this edition are cited parenthetically.

7. Hume, even as he argued against the partisan Whig account of history, endorsed the idea that this outcome represents "the complicated fabric of the most perfect government." Hume, *The History of England*, 3:327.

8. Young, "Scott and the Historians"; Trevor-Roper, "Sir Walter Scott and History"; Chandler, *England in 1819*.

9. Tosh, *The Pursuit of History*, 6–7.

10. My argument here underlines Anna Henchman's case for the important formal shift from the "single observer" to "constellations of characters" in nineteenth-century fiction. Henchman, *The Starry Sky Within*, 4–5.

11. Macaulay, *Critical, Historical and Miscellaneous Essays*, 428–9.

12. C. Hall, "Macaulay's 'History of England,'" 72.

13. Levine, *The Realistic Imagination*, 92–3.

14. In a previous study, I have argued that comparative history was a constitutive mode of historical analysis for nineteenth-century figures ranging from Charles Dickens to Karl Marx. The present chapter explores Scott's role in establishing that method. Griffiths, "The Comparative History of 'A Tale of Two Cities.'"

15. Garside, "Scott and the 'Philosophical' Historians"; Sutherland, "Fictional Economies."

16. Duncan, *Scott's Shadow*, 122–4.

17. Rigney, *The Afterlives of Walter Scott*, 65.

18. Croker, "The Antiquary," 126.

19. Frye, *Anatomy of Criticism*. Scott responded to Croker in his own "review" of *Tales of My Landlord*—posing as the *Quarterly*'s reviewer. Scott's main object is not to argue that Croker's individual valuations are wrong but rather to counter the accusation that Waverley novels lack "system." In his review, Scott is constrained by the precedential nature of the review format. Writing as the shared anonymous voice of the *Quarterly Review*, he can't very well come out and reverse the judgments he has, in effect, already made. Instead, he proceeds by arguing that, after a more careful evaluation, he perceives

"something of system" in the Waverley novels. Where Croker had seen the novel as falling into natural kinds, the reviewer of *Tales* now finds a form characterized by generic experiments between novel and drama and surprising continuities of character—what Ina Ferris terms the "hybrid nature" of Scott's historical novels. Ferris, *The Achievement of Literary Authority*.

20. Chandler, *England in 1819*, 106. See also Christensen, *Romanticism at the End of History*.

21. Duncan, *Modern Romance and Transformations of the Novel* and *Scott's Shadow*; Goodlad, *The Victorian Geopolitical Aesthetic*; Rigney, *The Afterlives of Walter Scott*; Trumpener, *Bardic Nationalism*; Chandler, *England in 1819*.

22. For a comprehensive study of Scott's global reception, see Rigney, *The Afterlives of Walter Scott*.

23. Crawford, "Walter Scott and European Union," 140.

24. Scott, *Waverley* (2007), 5. All further references to this edition are given parenthetically.

25. Duncan, *Modern Romance and Transformations of the Novel*, 23–6.

26. For a discussion of the limits of Humean sympathy, see Herdt, "Artificial Lives, Providential History, and the Apparent Limits of Sympathetic Understanding."

27. Hume, *The Life of David Hume*, 18–9. A similar sympathetic mode is engaged in Hume's description of the British subjects at the time of Charles I's trial; his "present distress" humanizes the king and so makes him available to their "generous" condescension. Hume, *History of England*, 1:453.

28. Hume, *Essays and Treatises on Several Subjects*, 429. Hume contrasts the universality of such passion with the plebian specificity of "the pleasantries of a waterman, the observations of a peasant, the ribaldry of a porter or hackney coachman, all of these are natural, and disagreeable. What an insipid comedy should we make of the chit-chat of the tea-table, copied faithfully and at full length?" (ibid., 116). On the relation between passion and sentiment in Hume's essays, see Costelloe, "Fact and Fiction."

29. See Potkay, *The Passion for Happiness*.

30. Braudy, *The Plot of Time*.

31. Scott, *The Antiquary*, 252. All further references to this edition are given parenthetically.

32. My account here parallels James Chandler's analysis of how Scott historicizes Smithean sympathy as placing oneself in the "case of another." Chandler, *England in 1819*, 226–8.

33. Buzard, *Disorienting Fiction*.

34. Duncan, *Scott's Shadow*, 131. "Double consciousness" evokes W. E. B. Du Bois's famous formulation of the racialized subject, torn between racial and national identity as this "two-ness,—an American, a Negro; two souls, two thoughts, two unreconciled strivings." Du Bois, *The Souls of Black Folk*, 3. Unlike the oscillation between absorption and reflection, or participation and observation, this implies a more immediate, more horizontal relation between the two perspectives, "two souls" lodged in one body. Du Bois, like Scott, emphasizes the immediate, serial relation between such identities.

35. Greiner, *Sympathetic Realism in Nineteenth-Century British Fiction*.

36. This is particularly true for historiography, as James Chandler has shown (*England in 1819*, 260–1). Mark Phillips's recent historiographical study serves as a strong

example. *On Historical Distance* provides a powerful mediation on the dimensionality of perspective in histories and opens its historical problem with the observation that many historians have seen the early nineteenth century as a turning point in the sense of the past. The first chapter begins with a quote from Macaulay's review of Henry Hallam's histories—the same review, discussed early in this chapter, that famously invokes Scott and trumpets the need for history to draw inspiration and method from Scott's novels—and Phillips' study closes with an epilogue subtitled, " 'Tis Forty Years Since," a wink at *Waverley*. This raises without addressing Scott's contribution to the more nuanced understanding of historical distance advocated by Phillips. Phillips, *On Historical Distance*. The reason, likely, is that Phillips had already given extended treatment to Scott's influence in *Society and Sentiment*.

37. Bruner, "The Narrative Construction of Reality," 56.

38. Sachs, *Romantic Antiquity*, 39.

39. Scott, "Tales of My Landlord."

40. Gieryn, "Boundary-Work and the Demarcation of Science from Non-Science," 781.

41. Manning, "Antiquarianism, the Scottish Science of Man, and the Emergence of Modern Disciplinarity," 67.

42. Ferris, "Bibliographic Romance."

43. Lynch, "Wedded to Books."

44. Gamer, *Romanticism and the Gothic.*

45. W. Smith, *Stratigraphical System of Organized Fossils*, ix–x.

46. See Winchester, *The Map That Changed the World.*

47. With the possible exception of ecology, as Susan Oliver argues. Oliver, "Sir Walter Scott's Transatlantic Ecology."

48. Anon., "Sir Walter Scott and William Laidlaw. Concluding Article," 73.

49. Buckland, *Novel Science*, 4, 70–2.

50. Lockhart, *Memoirs of the Life of Sir Walter Scott*, 53.

51. Heringman, *Sciences of Antiquity*, 4. See also Martin Rudwick's account of how these "erudite historians" contributed to geological theory in Rudwick, *Bursting the Limits of Time*, 181–7.

52. *Transactions of the Royal Society of Edinburgh* 4 (1798).

53. Arnaldo Momigliano wrote the foundational study for this line of thinking. Momigliano, "Ancient History and the Antiquarian."

54. These correspond to the two schools of antiquarianism that Daniel Woolf recognizes, one interested in philology and old manuscripts (philological antiquarianism), the other interested in ruins and artifacts (peripatetic antiquarianism). Woolf, *The Social Circulation of the Past*, 141–50. Whereas Woolf sees these as two different strains of antiquarianism, I am arguing that both served to sharpen the comparative methodologies of antiquarianism in general and, moreover, that the antiquities market prompted the development of forensic techniques that made these researches increasingly sophisticated. When seen in this light, what I call *forensic antiquarianism* should be recognized as an increasingly sophisticated methodology that applied to various domains of historical research.

55. See Trigger, *A History of Archaeological Thought*, 38–9.

56. Metzger and Ehrman, *The Text of the New Testament*, 101–2.

57. Ritson, "Observations on the Ancient English Minstrels."

58. Ibid., xxxi.

59. Heringman, *Sciences of Antiquity*, 2, 233–4, 274–6.

60. For a study of antiquarianism's impact on Victorian writing, see Zimmerman, *Excavating Victorians*.

61. Aghion, "Collecting Antiquities in Eighteenth-Century France."

62. Jobert, *The Knowledge of Medals*, 205.

63. Similarly, David Alvarez has argued that Addison uses coin collecting to theorize the productive relation between social virtue and antiquarianism. Alvarez, "Poetical Cash."

64. Aghion, "Collecting Antiquities in Eighteenth-Century France."

65. Pinkerton, *An Essay on Medals* (1808), 2:132, 145.

66. Rowlinson, *Real Money and Romanticism*.

67. Chandler, *England in 1819*, 277.

68. Jobert, *The Knowledge of Medals*, 183.

69. Pinkerton, *An Essay on Medals* (1808), 2:179–80.

70. Jobert, *The Knowledge of Medals*, 186–7.

71. Scott, "Essay on Imitation of the Ancient Ballad," 18.

72. Scott, *Minstrelsy of the Scottish Border*, 3:168. All future references are taken from this edition.

73. Ibid., 3:60.

74. Letter to Malcolm Laing, 14 April 1806. In Scott, *The Letters of Sir Walter Scott, 1787–1807*, 1:293–4.

75. Herd, *Ancient and Modern Scottish Songs, Heroic Ballads*.

76. Chambers, *The Romantic Scottish Ballads*.

77. Child, *The English and Scottish Popular Ballads*, 3:19. For a lively contemporary discussion of Chambers and the arguments against his position, see Clyne, *The Romantic Scottish Ballads and the Lady Wardlaw Heresy*.

78. For a discussion of how Francis James Child's own ballad collection served the purpose of American nationalism after the Civil War, see Cohen, *The Social Lives of Poems*, chap. 4.

79. Scott, "Ballads and Songs," 19r, v.

80. Buchan, *Ancient Ballads and Songs of the North of Scotland*.

81. McLane, *Balladeering, Minstrelsy, and the Making of British Romantic Poetry*, 34.

82. For more on the intermedial awareness of Romantic print culture, with particular reference to Scott, see Fielding, *Writing and Orality*; Piper, *Dreaming in Books*; Langan, "The Medium of Romantic Poetry."

83. McLane, *Balladeering, Minstrelsy, and the Making of British Romantic Poetry*, 68–70.

84. Scott's 1807 essay on ballad imitation footnotes a similar example of "some affected person [who] has struck in one or two factitious verses, which, like vulgar persons in a drawing-room, betray themselves by their over-finery." Scott, "Essay on Imitation of the Ancient Ballad," 19. Scott describes such "affected persons" as "vicious intromitters," a Scottish law term that describes "intermeddling with the effects of another . . . without legal authority." Discussed in "intromission, n." OED Online, June 2014, Oxford University Press, http://www.oed.com/view/Entry/98734?redirectedFrom=intromission. The "over-finery" of such verses makes such intermission "vicious." This lodges the "legal authority" to discriminate between good and bad interpolation, between collation and

intermission, in the office of aesthetic judgment; they are factitious because their manufacture has no bearing on historical fact. The "legal authority" is not the original author of the ballad but history itself.

85. Quoted in Ruthven, *Faking Literature*, 57–9.

86. Scott, "Essay on Imitations of the Ancient Ballad," 5.

87. Haywood, *The Making of History*; Haywood, *Faking It*; Groom, *Thomas Chatterton and Romantic Culture*. In Paul Saint-Amour's view, such Romantic forgeries remained important to modern writers like Oscar Wilde, who used them to confront distinctions between historicity, authenticity, and fabrication. Saint-Amour, *The Copywrights*.

88. Russett, *Fictions and Fakes*, 5.

89. Pinkerton, *An Essay on Medals* (1808), 2:164–5.

90. Simmel, *The Problems of the Philosophy of History*, 69.

91. Herder makes this case in "Über die Reichsgeschichte: ein historischer Spaziergang," excerpted in Beiser, *The German Historicist Tradition*, 119.

92. Ibid., 268–77. See also his discussion of Ranke, 268–77.

93. This famous statement comes in the preface to his *Geschichte der romanischen und germanischen Völker*, 274.

94. Scott, *Ivanhoe*, 9. All further references are given parenthetically for this edition.

95. For an account of its impact on rhetoric, see Copeland, *Rhetoric, Hermeneutics, and Translation in the Middle Ages*. For scientific discourse, see Crossgrove, "The Vernacularization of Science, Medicine, and Technology in Late Medieval Europe."

96. Copeland, *Rhetoric, Hermeneutics, and Translation in the Middle Ages*, 34.

97. Greene, *The Light in Troy*, 62.

98. Plutarchus and Wyatt, *T. Wyatis Translatyon of Plutarckes Boke Of the Quyete of Mynde*. For a discussion of this "clumsy" translation of *The Quyete of Mydne*, see Thomson, "Sir Thomas Wyatt."

99. Edmundson, *Lingua linguarum*.

100. Monboddo, *Of the Origin and Progress of Language*, 1:439. Volume 1, which focuses on grammatical analysis, is filled with this form of analogy, though its best definition occurs in volume 2.

101. "An Essay on Translation."

102. Tytler, *Essay on the Principles of Translation*, 6–7.

103. Ibid., 13–4.

104. Tytler's analysis is largely in line with the theorists of translation who came before, in the analysis of T. R. Steiner, *English Translation Theory, 1650–1800*, 32–9. I find something profoundly new in his reorientation of the perspective of the translator from the relation between content and style toward the relation between text and social context.

105. Apter, *The Translation Zone*.

106. Gamer, "Authors in Effect," 851.

107. Scott, "Essay on Imitation of the Ancient Ballad," 44–5.

108. Ibid., 43.

109. Scott, *The Letters of Sir Walter Scott, 1787–1807*, 166.

110. "A Grammar of the Sanskrita Language."

111. Crawford, *Devolving English Literature*.

112. Scott, *Guy Mannering*, xi.

113. Ibid., xii–xiii.

114. Scott, *Waverley* (1984), 498.

115. Giorgio Agamben argues that such "zones of indistinction"—subject positions explicitly excluded from the rule of law and civil armature by sovereign power—are central to modernity. Agamben, *Homo Sacer*, 64, 170, 181. Scott's novels suggest that any constellation of sovereign power stands among other competing systems (particularly at the borders) and that zones of indistinction in fact constitute historically specific zones of contact between those systems. For this reason, the exclusion of an individual subject from the rule of law has no necessary bearing on their access to alternative systems of justice and social regulation.

116. In contrast to Alison Lumsden's assertion that Scott used dialect to assert his "fundamental skepticism concerning the communicative potentialities of language," I see dialect as emphasizing Scott's confidence in the potentiality of language to communicate across history. Lumsden, *Walter Scott and the Limits of Language*, 6.

117. Underwood, "Linguistic Realism in 'Roderick Random.'"

118. Note that the relation between idiom and dialect is instable in Scott's fiction, in part because of the close attention to social gradation within regions as well as contact with those from outside. Here I take Elspeth's speech as idiomatic not because her Scots does not exhibit the regional distinction of a dialect, but because its contrast with her more standard English furnishes her language as an index into the contrast between patterns of speech within two differently classed social groups.

119. Galison, *Image and Logic*, 783.

120. In contrast, Maeve Adams has recently argued that, "in uniting the two dialects on the page, the novel unites two cultures in a single narrative that recognizes and transcends their differences." "'The Force of My Narrative,'" 950.

121. Chakrabarty, *Provincializing Europe*, 17–8.

122. Schleiermacher, "On the Different Modes of Translating," 142.

123. Alexander von Humboldt explained this as a depiction of the past that moves beyond the "skeleton of the event" that is contained in factual accounts, to "bring forth what is not present. . . . Like the poet, but in a different manner, he must take the scattered pieces he has gathered into himself and work them into a whole." Alexander von Humboldt, "On the Task of the Historian," in Mueller-Vollmer, *The Hermeneutics Reader*, 106. Mueller-Vollmer elaborates: "What is required is that the historian's 'investigative capability' (*forschende Kraft*) become assimilated with the object under investigation. Only when this takes place is he able to bridge the gap between himself and the historical phenomena." Ibid., 16.

124. Michelet, *The People*, 11, 14, 25.

125. Moretti, *Atlas of the European Novel, 1800–1900*, 40.

126. Rigney, *The Afterlives of Walter Scott*, 84, 103.

127. Ferris, *The Achievement of Literary Authority*, 227, 236.

128. Phillips, *On Historical Distance*, 6.

129. Jameson, *Marxism and Form*, 194.

130. By this I mean something different from what Chandler identifies as the "serial relation of cases" in the Waverley novels. Chandler, *England in 1819*, 223. Chandler identifies a secularizing shift from Christian casuistry to the higher-level "case of cases" in history. But this retains an emphasis upon history as a formal analogy, the second-order form identified through the analysis of events. I am arguing comparative

history added a crucially deformalizing impulse, producing an understanding of the past as a network of serial relationships defined by their inability to cohere in a unified form.

131. Here I'm thinking primarily about the novels themselves; Adams argues persuasively that their collection in the Magnum Opus edition of Scott's works coheres in a "new conception[] of rhetoric and persuasion [suited to] a new era of the imperial nation state." " 'The Force of My Narrative,' " 937.

132. Pittock, *The Reception of Sir Walter Scott in Europe*, 5.

133. Eliot, *The Mill on the Floss*, 270.

134. Developed by Joanna Drucker, Jerome McGann, and Bethany Nowviskie, "Ivanhoe" began as a bulletin board system (bbs), and began with a "ludic" rewriting of Scott's novel. http://www.ivanhoegame.org/.

135. See McGann's description of the genesis of the game, online at http://www.rc .umd.edu/pedagogies/commons/innovations/IVANHOE.html. Scott's response to the legions of readers who wished Rebecca and Ivanhoe otherwise, that "the prejudices of the age rendered such a union almost impossible," is both a defense and a thesis about the productive frustrations of historical difference. *Introductions*, 494. The "almost" carries a great deal of weight here, as the marker of a historical possibility produced through the novel. As Scott's contemporary Harriet Martineau would gloss this point, Scott disclosed both the obverse and reverse of history as it is given, "the ends for which freedom and power are desirable, as well as the disastrous effects of withholding them." Martineau, *Miscellanies*, 1:50.

136. Duncan, *Scott's Shadow*, 114.

137. Scott, "Biographical Memoir of Dr. Leyden," xlix, li.

138. Ibid., xlviii.

139. Scott, *Edinburgh Annual Register*, 1:209.

Chapter 3 • Spooky Action in Alfred Tennyson's In Memoriam A. H. H.

1. This point is finely elaborated by Stuart Curran for Percy Bysshe Shelley's "Adonais," his elegy for the fellow poet, John Keats. Curran, "Adonais in Context."

2. A. Tennyson, *In Memoriam* (2004), CXXVIII. All further references to this edition are given parenthetically by section number.

3. H. Tennyson, *Alfred Lord Tennyson*, 1:485.

4. "Notes."

5. H. Tennyson, *Alfred Lord Tennyson*, 2:469–78.

6. In addition to the volume of verse they planned to publish together (scuttled by Hallam's father), they composed poems together, as Cornelia Pearsall has noted. Pearsall, *Tennyson's Rapture*, 60.

7. Armstrong, *Victorian Poetry*, chaps. 2–3.

8. See my discussions of harmonic analogy in the prelude to this book.

9. Cameron, *Lyric Time*; Morgan, *Narrative Means, Lyric Ends*.

10. Tucker, *Epic*.

11. Harold Nicolson was possibly the first to observe the coordinate influence of Hallam's philosophical and religious writings, arguing that the "hint of evolution, here and there" in his posthumous *Remains in Prose and Verse*, provides "more precise indications of Tennysonian parallelisms." Nicolson, *Tennyson*, 71. Most recently, a transporting study

by Cornelia Pearsall has explored Tennyson's extensive engagement with Hallam's ideas regarding sympathy, religion, and oration in such poems as "St. Simeon Stylites" and "Tithonus." Pearsall, *Tennyson's Rapture*.

12. Tucker, *Tennyson and the Doom of Romanticism*, 185.

13. Hallam, *Remains, in Verse and Prose, of Arthur Henry Hallam*, 106. Further references to this edition are given parenthetically by page number.

14. As I note later, this line was changed to "the living soul" in the 1878 edition.

15. E. Gray, *The Poetry of Indifference*, 91.

16. Ricks, *Tennyson*, 216.

17. Gates, "Poetics, Metaphysics, Genre," 509–10.

18. My language here suggests how the *In Memoriam* stanza supports a more capacious dialectic of the kind that Andrew Cole has identified in Neoplatonic thought; a dialectic that unites "identity/difference" with "procession and return." Cole, *Birth of Theory*, 34–41. At the same time, as I will ultimately argue, Tennyson's emphasis upon return and consolidation into a singular "law" ultimately pushes *In Memoriam* away from the open-ended comparatism that characterizes harmonic analogy.

19. Ricks, *Tennyson*, 216.

20. Tucker, "The Fix of Form," 532.

21. Bradley, *A Commentary on Tennyson's In Memoriam*, 159.

22. Wolfson, *Formal Charges*, 3.

23. Any writer has experienced those moments of unbidden felicity where unified thought and language emerge on the page, as if unsummoned. Derek Attridge has characterized this as the "singularity" of literature—its moment of closest contact with the other, the alien. Attridge, *The Singularity of Literature*. In this vein, David Goslee has observed the "rhythm between [Tennyson's] openness to these intimations and their arrival" in the poem, and my point is that the *In Memoriam* stanza is built to evoke such moments, in the way the first two lines call out for an answer and must then wait for a response. Goslee, *Tennyson's Characters*, 97.

24. H. Tennyson, *Alfred Lord Tennyson*, 1:104.

25. Batchelor, *Tennyson*, 77.

26. A. Tennyson, *In Memoriam* (1982), 9.

27. Hallam, "The Papers of Arthur Henry Hallam."

28. Pearsall, *Tennyson's Rapture*, 12, 76–80.

29. "The Poetry of Sorrow."

30. Reproduced at Bonhams.com, auction 20922, http://www.bonhams.com/auctions /20922/lot/226/.

31. Hallam, *Remains*, 183. One side of the early Huntington manuscript clusters a group of poems (XLI, LXI, LXVI), that turn on images derived from Shakespeare ("thou art turn'd to something strange," "I love the, Spirit, and love, nor can / The soul of Shakespeare love thee more," "The shade by which my life was crost, / Which makes a desert in the mind, / Has made me kindly with my kind, / And like to him whose sight is lost"). And section LXI, with its short three-stanza structure and quick, two-line volta, stands as the nearest example of a Shakespearean sonnet within the larger poem.

32. Hallam, *The Letters of Arthur Henry Hallam*, 510.

33. Philip Flynn has explored the influence of Hallam's *Theodicaea* within the poem. Flynn, "Hallam and Tennyson."

34. Joseph, *Tennysonian Love*; Craft, "'Descend, Touch, and Enter'": Tennyson's Strange Manner of Address," 89.

35. The importance of this desire for communion can be clearly delineated if we juxtapose it to the rivalry and triangulation through heterosexual desire by which Eve Sedgewick characterizes homosocial relationships (and, particularly, Shakespeare's sonnets). Jeff Nunokawa extends Sedgwick's case in arguing the importance of a reciprocal tension between hetero- and homosexuality in *In Memoriam*. Sedgwick, *Between Men*; Nunokawa, "*In Memoriam* and the Extinction of the Homosexual." Tennyson's poem doesn't triangulate through third bodies so much as embrace the possibility of bringing Hallam and Tennyson together. The desire between the male figures is frank and acknowledged. If anything the desire for communion is triangulated through intangible questions in place of an orbiting object—problems of epistemology and faith. More basically, it is triangulated through the poem itself, which serves less as a substitute for Hallam than the place of his possibility. The poem is a firmament on which the desire to restore Hallam and its satisfaction can be realized, a place of their "meeting . . . love with love" (LXXXV).

36. If, as Joseph argues, Hallam serves as an "analogue of Christ" within *In Memoriam*, this is a feature of Tennyson's concern for Hallam's metaphysics. Joseph, *Tennysonian Love*, 68–9.

37. H. Tennyson, *Alfred Lord Tennyson*, 1:126.

38. Ibid., 1:xi.

39. Ibid., 1:268; Hallam, *The Letters of Arthur Henry Hallam*, 437.

40. H. Tennyson, *Alfred Lord Tennyson*, 1:60.

41. Hallam, "The Notebook of Arthur Henry Hallam." The *Remains* gives no indication of their extraction from this larger sequence.

42. Quoted in Shatto, "Tennyson's Revisions of 'In Memoriam.'"

43. Though the 2004 edition renders this line from section XCI as "Thy spirit in time among *they* peers" (emphasis added), I've revised the line to reflect the original spelling.

44. As Robert Pattison observes, the political ambition of these poems suggests an engagement with classical elegiac verse, which (unlike the genre of elegy itself), sought an ingenuous rhythm for more public address by truncating the heroic line. In contrast, Denise Gigante has argued the shorter line brushes into the less personal voice of the ballad, which "subsum[es] the speaker into one totalizing, collective voice." Pattison, *Tennyson and Tradition*, 108; Gigante, "Forming Desire," 488.

45. Goslee, *Tennyson's Characters*, 94.

46. Sendry, "*In Memoriam*," 56.

47. Kingsley, *Literary and General Lectures and Essays*, 20:81–2.

48. Kincaid, *Tennyson's Major Poems*, 83.

49. Kingsley, *Literary and General Lectures and Essays*, 20:90. We might say that such possibilities are opened up by the poet's doubt in the confident answers that religion and science try to offer to his questions. One striking feature of Kingsley's review is that Tennyson is so closely associated with the "poetry of doubt," in part through Kingsley's earlier analysis, that Tennyson himself is generally implicit in his criticism of skeptical verse. In place of doubtful verses, Kingsley demands a poetry of scientific (and muscular) Christian confidence.

50. Though the *Memoir* generally casts Tennyson as more consistent and more orthodox than *In Memoriam* would seem to suggest, Hallam recognizes how the poet's

assertion that "we have but faith, we cannot know" applies to modes of scientific knowledge as well as Christian faith. H. Tennyson, *Alfred Lord Tennyson*, 311.

51. Jason Rudy has argued that this debate finds a fuller political expression in the division between the idealism of conservative critics like John Wilson (a.k.a. "Christopher North") and more liberal reviewers like William Johnson Fox. Rudy, *Electric Meters*, 47–54.

52. Whewell's cordial relations with Tennyson are described in Martin, *Tennyson, the Unquiet Heart*, 54.

53. Whewell, *The Philosophy of the Inductive Sciences*, 242.

54. Mill, *A System of Logic*, 1:240.

55. Hacking, "Statistical Language, Statistical Truth and Statistical Reason."

56. In the syllogism, identity is constituted through the predicate's shared semantic content; in the stanza, the shared sound of the closing syllable. And yet, as we have seen, the stanzas often insist on a connection between shared sound and echoing sense.

57. See my discussion in chapter 1 and Snyder, *Reforming Philosophy*, 71–3.

58. Mill, *A System of Logic*, 2:213–4, 1:352.

59. Ibid., 2:219.

60. Hallam, *Remains*, 121.

61. Henchman, *The Starry Sky Within*, chap. 3.

62. Culler, *The Poetry of Tennyson*, 173–5.

63. For additional discussion of the larger plural worlds controversy, see Crowe, *The Extraterrestrial Life Debate*.

64. Whewell, *Astronomy and General Physics Considered with Reference to Natural Theology*, 280.

65. Holmes, *The Age of Wonder*, 61–3, 91–2. Herschel, in the introduction to his volume on astronomy for Lardner's *Cabinet Cyclopaedia* series (1833), put the matter simply: "The planets, which appear only as stars somewhat brighter than the rest, are to him spacious, elaborate, and habitable worlds; several of them vastly greater and far more curiously furnished than the earth he inhabits." Herschel, *A Treatise on Astronomy*, 2.

66. Chambers, *Vestiges of the Natural History of Creation*, 161.

67. Tucker, *Tennyson and the Doom of Romanticism*, 174.

68. See Snyder, *Reforming Philosophy*, chap. 2.

69. Peirce, *Collected Papers of Charles Sanders Peirce*, 7–8:137 (7.218).

70. Ibid., 1–2:53 (2.97).

71. Ibid., 7–8:137 (7.218).

72. See the discussion in Merrell, *Entangling Forms within Semiosic Processes*, 22, 66–7.

73. Daston and Galison, *Objectivity*; Levine, *Dying to Know*.

74. E. Gray, *The Poetry of Indifference*, 73–6; Joseph, *Tennysonian Love*, 18.

75. Henchman, "The Globe We Groan In," 37.

76. A. Culler, *The Poetry of Tennyson*, 175.

77. Kincaid, *Tennyson's Major Poems*, 109.

78. Rudy, *Electric Meters*, 16.

79. Hoffman, "Arthur Hallam's Spirit Photograph and Tennyson's Elegiac Trace," 613.

80. From the 1888 diary of Ella Coltman, quoted in ibid., 617.

81. Tucker, *Tennyson and the Doom of Romanticism*, 378.

82. Armstrong, *Victorian Poetry*, 252.

83. For more information on Sloane, see Walker, MacGregor, and Hunter, *From Books to Bezoars*. While Buffon was not an explorer, in his capacity as director of the Jardin du Roy, later to become the *Muséum national d'histoire naturelle*, he coordinated the vast collection apparatus of the French empire (Roger and Williams, *Buffon*). With regard to nineteenth-century specimen collection in Britain, see Raby, *Bright Paradise*.

84. Lamarck, *Zoological Philosophy*, 29. All subsequent references are given parenthetically by page number.

85. Cuvier and Rudwick, *Georges Cuvier, Fossil Bones, and Geological Catastrophes*, 185. All further citations are given parenthetically by page number.

86. Rudwick, *Bursting the Limits of Time*, 181–7; Heringman, *Sciences of Antiquity*.

87. "La voie de l'analogie est loin de pouvoir fournier des indications précis." In Saint-Hilaire, "Mémoires sur la structure et les usages de l'appareil olfactif dans les poissons," 342.

88. For more on analogy and biblical hermeneutics, see chapter 4; for analogy's relation to natural theology, see chapter 5.

89. For more discussion of homology, particularly the work of J. V. Poncelet, see the prelude to this volume.

90. He is said to have remarked, "I wish they would be content to let me be the Owen of England." In Owen, *The Life of Richard Owen*, 1:327.

91. For an extended survey of Owen's career that disentangles his thinking from the retrospect of Darwinian science, see Rupke, *Richard Owen*.

92. Owen, *On the Archetype and Homologies of the Vertebrate Skeleton*, 668, 674.

93. In modern cladistics (the study of taxonomic relations), homology and analogy are now reciprocal; to say that a given creature has 60 percent homology to another means there is a 60 percent chance that their shared patterns are part of common descent, and a 40 percent chance it is analogy. For a discussion of how Owens's different categories of "formal" and "serial" homology continue to develop, see Pinna, "Concepts and Tests of Homology in the Cladistic Paradigm."

94. See Gordon L. Miller's introduction to Goethe, *The Metamorphosis of Plants*; and Koerner, "Goethe's Botany."

95. Owen, *Report on the Archetype and Homologies of the Vertebrate Skeleton*, 3–6.

96. The relationship between formal analogy and theories of scientific modeling is taken up in chapter 5.

97. Owen, "Darwin 'On the Origin of Species,'" 499–500.

98. Reprinted in Ricks, "Tennyson's 'Hail, Briton!' and 'Tithon,'" 53–5.

99. For discussions of climate change in British Romanticism, see Carroll, "Crusades against Frost"; Carroll, "Mary Shelley's Global Atmosphere." For similar discussions in the Victorian and modernist novel, see Taylor, *The Sky of Our Manufacture*.

100. Owen, *The Life of Richard Owen*, 1:388.

Chapter 4 · Falsifying George Eliot

1. Leavis, *The Great Tradition*.

2. See Beaty, *Middlemarch from Notebook to Novel*.

3. Eliot, *Middlemarch*, 20. All further references are given parenthetically by page number.

4. Hertz, *George Eliot's Pulse*, 17.

5. Later in this chapter I explore the relation between Eliot's comparative method and the "contrapuntal" reading that Edward Said has proposed as one alternative to the comparative methodologies of orientalism. Said, *Culture and Imperialism*.

6. G. Levine, *The Realistic Imagination*, 5.

7. Herbert, *Culture and Anomie*; A. Miller, *The Burdens of Perfection*; Kurnick, "An Erotics of Detachment."

8. Eliot, *Felix Holt, the Radical*, 111. All further references given parenthetically by page number.

9. See Hardy, *The Novels of George Eliot*. Also valuable is her ever-rewarding *Particularities*, which gives a "slow" surface reading of *Middlemarch* avant la lettre, cautioning, "The act of comparison is a dangerous tool in criticism. We may too easily select the material for comparison in order to back up our prejudices and preferences." Hardy, *Particularities*, 15.

10. Here I draw on work by Deidre Lynch and Alex Woloch that examines how characters operate systematically within the space of the novel and within social and economic codes that structure how they are read and understood. Comparative historicism is important to the novel because it offered one way of reading what Woloch terms individual "character-spaces" as part of a larger system. As Woloch puts this, "the character-system offers not simply many *intersecting* individuals but many *intersecting* character-spaces, each of which encompasses an *embedded* interaction between the discretely implied person and the dynamically elaborated narrative form." Lynch, *The Economy of Character*; Woloch, *The One vs. the Many*, 17–8.

11. Apter, *Against World Literature*, 4.

12. Eliot, *Adam Bede*, 10. All further references are given parenthetically by page number.

13. In an influential reading, Raymond Williams understands Eliot's idiomatic focus as an awkward condescension, arguing that the "known community" in Eliot's fiction is provided in "an uneasy contact, in language, with another interest and another sensibility." Williams, "The Knowable Community in George Eliot's Novels," 261.

14. Beer, *Darwin's Plots*, 148.

15. Eliot, *Middlemarch*, 264; Eliot and Cave, *Daniel Deronda*, 7; Eliot, *The Impressions of Theophrastus Such*, 8.

16. Printed in Collins, "Questions of Method," 389. Additional references are given parenthetically by page number.

17. For a summary of *harmonic analogy*, in contrast to *formal analogy*, see the prelude to this book.

18. Shaw, *Narrating Reality*, 4; Greiner, *Sympathetic Realism in Nineteenth-Century British Fiction*.

19. Gallagher, "George Eliot," 66.

20. Eliot, *Essays of George Eliot*, 435; Eliot, *Poems, Essays, and Leaves from a Note-Book*, 192–4.

21. Eliot, *Poems, Essays, and Leaves from a Note-Book*, 193.

22. Michael Carignan notes that Eliot's theory may also draw from Edward Bulwer Lytton's discussion of the "analogical hypothesis" in *The Last of the Barons*. Carignan, "Analogical Reasoning in Victorian Historical Epistemology." Bulwer Lytton argues that

the analogical hypothesis helps the writer "to solve the disputes and difficulties of contradictory evidence by the philosophy of the human heart," a thesis about the unity of human nature which is a legacy of Enlightenment thought. Lytton, *The Last of the Barons*, 1:vii.

23. "The Future of Geology"; "Ruth and Villette"; "Thackeray's Works"; "Iconoclasm in Philosophy"; "Balzac and His Writings"; "The Universal Postulate"; "Results of the Census of 1851"; "Goethe as a Man of Science."

24. "Science"; "Life and Doctrine of Geoffroy St. Hilaire."

25. "Contemporary Literature."

26. I do not mean to ignore the important distinction between comparative literary history (*Geschichte*) and comparative literary science (*Wissenschaft*) that later emerged in German scholarship but to take up an earlier moment in which the boundaries were less clearly demarcated.

27. Dillane, *Before George Eliot*. In the most extensive study to date of Eliot's editorial work, Dillane cautions against the critical tendency to read that work as a careful preparation for the novels that followed, a tendency so strong that literary historians refer to her exclusively *as* George Eliot.

28. G. Levine, *Darwin and the Novelists*, 226.

29. Beer, *Darwin's Plots*, 149.

30. Shuttleworth, *George Eliot and Nineteenth-Century Science*.

31. Kuhn, *The Structure of Scientific Revolutions*.

32. G. Levine, *The Realistic Imagination*, 95.

33. Shaw, *Narrating Reality*; McCaw, *George Eliot and Victorian Historiography*. Eliot was perhaps the most perceptive nineteenth-century reader of Scott, not least because she came to recognize his deep engagement with the relational turn in history. For a further discussion of Scott's historicist influence on both Eliot and on Honoré de Balzac, see Goodlad, *The Victorian Geopolitical Aesthetic*, chap. 6. Both Gordon Haight and Joseph Nicholes have explored the careful imitation of Scott in Eliot's earliest juvenile fictions. Haight, *George Eliot*, 652; Nicholes, "Vertical Context in Middlemarch." Her essay on the "Historic Imagination" similarly takes Scott to heart. There we find many of the ingredients of Scott's fiction: an emphasis on historical translation, a careful coordination of historical distance and connection, and the relation between major historical events and the "view from below" of social history. Eliot, *Poems, Essays, and Leaves from a Note-Book*, 192–4.

34. Quoted in Karl, *George Eliot, Voice of a Century*, 330.

35. This includes duplicate works. Baker, *The Libraries of George Eliot and George Henry Lewes*, 104–6.

36. G. Levine, *The Realistic Imagination*, 95.

37. Norman Vance provided the most extended study of Victorian interest in Roman history in *The Victorians and Ancient Rome*. Virginia Hoselitz focuses on how Victorians interpreted their own history through the lens of classical Rome in *Imagining Roman Britain*.

38. Gibbon, *The History of the Decline and Fall of the Roman Empire*.

39. The third book of John Ruskin's *Modern Painters* (which Eliot reviewed) takes up the significance accorded to "historical" painting in the British Academy (under the influence of John Reynolds), though it broadens the category of history considerably.

40. Gilmore, *The Victorian Novel and the Space of Art*, 100–1.

41. Siegel, *Haunted Museum*.

42. Joseph Wiesenfarth makes this point for art generally in "Middlemarch: The Language of Art." Abigail S. Rischin sees this scene as an example of the importance of statuary, particularly the Ariadne statue she reclines against, as a major object of such discussions in "Beside the Reclining Statue." For a discussion of the place of portraiture in *Middlemarch* that argues for its realist engagement, see Hollander, "Ariadne and the Rippled Nose."

43. Flint, "The Materiality of Middlemarch," 66.

44. Yeazell, *Art of the Everyday*, chap. 4.

45. This is from her review of Wilhelm Heinrich Riehl's *Natural History of the German People*, discussed in the next section. Eliot, "The Natural History of German Life," 144. All further references are given parenthetically by page number.

46. I discuss *harmonic analogy* in relation to logic and Charles Sanders Peirce's theory of "abduction" in chapter 3.

47. Shaw, *Narrating Reality*, 91.

48. Ibid., 104, 99, 94.

49. The problem is apparent in Shaw's analysis of the second chapter of Scott's *Waverley*, which reads the characteristic news media of the eighteenth and nineteenth centuries as an example of historicist metonymy. Ibid., 104–6. The problem is that the passage contains at least one explicit metaphor and no explicit examples of metonymy. Shaw works around this problem, arguing that the metaphor in which news is "distilled" is displaced into the metonymic social relations explored in the passage, but it is typical in the analysis of metonymic realism.

The extraordinarily long life of the metaphor-metonymy distinction seems to be due to its fitness for the paradigmatic-syntagmatic distinction of Prague-school structural linguistics and to Roman Jakobson's claim that this distinction is supported empirically by a split between two kinds of aphasic disorders. Now that these two kinds of disorders have fallen out of the *Diagnostic and Statistical Manual of Mental Disorders*, 5th ed. (DSM-V) and clinical use, perhaps it's time to take Jonathan Culler's advice: "One frequently wishes, when reading and writing about figures, to put an end to the tropological inflation of tropes. Could we not avoid all these problems if we restricted *metaphor* and *metonymy* to their literal meanings?" Culler adds, however, that "a certain austerity in their use might indeed avoid some problems, but in fact the issues that have emerged in the swings and reversals of *metaphor* and *metonymy* have an uncanny way of reappearing everywhere in this domain, particularly when one wishes to distinguish the literal from the metaphorical." Culler, *The Pursuit of Signs*, 225.

50. McCaw, *George Eliot and Victorian Historiography*; J. Miller, *Reading for Our Time*, 51.

51. Quoted in McCaw, *George Eliot and Victorian Historiography*, 7.

52. Ricœur, *Rule of Metaphor*, 154, 200.

53. Metaphor has received far more critical attention that metonymy, and I will focus on it here. In Paul Ricœur's classic formulation of metaphor, the "semantic impertinence," precipitated by the contrast between semantic error and syntactic appropriateness, invites us to "see" one thing "as" another. Ibid., 154. This expands on Max Black's description of metaphor as mapping the attributes of the "vehicle" on to the "tenor." Black, *Models and Metaphors*. In both metaphor and metonymy, one thing is framed in terms of the other. In the prelude, I argue that, while *formal analogy* can be seen as a

similar "mapping" of one relation to another, *harmonic analogy* points to a more equal interaction between the objects of comparison. While Ricœur's emphasis on "seeing" the tenor "as" the vehicle works to keep some of that resonance alive for metaphor, its paradigmatic substitution pushes metaphor toward formalization/mapping. The power of metaphor consists in the irresistible way that the new perspective takes precedence.

54. The only extended study is Carpenter, *George Eliot and the Landscape of Time*. The first chapter of that study gives a useful summary of contemporary millennialism and the "continuous-historical" method.

55. Eliot began by largely rewriting a manuscript translation first tackled by Rufa Brabant. Haight, *George Eliot*, 53.

56. Anger, *Victorian Interpretation*; Hill, "Translating Feuerbach, Constructing Morality."

57. Strauss and Eliot, *The Life of Jesus*, 1:31. All further references are given parenthetically by volume and page.

58. His most important gleaning from comparative mythology was this distinction between *myth* (the invention of historical events from an "idea") and *legend* (finding a higher "idea" in actual historical events).

59. Feuerbach, *The Essence of Christianity*, xvii.

60. Cf. "For her conception of form is in the end self-defeating. To the extent that she is able in her last novels to achieve a high degree of form by showing very intricate relations within a novel, the beginning and ending become increasingly false in that they artificially cut off relations which the novel itself sends outward, as it were, from its complexity to the rest of the universe." Mansell, "George Eliot's Conception of 'Form.'"

61. I'm not fully persuaded that Eliot really intends to signal "her disquietude in the face of Riehl's approach." Dillane, "Re-reading George Eliot's 'Natural History,'" 248, 57. As I argue, the corrective seems calculated not to resist Riehl as much as to adapt him to her own (considerably distinct) ends. Indeed, Eliot reads Riehl's history against the grain in key respects, in part because (as she noted in her journal) she only read half, preferring to spend her time plashing along the Ilfracombe coast as Lewes worked on his seaside studies. Eliot, "Journal."

62. Eliot, *George Eliot's Life*, 1:39.

63. Riehl, *The Natural History of the German People*, 107.

64. See Pinney, "The Authority of the Past in George Eliot's Novels," 134; Wohlfarth, "Daniel Deronda and the Politics of Nationalism."

65. Eliot, "The Natural History of German Life," 164. All further references are given parenthetically by page.

66. Ironically, for all the ethnographic sophistication of his enquiry into the ethnic makeup of the German state, Riehl misses even the broadest distinctions that delineate the components of the British Union, insofar as he insists that Scott's novels are about the unified "social core" of the "English." Riehl, *The Natural History of the German People*, 23–4. Scott's novels are predicated on the *lack* of a "social core," in Riehl's sense, insofar as the various social and historical formations of Scottish society are played against the (similarly variegated) "English." As I argue in chapter 2, this is especially evident in his nuanced treatment of dialect, which projects complex understanding of the role of linguistic mediation in constructing social and historical difference.

67. A. Anderson, *The Powers of Distance*, 3.

68. The nuanced sense of language's past includes an allusion to previous failed efforts to "construct a universal language" that would provide a key to all philologies, for example, John Wilkins's *Essay toward a Real Character, and a Philosophical Language* (1668).

69. Quoted in J. Taylor, "The Novel as Climate Model," 21.

70. B. Gray, *George Eliot and Music*; Clapp-Itnyre, *Angelic Airs, Subversive Songs*; Sousa Correa, *George Eliot, Music and Victorian Culture*.

71. Capuano, "An Objective Aural-Relative in Middlemarch," 924.

72. Gallagher, "Formalism and Time," 251; Dames, *The Physiology of the Novel*, 10-1. For a wide-ranging account of the importance of momentary time to the Victorian novel, see Zemka, *Time and the Moment*.

73. For a more extended discussion of allegory in relation to analogy, see chapter 1.

74. Eliot, *The Legend of Jubal and Other Poems*, 11, 13. Further references are given parenthetically by page number.

75. Macfarren, *Six Lectures on Harmony*, 3–4.

76. Ibid., 3.

77. For a summary of Eliot's sympathetic theories, see Ermarth, "George Eliot's Conception of Sympathy."

78. Greiner, *Sympathetic Realism in Nineteenth-Century British Fiction*, 1, 16.

79. By taking Eliot's claims for sympathy seriously on their own terms, I find further support for the model of distributed encounter offered by comparative historicism, as organized by harmonic analogy. See the discussions of both in the introduction and prelude to this book. For another extended discussion of Eliot's sympathetic imagination as a kind of rhetoric, see Doyle, *The Sympathetic Response*.

80. Gallagher, "George Eliot," 70.

81. G. Levine, *The Realistic Imagination*, 308. I note that Levine's comparative method in that essay deploys the pattern analysis of the realist novel itself, arguing that "striking but apparently accidental similarities . . . reflect certain historically important similarities in the world views" of Eliot and Joseph Conrad, as well as their characters.

82. Shaw explains that "knowing the Other while remaining a part of one's own culture is a problem for historicism." Shaw, *Narrating Reality*, 237.

83. Said, *Beginnings*, xxiii.

84. Said, *Culture and Imperialism*, 32.

85. Said characterizes this as an example of a "regrasping-of-life scene," in which Dorothea gets a limited glimpse the complex world beyond the Victorian novel. Ibid., 143.

86. Matz, *Satire in an Age of Realism*.

87. Despite the report of John Cross, Eliot's second husband, that she "kept the idea resolutely out of her mind" until, "abandoning herself to the inspiration of the moment, she wrote the whole scene exactly as it stands," Jerome Beaty has shown that Eliot began plotting the scene about halfway through writing the novel. Quoted in Beaty, "Visions and Revisions," 663.

88. Quoted in Hughes, "Constructing Fictions of Authorship in George Eliot's Middlemarch, 1871–1872," 159. Hughes's rich exploration of the advertising and serial runs of *Middlemarch* follows Flint's *Victorian Woman Reader* in seeing Rosamond as Eliot's reflection on the contemporary literary market.

89. Marcus, *Between Women*, 81–2. In order to recognize this more complex, more harmonious relationship, Marcus formulates her method of "just reading," later charac-

terized with Steven Guest, as "surface reading." Best and Marcus, "Surface Reading." As I note in the prelude, we might recognize this contrast between Jamesonian "depth" and surface reading as an attempt to get at the difference between the formal and harmonic relationships that are often organized through analogies (see p. 270, n. 49).

90. Rebecca Mitchell has made the reverse case, observing that Eliot's "Rosamond Plot" argues for the unknowability of alterity in a substantial revision of the analogous plot within *Jane Eyre*. Mitchell, "The Rosamond Plots."

91. Schor, *Curious Subjects*.

92. Kurnick, *Empty Houses*, 67–8.

93. Eliot, *George Eliot's Life*, 64.

94. The notion of form as a universal received its most influential early exposition in Plato's *Republic* and his allegory of the cave. In Jerome McGann's analysis, this theory of form received its influential modern interpretation as part of a "Romantic ideology," pronounced in the poetry of Percy Bysshe Shelley and John Keats, that saw form as "a universal rather than historical phenomenon" and shaped twentieth-century criticism. McGann, *The Beauty of Inflections*, 62. See also McGann, *The Romantic Ideology*. Gallagher has further explored how this notion of "timeless" form shaped Pater and the new critics but advocates instead an idea of form that can address "the temporal nature" of literature. Gallagher, "Formalism and Time," 231. Yet, as Susan Wolfson notes in the special issue of *Modern Language Quarterly* that includes Gallagher's essay, there are many other ways of "reading for form" that emphasize form's eventful, temporal, and changeful nature—alternatives that, in the long view, take inspiration from Marxist formalism and its emphasis on the historical and cultural determinations of form. Wolfson, "Reading for Form."

95. C. Levine, *Forms*, 8–9. All further citations are given parenthetically by page number.

96. Lewes, *Problems of Life and Mind: First Series*, 1:39–40. All further references are given parenthetically by volume and page number.

97. See the prelude to this book, as well as Griffiths, "The Intuitions of Analogy in Erasmus Darwin's Poetics."

98. It should be noted that Lewes emphasizes the language of comparison over analogy, because he generally sees comparison as an empirical method, rather than a mode of philosophical speculation. In this way, Lewes's relational nominalism—his conviction that comparison disclosed real natural patterns—stands in contrast to contemporary efforts, discussed in chapter 3, to come to terms with analogy as a form of imperfect induction.

99. Herbert, *Victorian Relativity*.

100. Lewes, *Problems of Life and Mind: Third Series*, 1:118–26. Further citations are given parenthetically by volume and page number.

101. Duncan, "George Eliot's Science Fiction," 17.

102. The novelist Anthony Trollope complained of her "dissection of mind," in which "everything that comes before her is pulled to pieces so that the inside of it shall be seen." Trollope, *An Autobiography*, 2:66–7.

103. C. Darwin, *On the Origin of Species by Means of Natural Selection* (1859), 489.

104. Ibid., 490.

105. Eliot, *George Eliot's Life*, 263.

106. Tennyson, "Darwin's Gemmule."

107. C. Darwin, *The Variation of Plants and Animals under Domestication*, 1:7.

108. See Collins's discussion in "Questions of Method," 399.

109. Lewes, "Mr Darwin's Hypotheses—Part 3," 67.

110. Lewes, "Mr Darwin's Hypotheses—Part 2," 626.

111. Lewes, "Mr Darwin's Hypotheses—Part 4," 500. For a discussion of Darwin's "lumpers" and "splitters," see the introduction to this book.

112. Lewes, "Mr Darwin's Hypotheses—Part 3," 79.

113. Eliot, *Middlemarch*, 211.

Chapter 5 · *The* Origin *of Charles Darwin's* Orchids

This chapter is derived, in part, from an article published in *Nineteenth-Century Contexts*, September 24, 2015, available at http://www.tandfonline.com/doi/pdf/10.1080/08905495.2015.1080727.

1. C. Darwin, "Notebook M," 35–6.

2. C. Darwin, *On the Origin of Species by Means of Natural Selection* (1859), 91. All further citations are given parenthetically from this edition.

3. For a definition of *harmonic analogy*, in relation to what I have termed *formal analogy*, see the introduction and prelude to this book.

4. Reiss, "Natural Selection and the Conditions for Existence."

5. Fodor, "Against Darwinism."

6. Olsen et al., "Function-Based Isolation of Novel Enzymes from a Large Library."

7. Shapin and Schaffer, *Leviathan and the Air-Pump*.

8. Galison, *Image and Logic*, 62.

9. Daston and Galison. *Objectivity*. See my discussion in the introduction.

10. For an evaluation of the contrast between "historical" and "experimental" science, see Cleland, "Historical Science, Experimental Science, and the Scientific Method." For a tendentious account of the "descriptive" versus "hypothesis-driven" contrast, see Casadevall and Fang, "Descriptive Science."

11. Beer, *Darwin's Plots*, xxv. Further references are given parenthetically by page number.

12. Ibid.; G. Levine, *Darwin and the Novelists*; R. Richards, *The Romantic Conception of Life*. Between *Darwin's Plots* and *Darwin and the Novelists*, Beer and Levine effectively framed the problem of Darwin and literature in terms of poetic influence and novelistic impact, with the literary character of Darwin's language serving as a bridge. This chapter asks what Darwin studies might have looked like had Beer taken the novel as her primary concern, or Levine worked from the novelists into Darwin. Beer has provided the most extended account of how poets like John Milton and William Wordsworth (also a favorite of Darwin in his younger years) helped to structure that theory. Yet her hugely influential study in *Darwin's Plots* finds few examples of direct poetic influence. Her most powerful readings come as she breaks literary influence down into more basic formal categories: Darwin's way of using devices like metaphor, personification, and analogy; the importance of mythic narratives of decline and redemption; and Darwin's influential (and deferential) narrative persona. For its part, Levine's *Darwin and the Novelists* works out from Darwin to explore his impact on the nineteenth-century novel. Levine's other studies of

Darwin's language have centered our attention on the imaginative, ethical, and affective range of his works. G. Levine, *Darwin Loves You*; G. Levine, *Darwin the Writer*.

13. G. Levine, "Reflections on Darwin and Darwinizing," 226.

14. Darwin, "Letter to John Murray," March 31, 1859.

15. As discussed in the introduction and prelude to this book, the complicated transition between analogy and comparison is central to this strategy. This intermixture of analogy and comparison, organized through a newly relativized sense of historical relationships, constituted the new comparative history.

16. C. Darwin, "Notebook B: [Transmutation of Species (1837–1838)]," 129–30.

17. Ibid., 43.

18. C. Darwin, "Notebook C: [Transmutation of Species (1838.02–1838.07)]," 60.

19. Prout, *Chemistry, Meteorology, and the Function of Digestion Considered with Reference to Natural Theology*, 79.

20. C. Darwin, "Notebook C: [Transmutation of Species (1838.02–1838.07)]," 61.

21. Quoted in ibid., 61 n. 3.

22. Ibid., 139–40.

23. C. Darwin, "Notebook E: [Transmutation of Species (1838–1839)]," 128.

24. Mill, *A System of Logic, Ratiocinative and Inductive*, 96.

25. Bopp, *A Comparative Grammar of the Sanscrit, Zend, Greek, Latin, Lithuanian, Gothic, German and Sclavonic Languages*; Macaulay, "Neale's Romance of History: England." Chapter 3 discusses Owen, *On the Archetypes and Homologies of the Vertebrate Skeleton*.

26. Grosz, *The Nick of Time*, 8.

27. Wallace, "On the Tendency of Species to Form Varieties," 61.

28. DeLanda, *Intensive Science and Virtual Philosophy*, 16.

29. Jauss, *Toward an Aesthetic of Reception*. In *The Structure of Scientific Revolutions*, Thomas Kuhn described this dilemma in terms of paradigms, arguing that "each group must use its own paradigm to argue in that paradigm's defense. . . . It cannot be made logically or even probabalistically compelling for those who refuse to step into the circle." Kuhn, *The Structure of Scientific Revolutions*, 94.

30. Latour, *Pandora's Hope*.

31. C. Darwin, *The Life and Letters of Charles Darwin*, 1:149.

32. For a longer discussion of Levine's theory of formal affordance, see chapter 4. C. Levine, *Forms*.

33. For a discussion of entangled reference, see the prelude.

34. Jörg Zinken, in a study of newspaper accounts, has argued for a similar phenomenon in which "discourse metaphors" are organized through their reference to regular, form-specific analogies. Zinken, "Discourse Metaphors."

35. Beer argues that "natural selection is a pithy rejoinder to 'natural theology'" (xviii), but it is more immediately a reflection of this basic analogy between the natural process he describes and domestic selection. Moreover, as I argue at the close of this chapter, I think Darwin was less intent to throw elbows than to draft natural theology into an evolutionary understanding.

36. G. Levine, *Darwin and the Novelists*.

37. Dawson, "Literature," 436.

38. Martineau, *Life in the Wilds*, ix–xiii.

39. It should be noted that Martineau does not want to think of her *Illustrations* as historical fictions, in which history is a negative presence. To her way of thinking, history clouds the picture instead of clarifying it, introducing those accidents of circumstance that make the universal features of moral science harder to discern. Yet her comparative method and investment in carefully localized communities and customs indicate Scott's influence.

40. C. Darwin, *The Life and Letters of Charles Darwin*, 1:112–25.

41. For a discussion of the reading lists, and the problem of Darwin's literary tastes, see Stevens, "Darwin's Humane Reading."

42. C. Darwin, *The Life and Letters of Charles Darwin*, 1:102.

43. Ibid., 1:145.

44. Beer, "Darwin's Reading and Fictions of Development," 543.

45. Buckland, "Losing the Plot."

46. Margaret Russett and Joseph Dane see both Horace Walpole's *Castle of Otranto* and Thomas Chatterton's forgeries as key antecedents for this trope. Russett and Dane, "Everlastinge to Posterytie."

47. Meinhold, *Mary Schweidler*, iv.

48. Ibid., vii.

49. C. Darwin, *The Life and Letters of Charles Darwin*, 1:116.

50. C. Darwin, "Letter to Joseph Hooker," July 27, 1861.

51. C. Darwin, "Letter to Joseph Hooker," August 30, 1861.

52. C. Darwin, "Letter to John Lindley," October 18, 1861.

53. C. Darwin, "Letter to A. G. More," June 2, 1861.

54. C. Darwin, "Letter to John Murray," September 24, 1861.

55. Voskuil, "Victorian Orchids and the Forms of Ecological Society."

56. My argument here expands on my previous discussion in Griffiths, "Flattening the World."

57. I am indebted to discussions with botanists Joseph Arditti, Tim Yam, John Elliott, Ken Cameron, and John van Wyhe for their extraordinary help in learning more about the Darwin's work on orchids and the history of their cultivation.

58. "Floriculture."

59. See Reinikka, *A History of the Orchid*.

60. Leifchild, "On the Various Contrivances by Which British and Foreign Orchids Are Fertilised by Insects."

61. C. Darwin, "Letter to John Murray," June 13, 1862.

62. C. Darwin, "Letter to John Murray," September 21, 1861.

63. Kelley, *Clandestine Marriage*.

64. E. Darwin, *The Loves of the Plants*, 2:26.

65. Grosz, *The Nick of Time*, 91. Grosz places particular emphasis on Darwin's *The Descent of Man* (1871).

66. C. Darwin, *On the Various Contrivances by Which British and Foreign Orchids Are Fertilised by Insects*, 45–6. All further references are given parenthetically by page number.

67. My thanks to Arditti, Yam, Elliott, and Cameron for their help in learning more about the *Cycnoches*.

68. J. Smith, *Charles Darwin and Victorian Visual Culture*.

69. Bateman and Cruikshank, *The Orchidaceae of Mexico & Guatemala*.

70. Richard Doyle has made this argument with respect to psychedelic plants. Doyle, *Darwin's Pharmacy*.

71. "Floriculture."

72. Secord, "Darwin and the Breeders: A Social History," 519.

73. Figure 1 appears several pages later, interleaved between pages 18 and 19.

74. E. Darwin, *The Loves of the Plants*, 2:26.

75. I use "intentive" rather than "intentional" in order to emphasize the flexibility of Darwin's treatment of intent in nature. As I will shortly argue, while Darwin felt confident that such apparent intent was an emergent phenomenon of natural selection, he also wanted to create space for those naturalists who ascribed such seeming intent to a creator.

76. C. Darwin, "Letter to John Murray," September 21, 1861.

77. See Ruse, *Darwinism and Its Discontents*.

78. Leifchild, *The Higher Ministry of Nature*, 229–30.

79. See, for instance, Ayala, *Darwin's Gift to Science and Religion*.

80. An exception is Dilley, "Charles Darwin's Use of Theology in the Origin of Species."

81. Kant, *Critique of Pure Reason*, A687/B715–A689/B717.

82. Bruner, *Actual Minds, Possible Worlds*.

83. Recently, Bruce Clarke has argued that the interplay between narratives is "a systematic response to communicative demands crucial to [both] the self-maintenance of social groups" and, also, to the apprehension of organized systems in general. Clarke, *Posthuman Metamorphosis*, 47.

84. "Fertilization of Orchids."

85. "On the Various Contrivances by Which British and Foreign Orchids . . . "

86. Campbell, "On the Supernatural," 392. Further references are given parenthetically by page.

87. C. Darwin, *On the Various Contrivances by Which British and Foreign Orchids Are Fertilised by Insects*, 340.

88. A. Gray, "Botany [Darwin's Fertilisation of Orchids Pt. 1]," 139.

89. A. Gray, "Botany [Darwin's Fertilisation of Orchids Pt. 2]," 428–9.

90. See Dawson, *Darwin, Literature and Victorian Respectability*, 1–3.

91. Ultimately, the *Orchids* challenges the strong thesis of the nineteenth century as an age of secularization and Darwin's assumed place in it. Ongoing scrimmages, like those between the Discovery Institute and Richard Dawkins, line up within the bounds that Darwin chalked out. I suspect he'd be surprised how little either side, when playing near its respective sideline, concerns itself with the substantial common ground of fact.

92. G. Levine, *Darwin Loves You*.

93. C. Taylor, *A Secular Age*.

94. Gray, "Botany [Darwin's Fertilisation of Orchids Pt. 1]," 139.

95. See Turner, *Between Science and Religion*.

96. Ayala, "Teleological Explanations."

97. Barthes, "The Death of the Author," 147.

98. Séan Burke, in his extended study of the death and "return" of the author, argues that Barthes and Foucault never intended to evacuate the author as an important category

of analysis, intending rather, Burke argues, a critique of Edmund Husserl's transcendental ego. As he puts it, "The *death or disappearance* of the author was not at issue but rather the incompatibility of authorial categories with immanent analyses." Burke, *The Death and Return of the Author*, 16.

99. Amigoni, *Cults, Colonies, and Evolution*, 9.

100. Bennett, *Vibrant Matter*, 8.

101. Haraway, *When Species Meet*; Doyle, *Darwin's Pharmacy*; Hall, *Plants as Persons*; Kohn, *How Forests Think*; Nealon, *Plant Theory*.

102. C. Darwin and King-Hele, *The Life of Erasmus Darwin*, 16. All further citations to this edition are given parenthetically by page.

Coda · *Climate Science and the "No-Analog Future"*

1. US EPA, "Future Climate Change."

2. For recent work on climate change in nineteenth-century literature, see Carroll, "Crusades against Frost"; Carroll, "Mary Shelley's Global Atmosphere"; J. Taylor, *The Sky of Our Manufacture*.

3. Williams and Jackson, "Novel Climates, No-Analog Communities, and Ecological Surprises."

4. This phrase was coined by Douglas Fox in his report on the proposal by Williams and Jackson. Fox, "Back to the No-Analog Future?"

Abrams, M. H. *The Mirror and the Lamp: Romantic Theory and the Critical Tradition*. New York: Oxford University Press, 1953.

Adams, Maeve. "'The Force of My Narrative': Persuasion, Nation, and Paratext in Walter Scott's Early Waverley Novels." *ELH* 82, no. 3 (Fall 2015): 937–67.

Agamben, Giorgio. *Homo Sacer*. Translated by Daniel Heller-Roazen. Stanford, Calif.: Stanford University Press, 1998.

———. *The Coming Community*. Minneapolis: University of Minnesota Press, 1993.

Aghion, Irène. "Collecting Antiquities in Eighteenth-Century France." *Journal of the History of Collections* 14, no. 2 (2002): 193–203.

Allewaert, Monique. *Ariel's Ecology: Plantations, Personhood, and Colonialism in the American Tropics*. Minneapolis: University of Minnesota Press, 2013.

Alvarez, David. "'Poetical Cash': Joseph Addison, Antiquarianism, and Aesthetic Value." *Eighteenth-Century Studies* 38, no. 3 (2005): 509–31.

Amigoni, David. *Cults, Colonies, and Evolution: Literature, Science and Culture in Nineteenth-Century Writing*. New York: Cambridge University Press, 2007.

Amigoni, David, and James Elwick. *The Evolutionary Epic*. Vol. 4 of *Victorian Science and Literature*, edited by Gowan Dawson and Bernard Lightman. London: Pickering & Chatto, 2011.

Anderson, Amanda. *The Powers of Distance: Cosmopolitanism and the Cultivation of Detachment*. Princeton, N.J.: Princeton University Press, 2001.

Anderson, Benedict. *Imagined Communities: Reflections on the Origin and Spread of Nationalism*. New York: Verso, 1991.

Anger, Suzy. *Victorian Interpretation*. Ithaca, N.Y.: Cornell University Press, 2005.

Anon. "'The Botanic Garden.'" *Monthly Review, or Literary Journal* 80 (July 1789): 337–43.

———. "'The Botanic Garden.'" *English Review* 14 (July 1789): 1–8.

———. "'The Botanic Garden.'" *Critical Review, or Annals of Literature* 68 (November 1789): 375–81.

———. "The Botanic Garden." *European Magazine and London Review*, no. 27 (February 1795): 74–77.

———. "Darwin's Zoonomia." *British Critic and Quarterly Review* 5 (1795): 113–22.

———. "Sir Walter Scott and William Laidlaw. Concluding Article." *Chamber's Edinburgh Journal* 82 (July 26, 1845): 72–74.

Apter, Emily S. *Against World Literature: On the Politics of Untranslatability*. New York: Verso, 2013.

———. "'Je ne crois pas beaucoup à la littérature comparée': Universal Poetics and Postcolonial Comparatism." In *Comparative Literature in an Age of Globalization*, edited by Haun Saussy, 54–62. Baltimore: Johns Hopkins University Press, 2006.

———. *The Translation Zone: A New Comparative Literature*. Princeton, N.J.: Princeton University Press, 2006.

Aristotle. *The Complete Works of Aristotle: The Revised Oxford Translation*. Edited by Jonathan Barnes. 2 vols. Princeton, N.J.: Princeton University Press, 1995.

———. *Posterior Analytics*. Translated by E. S. Forster and Hugh Tredennick. Cambridge, Mass.: Harvard University Press, 1989.

Armstrong, Isobel. *Victorian Poetry: Poetry, Poetics, and Politics*. London: Routledge, 1993.

Arnold, Matthew. "Literature and Science." *Nineteenth Century*, no. 19 (August 1882): 216–30.

———. *Literature & Dogma*. London: Macmillan, 1873.

Attridge, Derek. *The Singularity of Literature*. London: Routledge, 2004.

Ayala, Francisco José. *Darwin's Gift to Science and Religion*. Washington, D.C.: Joseph Henry Press, 2007.

———. "Teleological Explanations." In *Evolution*, edited by Theodosius Grigorievich Dobzhansky, Francisco José Ayala, and G. Ledyard Stebbins, 497–504. San Francisco, Calif.: Freeman, 1977.

Bacon, Francis. *The New Organon*. Edited by Lisa Jardine and Michael Silverthorne. Cambridge: Cambridge University Press, 2002.

Badiou, Alain. *Manifesto for Philosophy; Followed by Two Essays: "The (Re)turn of Philosophy Itself" and "Definition of Philosophy."* Edited by Norman Madarasz. Albany: State University of New York Press, 1999.

Baker, William. *The Libraries of George Eliot and George Henry Lewes*. English Literary Studies. Victoria, B.C.: University of Victoria, 1981.

Bakhtin, Mikhail Mikhailovich. *The Dialogic Imagination: Four Essays by M. M. Bakhtin*. Edited by Michael Holquist. Translated by Caryl Emerson. Austin: University of Texas Press, 1981.

"Balzac and His Writings." *Westminster Review* 60, no. 117 (July 1853): 199–214.

Barrett, Paul H., and Howard E. Gruber. *Darwin on Man: A Psychological Study of Scientific Creativity*. Edited by Charles Darwin. New York: E. P. Dutton, 1974.

Barthes, Roland. "The Death of the Author." In *Image, Music, Text*, 142–8. New York: Hill and Wang, 1978.

Batchelor, John. *Tennyson: To Strive, to Seek, to Find*. New York: Pegasus Books, 2013.

Bateman, James, and George Cruikshank. *The Orchidaceae of Mexico & Guatemala*. London: Ackermann, 1837.

Beaty, Jerome. *"Middlemarch" from Notebook to Novel: A Study of George Eliot's Creative Method*. Urbana: University of Illinois Press, 1960.

———. "Visions and Revisions: Chapter LXXXI of Middlemarch." *PMLA* 72, no. 4 (1957): 662–79.

Beer, Gillian. *Darwin's Plots: Evolutionary Narrative in Darwin, George Eliot, and Nineteenth-Century Fiction*. Cambridge: Cambridge University Press, 2000.

———. "Darwin's Reading and Fictions of Development." In *The Darwinian Heritage*, edited by David Kohn and Malcolm J. Kottler, 543–88. Princeton, N.J.: Princeton University Press, in association with Nova Pacifica, 1985.

Beiser, Frederick C. *The German Historicist Tradition*. Oxford: Oxford University Press, 2011.

Bender, John B., and David E. Wellbery. *The Ends of Rhetoric: History, Theory, Practice*. Stanford, Calif.: Stanford University Press, 1990.

Benedict, Barbara. *Curiosity: A Cultural History of Early Modern Inquiry*. Chicago: University of Chicago Press, 2001.

Benjamin, Walter. "Theses on the Philosophy of History." In *Illuminations*, edited by Hannah Arendt and translated by Harry Zohn, 253–64. New York: Schocken, 1969.

Bennett, Jane. *Vibrant Matter: A Political Ecology of Things*. Durham, N.C.: Duke University Press, 2010.

Best, Stephen, and Sharon Marcus. "Surface Reading: An Introduction." *Representations* 108, no. 1 (2009): 1–21.

Bewel, Alan. "Erasmus Darwin's Cosmopolitan Nature." *ELH* 76, no. 1 (Spring 2009): 19–48.

Black, Max. *Models and Metaphors: Studies in Language and Philosophy*. Ithaca, N.Y.: Cornell University Press, 1962.

Blanton, C. D. "Theory by Analogy." *PMLA* 130, no. 3 (May 2015): 750–8.

Bohr, Niels. "On the Constitution of Atoms and Molecules. Pt. 1." *Philosophical Magazine*, ser. 6, 26, no. 151 (July 1913): 1–25.

Bopp, Franz. *A Comparative Grammar of the Sanscrit, Zend, Greek, Latin, Lithuanian, Gothic, German and Sclavonic Languages*. Edited by Edward B. Eastwick. London: Madden and Malcolm, 1845.

Bradley, A. C. *A Commentary on Tennyson's "In Memoriam."* 3rd ed. Hamden, Conn.: Archon Books, 1966.

Brassier, Ray. "Concepts and Objects." In *The Speculative Turn: Continental Materialism and Realism*, edited by Levi R. Bryant, Nick Srnicek, and Graham Harman, 47–55. Melbourne: re.press, 2011.

Braudy, Leo. *The Plot of Time: Narrative Form in Hume, Fielding, and Gibbon*. Los Angeles, Calif.: Figueroa Press, 2003.

Brontë, Charlotte. *Shirley*. Oxford: Oxford University Press, 1998.

Bruner, Jerome S. *Actual Minds, Possible Worlds*. Cambridge, Mass.: Harvard University Press, 1986.

———. "The Narrative Construction of Reality." In *Narrative Intelligence*, edited by Michael Mateas and Phoebe Sengers, 41–62. Amsterdam: J. Benjamins, 2003.

Bryant, Levi R., Nick Srnicek, Graham Harman. *The Speculative Turn: Continental Materialism and Realism*. Melbourne: re.press, 2011.

Brylowe, Thora. "Romantic Arts & Letters: British Print, Paint, Engraving, 1760–1830." Ph.D. diss., Carnegie Mellon University, 2009.

Buchan, Peter. *Ancient Ballads and Songs of the North of Scotland*. Edinburgh: Printed for W. & D. Laing, and J. Stevenson, 1828.

Buckland, Adelene. "Losing the Plot: The Geological Anti-Narrative." *19: Interdisciplinary Studies in the Long Nineteenth Century* 11 (2010).

———. *Novel Science: Fiction and the Invention of Nineteenth-Century Geology*. Chicago: University of Chicago Press, 2013.

Budge, Gavin. "Erasmus Darwin and the Poetics of William Wordsworth: 'Excitement without the Application of Gross and Violent Stimulants.'" *Journal for Eighteenth-Century Studies* 30, no. 2 (2007): 279–308.

———. "Introduction: Science and Soul in the Midlands Enlightenment." *Journal for Eighteenth-Century Studies* 30, no. 2 (2007): 157–60.

Bulwer Lytton, Edward. *The Last of the Barons*. New York: Charles Scribner's Sons, 1903.

Burguière, André. *The Annales School: An Intellectual History*. Translated by Jane Marie Todd. Ithaca, N.Y.: Cornell University Press, 2009.

Burke, Séan. *The Death and Return of the Author*. 2nd ed. Edinburgh: Edinburgh University Press, 1998.

Burnett, James (Lord Monboddo). *Of the Origin and Progress of Language.* Vol. 1. Edinburgh: Printed for J. Balfour; London: T. Cadell, 1774.

Burwick, Frederick. "John Boydell's Shakespeare Gallery and the Stage." *Shakespeare-Jahrbuch* 133 (1997): 54–76.

Butterfield, Herbert. *The Whig Interpretation of History.* New York: W. W. Norton, 1965.

Buzard, James. *Disorienting Fiction: The Autoethnographic Work of Nineteenth-Century British Novels.* Princeton, N.J.: Princeton University Press, 2005.

Cameron, Sharon. *Lyric Time: Dickinson and the Limits of Genre.* Baltimore: Johns Hopkins University Press, 1979.

Campbell, Joseph. "On the Supernatural." *Edinburgh Review* 116, no. 236 (October 1862): 378–97.

Canning, G., and J. H. Frere. "Loves of the Triangles." In *Poetry of the-Anti Jacobin*, 4th ed., 125–39. London: J. Wright, 1801.

Capuano, Peter J. "An Objective Aural-Relative in *Middlemarch.*" *Studies in English Literature, 1500–1900* 47, no. 4 (2007): 921–41.

Carignan, Michael. "Analogical Reasoning in Victorian Historical Epistemology." *Journal of the History of Ideas* 64, no. 3 (2003): 445–64.

Carlyle, Thomas. "The Memoirs of the Life of Sir Walter Scott, Baronet." *Westminster Review* 6, no. 2 (January 1838): 293–345.

Carpenter, Mary Wilson. *George Eliot and the Landscape of Time: Narrative Form and Protestant Apocalyptic History.* Chapel Hill: University of North Carolina Press, 1986.

Carroll, Siobhan. "Crusades against Frost: *Frankenstein*, Polar Ice, and Climate Change in 1818." *European Romantic Review* 24, no. 2 (2013): 211–30.

———. "Mary Shelley's Global Atmosphere." *European Romantic Review* 25, no. 1 (February 2014): 3–17.

Casadevall, Arturo, and Ferric C. Fang. "Descriptive Science." *Infection and Immunity* 76, no. 9 (2008): 3835–6.

Chakrabarty, Dipesh. "The Climate of History: Four Theses." *Critical Inquiry* 35, no. 2 (2009): 197–222.

———. *Provincializing Europe: Postcolonial Thought and Historical Difference.* Princeton, N.J.: Princeton University Press, 2000.

———. "Radical Histories and the Question of Enlightenment Rationalism: Some Recent Critiques of Subaltern Studies." In *Mapping Subaltern Studies and the Postcolonial*, edited by Vinayak Chaturvedi, 256–80. London: Verso, 2000.

Chambers, Robert. *The Romantic Scottish Ballads: Their Epoch and Authorship.* Edinburgh: W. and R. Chambers, 1849.

———. *"Vestiges of the Natural History of Creation" and Other Evolutionary Writings.* Edited by James A. Secord. Chicago: University of Chicago Press, 1994.

Chandler, James. *England in 1819: The Politics of Literary Culture and the Case of Romantic Historicism.* Chicago: University of Chicago Press, 1998.

Chandler, James, Arnold I. Davidson, and Harry D. Harootunian. *Questions of Evidence: Proof, Practice, and Persuasion across the Disciplines.* Chicago: University of Chicago Press, 1994.

Child, Francis James. *The English and Scottish Popular Ballads.* 5 vols. Boston: Houghton, Mifflin, 1882.

Christensen, Jerome. *Romanticism at the End of History.* Baltimore: Johns Hopkins University Press, 2000.

Clapp-Itnyre, Alisa. *Angelic Airs, Subversive Songs: Music as Social Discourse in the Victorian Novel.* Athens: Ohio University Press, 2002.

Clarke, Bruce. *Posthuman Metamorphosis: Narrative and Systems.* New York: Fordham University Press, 2008.

Cleland, Carol E. "Historical Science, Experimental Science, and the Scientific Method." *Geology* 29 (2001): 987–90.

Clyne, Norval. *The Romantic Scottish Ballads and the Lady Wardlaw Heresy.* Norwood, Pa.: Norwood Editions, 1974.

Cohen, Michael. *The Social Lives of Poems in Nineteenth-Century America.* Philadelphia: University of Pennsylvania Press, 2015.

Cole, Andrew. *The Birth of Theory.* Chicago: University of Chicago Press, 2014.

Coleridge, Samuel Taylor. *Biographia Literaria, Chapters I–IV, XIV–XXII.* Edited by George Sampson and Arthur Quiller-Couch. Cambridge: University Press, 1920.

———. "Letter to Robert Southey." In *Collected Letters of Samuel Taylor Coleridge,* edited by Earl Leslie Griggs, vol. 2. Oxford: Clarendon Press, 1956.

———. *Letters of Samuel Taylor Coleridge.* Edited by Ernest Hartley Coleridge. 2 vols. Boston: Houghton, Mifflin, 1895.

Collingwood, R. G. *The Idea of History.* Edited by T. M. Knox. Mansfield Centre, Conn.: Martino Publishing, 2014.

Collins, K. K. "Questions of Method: Some Unpublished Late Essays." *Nineteenth-Century Fiction* 35, no. 3 (1980): 385–405.

"Contemporary Literature." *Westminster Review* 61, no. 120 (April 1854): 564.

Copeland, Rita. *Rhetoric, Hermeneutics, and Translation in the Middle Ages: Academic Traditions and Vernacular Texts.* Cambridge: Cambridge University Press, 1991.

Costelloe, Timothy M. "Fact and Fiction: Memory and Imagination in Hume's Approach to History." In *David Hume: Historical Thinker, Historical Writer,* edited by Mark G. Spencer, 181–200. University Park: Pennsylvania State University Press, 2013.

Craft, Christopher. "'Descend, Touch, and Enter': Tennyson's Strange Manner of Address." *Genders* 1 (1988): 83–101.

Crawford, Robert. *Devolving English Literature.* New York: Clarendon Press, 1992.

———. "Walter Scott and European Union." *Studies in Romanticism* 40, no. 1 (2001): 137–52.

Croker, John Wilson. "The Antiquary." *Quarterly Review* 40 (April 1816): 125–39.

Crossgrove, William. "The Vernacularization of Science, Medicine, and Technology in Late Medieval Europe: Broadening Our Perspectives." *Early Science and Medicine* 5, no. 1 (2000): 47–63.

Crowe, Michael J. *The Extraterrestrial Life Debate, 1750–1900: The Idea of a Plurality of Worlds from Kant to Lowell.* Cambridge: Cambridge University Press, 1986.

Culler, A. Dwight. *The Poetry of Tennyson.* New Haven: Yale University Press, 1977.

———. *The Victorian Mirror of History.* New Haven: Yale University Press, 1985.

Culler, Jonathan D. *The Pursuit of Signs: Semiotics, Literature, Deconstruction.* Ithaca, N.Y.: Cornell University Press, 1981.

Curran, Stuart. "'Adonais' in Context." In *Shelley Revalued: Essays from the Gregynog Conference,* edited by Kelvin Everest, 165–82. Totowa, N.J.: Barnes & Noble, 1983.

Cuvier, Georges, and M. J. S. Rudwick. *Georges Cuvier, Fossil Bones, and Geological Catas-trophes.* Translated by M. J. S. Rudwick. Chicago: University of Chicago Press, 1997.

Dames, Nicholas. *The Physiology of the Novel: Reading, Neural Science, and the Form of Victorian Fiction.* Oxford: Oxford University Press, 2007.

Darnton, Robert. "What Is the History of Books?" *Daedalus* 111, no. 3 (1982): 65–83.

Darrigol, Olivier. *A History of Optics: From Greek Antiquity to the Nineteenth Century.* Oxford: Oxford University Press, 2012.

Darwin, Charles. *The Autobiography of Charles Darwin, 1809–1882. With Original Omis-sions Restored.* Edited by Nora Barlow. New York: Harcourt, Brace, 1959.

———. "Letter to A. G. More." MS, June 2, 1861. Letter 3174. Darwin Correspondence Project. http://www.darwinproject.ac.uk/entry-3174.

———. "Letter to John Lindley." MS, October 18, 1861. Letter 3289. Darwin Correspon-dence Project. http://www.darwinproject.ac.uk/entry-3289.

———. "Letter to John Murray." MS, March 31, 1859. Letter 2441. Darwin Correspondence Project. http://www.darwinproject.ac.uk/entry-2441.

———. "Letter to John Murray." MS, September 21, 1861. Letter 3259. Darwin Corre-spondence Project. http://www.darwinproject.ac.uk/entry-3259.

———. "Letter to John Murray." MS, September 24, 1861. Letter 3264. Darwin Correspon-dence Project. http://www.darwinproject.ac.uk/entry-3264.

———. "Letter to John Murray." MS, June 13, 1862. Letter 3602. Darwin Correspondence Project. http://www.darwinproject.ac.uk/entry-3602.

———. "Letter to Joseph Hooker." MS, July 27, 1861. Letter 3220. Darwin Correspon-dence Project. http://www.darwinproject.ac.uk/entry-3220.

———. "Letter to Joseph Hooker." MS, August 30, 1861. Letter 3238. Darwin Correspon-dence Project. http://www.darwinproject.ac.uk/letter/entry-3238.

———. "Letter to Joseph Hooker," MS, August 1, 1857. Letter 2130. Darwin Correspon-dence Project. http://www.darwinproject.ac.uk/entry-2130.

———. *The Life and Letters of Charles Darwin.* 3 vols. London: John Murray, 1887.

———. "Notebook B: [Transmutation of Species (1837–1838)]," August 1837. DAR 121. Cambridge University Library. In *The Complete Work of Charles Darwin Online,* ed-ited by John Van Wyhe. 2002–. http://darwin-online.org.uk/.

———. "Notebook C: [Transmutation of Species (1838.02–1838.07)]," 1838. DAR 122. Cambridge University Library. In *The Complete Work of Charles Darwin Online,* edi-ted by John Van Wyhe. 2002–. http://darwin-online.org.uk/.

———. "Notebook E: [Transmutation of Species (1838–1839)]," September 1838. DAR 124. Cambridge University Library. In *The Complete Work of Charles Darwin Online,* edited by John Van Wyhe. 2002–. http://darwin-online.org.uk/.

———. "Notebook M: [Metaphysics on morals and speculations on expression (1838)]," 1838. Cambridge University Library. In *The Complete Work of Charles Darwin Online,* edited by John Van Wyhe. 2002–. http://darwin-online.org.uk/.

———. *On the Origin of Species by Means of Natural Selection.* 1st ed. London: John Mur-ray, 1859. In *The Complete Work of Charles Darwin Online,* edited by John Van Wyhe. 2002–. http://darwin-online.org.uk/.

———. *On the Origin of Species by Means of Natural Selection.* 3rd ed. London: John Mur-ray, 1861. In *The Complete Work of Charles Darwin Online,* edited by John Van Wyhe. 2002–. http://darwin-online.org.uk/.

————. *On the Various Contrivances by Which British and Foreign Orchids Are Fertilised by Insects, and on the Good Effects of Intercrossing.* London: John Murray, 1862. In *The Complete Work of Charles Darwin Online*, edited by John Van Wyhe. 2002–. http://darwin-online.org.uk/.

————. *The Variation of Plants and Animals under Domestication.* 2 vols. London: John Murray, 1868. In *The Complete Work of Charles Darwin Online*, edited by John Van Wyhe. 2002–. http://darwin-online.org.uk/.

Darwin, Charles, and Desmond King-Hele. *Charles Darwin's "The Life of Erasmus Darwin."* Cambridge: Cambridge University Press, 2003.

Darwin, Erasmus. *The Collected Writings of Erasmus Darwin.* Edited by Martin Priestman. 9 vols. Bristol: Thoemmes Continuum, 2004.

————. "Letter to Joseph Johnson." May 23, 1784. In *The Collected Letters of Erasmus Darwin*, edited by Desmond King-Hele, 287. Cambridge: Cambridge University Press, 2007.

————. *The Loves of the Plants.* 2nd ed. 2 vols. London: J. Johnson, 1791.

Darwin, Erasmus, and Stuart Harris. *Cosmologia.* Sheffield: S. Harris, 2002.

da Sousa Correa, Delia. *George Eliot, Music and Victorian Culture.* Basingstoke: Palgrave Macmillan, 2003.

Daston, Lorraine. "Historical Epistemology." In *Questions of Evidence: Proof, Practice, and Persuasion across the Disciplines*, edited by James Chandler, Arnold I. Davidson, and Harry D. Harootunian, 282–89. Chicago: University of Chicago Press, 1994.

Daston, Lorraine, and Peter Galison. *Objectivity.* New York: Zone Books, 2007.

Daston, Lorraine, and Katherine Park. *Wonders and the Order of Nature, 1150–1750.* New York: Zone Books, 1998.

Davis, Michael T. *George Eliot and Nineteenth-Century Psychology: Exploring the Unmapped Country.* Aldershot: Ashgate, 2006.

————. *London Corresponding Society, 1792–1799.* London: Pickering & Chatto, 2002.

Dawson, Gowan. *Darwin, Literature and Victorian Respectability.* Cambridge: Cambridge University Press, 2007.

————. "Literature." In *The Cambridge Encyclopedia of Darwin and Evolutionary Thought*, edited by Michael Ruse. Cambridge: Cambridge University Press, 2013.

Dawson, Gowan, and Bernard Lightman. *Victorian Science and Literature.* 4 vols. London: Pickering & Chatto, 2010.

DeLanda, Manuel. *Intensive Science and Virtual Philosophy.* London: Continuum, 2002.

de Man, Paul. "The Rhetoric of Temporality." In *Interpretation: Theory and Practice*, edited by Charles S. Singleton, 173–209. Baltimore: Johns Hopkins Press, 1969.

DeWitt, Anne. *Moral Authority, Men of Science, and the Victorian Novel.* Cambridge: Cambridge University Press, 2013.

Dickens, Charles. *A Tale of Two Cities.* Edited by Richard Maxwell. London: Penguin Books, 2003.

Dillane, Fionnuala. *Before George Eliot: Marian Evans and the Periodical Press.* Cambridge: Cambridge University Press, 2013.

————. "Re-reading George Eliot's 'Natural History': Marian Evans, 'the People,' and the Periodical." *Victorian Periodicals Review* 42, no. 3 (2009): 244–66.

Dilley, Stephen. "Charles Darwin's Use of Theology in the *Origin of Species.*" *British Journal for the History of Science* 45 (2012): 29–56.

D'Israeli, Isaac. *Curiosities of Literature.* 2 vols. London: H. Murray, 1794.

———. *Curiosities of Literature.* 7th ed. 5 vols. London: J. Murray, 1823.

Doyle, Mary Ellen. *The Sympathetic Response: George Eliot's Fictional Rhetoric.* Rutherford, N.J.: Fairleigh Dickinson University Press, 1981.

Doyle, Richard. *Darwin's Pharmacy: Sex, Plants, and the Evolution of the Noösphere.* Seattle: University of Washington Press, 2011.

Du Bois, W. E. B. "Races." *The Crisis* 4, no. 2 (August 1911): 157–8.

———. *The Souls of Black Folk: Essays and Sketches.* Chicago: A. C. McClurg, 1903.

Duff, David. *Romanticism and the Uses of Genre.* Oxford: Oxford University Press, 2009.

Duncan, Ian. "George Eliot's Science Fiction." *Representations* 125 (2014): 15–39.

———. *Modern Romance and Transformations of the Novel: The Gothic, Scott, Dickens.* Cambridge: Cambridge University Press, 1992.

———. *Scott's Shadow: The Novel in Romantic Edinburgh.* Princeton, N.J.: Princeton University Press, 2007.

Dutilh Novaes, Catarina. "The Different Ways in Which Logic Is (Said to Be) Formal." *History and Philosophy of Logic* 32, no. 4 (2011): 303–32.

———. *Formal Languages in Logic: A Philosophical and Cognitive Analysis.* Cambridge: Cambridge University Press, 2012.

Edmundson, Henry. *Lingua Linguarum, The Natural Language of Languages in a Vocabulary . . . Contrived and Built upon Analogy, a Designe Further Improvable, and Applicable to the Gaining of any Language.* London: n.p., 1658.

Eliot, George. *Adam Bede.* Edited by Margaret Reynolds. London: Penguin, 2008.

———. *Daniel Deronda.* Edited by Terence Cave. London: Penguin Books, 1995.

———. *Essays of George Eliot.* Edited by Thomas Pinney. New York: Columbia University Press, 1963.

———. *Felix Holt, the Radical.* Edited by Lynda Mugglestone. London: Penguin Books, 1995.

———. *George Eliot's Life: As Related in Her Letters and Journals.* Edited by John W. Cross. 3 vols. Edinburgh: Blackwood, 1885.

———. *The Impressions of Theophrastus Such.* Edited by D. J. Enright. London: Everyman, J. M. Dent, 1995.

———. "Journal." MS, 61 1854. George Eliot and George Henry Lewes Collection. Beinecke Library, Yale University.

———. *The Legend of Jubal and Other Poems.* Edinburgh: Blackwood, 1874.

———. *Middlemarch.* Edited by W. J. Harvey. London: Penguin Books, 1994.

———. *Poems, Essays, and Leaves from a Notebook.* New York: Doubleday, Page, 1904.

———. *The Mill on the Floss.* Edited by Oliver Lovesey. Peterborough, Ont.: Broadview Press, 2007.

———. "The Natural History of German Life." In *The Essays of "George Eliot,"* edited by Nathan Sheppard, 141–77. New York: Funk & Wagnalls, 1883.

Endersby, Jim. *Imperial Nature: Joseph Hooker and the Practices of Victorian Science.* Chicago: University of Chicago Press, 2008.

————. "Lumpers and Splitters: Darwin, Hooker, and the Search for Order." *Science* 326, no. 5959 (2009): 1496–9.

Engelstein, Stefani. "The Allure of Wholeness: The Eighteenth-Century Organism and the Same-Sex Marriage Debate." *Critical Inquiry* 39, no. 4 (2013): 754–76.

Ermarth, Elizabeth Deeds. "George Eliot's Conception of Sympathy." *Nineteenth-Century Fiction* 40, no. 1 (1985): 23–42.

"An Essay on Translation." *Gentleman's Magazine* 41 (August 1771): 349–50.

Febvre, Lucien. "Albert Mathiez: Un tempérament, une éducation." *Annales d'histoire économique et Sociale* 18, no. 4 (1932): 573–6.

Felski, Rita, and Susan Stanford Friedman. Introduction to *Comparison: Theories, Approaches, Uses*, edited by Rita Felski and Susan Stanford Friedman, 1–12. Baltimore: Johns Hopkins University Press, 2013.

Ferguson, Adam. *An Essay on the History of Civil Society*. 6th ed. London: T. Cadell, 1793.

Ferris, Ina. *The Achievement of Literary Authority: Gender, History, and the Waverley Novels*. Ithaca, N.Y.: Cornell University Press, 1991.

————. "Bibliographic Romance: Bibliophilia and the Book-Object." *Romantic Circles*, February 1, 2004. http://www.rc.umd.edu/praxis/libraries/ferris/ferris.html.

"Fertilization of Orchids." *London Review and Weekly Journal of Politics, Literature, Art and Society* 4, no. 102 (June 14, 1862): 553–4.

Feuerbach, Ludwig. *The Essence of Christianity*. 2nd ed. Amherst, Mass.: Prometheus Books, 1989.

Fielding, Penny. *Writing and Orality: Nationality, Culture, and Nineteenth-Century Scottish Fiction*. Oxford: Oxford University Press, 1996.

Flint, Kate. "The Materiality of *Middlemarch*." In *Middlemarch in the Twenty-First Century*, edited by Karen Chase, 65–86. Oxford: Oxford University Press, 2006.

"Floriculture." *London Quarterly Review* 24 (1865): 50–79.

Flynn, Philip. "Hallam and Tennyson: The 'Theodicaea Novissima' and *In Memoriam*." *Studies in English Literature, 1500–1900* 19, no. 4 (Autumn 1979): 705–20.

Flynn, Thomas R. *Sartre, Foucault, and Historical Reason*. 2 vols. Chicago: University of Chicago Press, 2005.

Fodor, Jerry A. "Against Darwinism." *Mind and Language* 23, no. 1 (2008): 1–24.

————. "Special Sciences (or: The Disunity of Science as a Working Hypothesis)." *Synthese: An International Journal for Epistemology, Methodology and Philosophy of Science* 28, no. 2 (1974): 97–115.

Foucault, Michel. *The Archaeology of Knowledge; and, The Discourse on Language*. New York: Pantheon, 1982.

————. *The Order of Things: An Archaeology of the Human Sciences*. New York: Pantheon Books, 1971.

————. "What Is an Author?" In *Language, Counter-Memory, Practice*, 113–38. Ithaca, N.Y.: Cornell University Press, 1980.

Fox, Douglas. "Back to the No-Analog Future?" *Science* 316, no. 5826 (May 2007): 823–5.

French, Robert, and Douglas Hofstadter. "Tabletop: An Emergent, Stochastic Model of Analogy-Making." In *Proceedings of the Thirteenth Annual Conference of the Cognitive*

Science Society, edited by Kristian J. Hammond and Dedre Gentner, 708–13. Hillsdale, N.J.: LEA, 1991.

Frye, Northrop. *Anatomy of Criticism*. Princeton, N.J.: Princeton University Press, 1957.

Fulford, Tim. "Darwin in the Air." Paper Presented at "The Darwins Reconsidered," University of Roehampton, London, September 4, 2015.

Furet, François. *In the Workshop of History*. Translated by Jonathan Mandelbaum. Chicago: University of Chicago Press, 1984.

Furniss, Tom. "A Romantic Geology: James Hutton's 1788 'Theory of the Earth.'" *Romanticism* 16, no. 3 (Fall 2010): 305–21.

"The Future of Geology." *Westminster Review* 58, no. 113 (July 1852): 67–94.

Galison, Peter. *Image and Logic: A Material Culture of Microphysics*. Chicago: University of Chicago Press, 1997.

Gallagher, Catherine. "Formalism and Time." *Modern Language Quarterly* 61, no. 1 (2000): 229–51.

———. "George Eliot: Immanent Victorian." *Representations* 90, no. 1 (2005): 61–74.

———. *Nobody's Story: Vanishing Acts of Women Writers in the Marketplace, 1670–1820*. Oxford: Oxford University Press, 1995.

Gallagher, Catherine, and Stephen Greenblatt. *Practicing New Historicism*. Chicago: University of Chicago Press, 2000.

Gamer, Michael. "Authors in Effect: Lewis, Scott, and the Gothic Drama." *ELH* 66, no. 4 (1999): 831–61.

———. *Romanticism and the Gothic: Genre, Reception, and Canon Formation*. Cambridge: Cambridge University Press, 2000.

Garside, Peter D. "Scott and the 'Philosophical' Historians." *Journal of the History of Ideas* 36, no. 3 (September 1975): 497–512.

Gates, Sarah. "Poetics, Metaphysics, Genre: The Stanza Form of *In Memoriam*." *Victorian Poetry* 37, no. 4 (1999): 507–20.

Gentner, Dedre. "Structure-Mapping: A Theoretical Framework for Analogy." *Cognitive Science* 7 (1983): 155–70.

Gernsheim, Helmut, and Alison Gernsheim. *The History of Photography from the Earliest Use of the Camera Obscura in the Eleventh Century up to 1914*. London: Oxford University Press, 1955.

Gibbon, Edward. *The History of the Decline and Fall of the Roman Empire*. 6 vols. London: Strahan & Cadell, 1776.

Gieryn, Thomas F. "Boundary-Work and the Demarcation of Science from Non-Science: Strains and Interests in Professional Ideologies of Scientists." *American Sociological Review* 48 (December 1983): 781–95.

Gigante, Denise. "Forming Desire: On the Eponymous *In Memoriam* Stanza." *Nineteenth-Century Literature* 53, no. 4 (1999): 480–504.

Gilmartin, Kevin. *Writing against Revolution: Literary Conservatism in Britain, 1790–1832*. Cambridge: Cambridge University Press, 2007.

Gilmore, Dehn. *The Victorian Novel and the Space of Art: Fictional Form on Display*. New York: Cambridge University, 2013.

Glissant, Édouard. *The Poetics of Relation*. Translated by Betsy Wing. Ann Arbor: University of Michigan, 1997.

"Goethe as a Man of Science." *Westminster Review* 58, no. 114 (October 1852): 479–506.

Goethe, Johann Wolfgang von. *The Metamorphosis of Plants.* Edited by Gordon L. Miller. Cambridge, Mass.: MIT Press, 2009.

Goodlad, Lauren. *The Victorian Geopolitical Aesthetic: Realism, Sovereignty, and Transnational Experience.* Oxford: Oxford University Press, 2015.

Goslee, David. *Tennyson's Characters: Strange Faces, Other Minds.* Iowa City: University of Iowa Press, 1989.

Grady, Hugh. *Shakespeare, Machiavelli, and Montaigne: Power and Subjectivity from "Richard II" to "Hamlet."* Oxford: Oxford University Press, 2002.

"A Grammar of the Sanskrita Language." *Edinburgh Review* 13, no. 26 (January 1809): 366–81.

Gray, Asa. "Botany [Darwin's Fertilisation of Orchids Pt. 1]." *American Journal of Science and Arts* 34 (July 1862): 138–44.

———. "Botany [Darwin's Fertilisation of Orchids Pt. 2]." *American Journal of Science and Arts* 34 (November 1862): 420–9.

Gray, Beryl. *George Eliot and Music.* New York: St. Martin's, 1989.

Gray, Erik Irving. *The Poetry of Indifference: From the Romantics to the Rubáiyát.* Amherst: University of Massachusetts Press, 2005.

Greenblatt, Stephen. *Renaissance Self-Fashioning: From More to Shakespeare.* Chicago: University of Chicago Press, 1980.

Greene, Thomas M. *The Light in Troy: Imitation and Discovery in Renaissance Poetry.* New Haven: Yale University Press, 1982.

Greiner, Rae. *Sympathetic Realism in Nineteenth-Century British Fiction.* Baltimore: Johns Hopkins University Press, 2012.

Griffiths, Devin. "The Comparative History of *A Tale of Two Cities.*" *ELH* 80, no. 3 (2013): 811–38.

———. "Flattening the World: Natural Theology and the Ecology of Darwin's *Orchids.*" *Nineteenth-Century Contexts* 37, no. 5 (2015): 431–52.

———. "The Intuitions of Analogy in Erasmus Darwin's Poetics." *Studies in English Literature, 1500–1900* 51, no. 3 (2011): 645–65.

———. "The Radical's Catalogue: Antonio Panizzi, Virginia Woolf, and the British Museum Library's *Catalogue of Printed Books.*" *Book History* 18 (2015): 134–65.

Grimaldi, David A., and Michael S. Engel. "Why Descriptive Science Still Matters." *BioScience* 57, no. 8 (2007): 646.

Groom, Nick. *Thomas Chatterton and Romantic Culture.* New York: St. Martin's, 1999.

Grosz, Elizabeth A. *The Nick of Time: Politics, Evolution, and the Untimely.* Durham, N.C.: Duke University Press, 2004.

Guillory, John. "Enlightening Mediation." In *This Is Enlightenment,* edited by Clifford Siskin and William Warner, 37–63. Chicago: University of Chicago Press, 2010.

Hacking, Ian. "Statistical Language, Statistical Truth and Statistical Reason: The Self-Authentication of a Style of Scientific Reasoning." In *The Social Dimensions of Science,* edited by Ernan McMullin, 130–57. Notre Dame, Ind.: University of Notre Dame Press, 1992.

———. "Styles of Scientific Reasoning." In *Post-Analytic Philosophy,* edited by John Rajchman and Cornel West, 145–65. New York: Columbia University Press, 1985.

Haight, Gordon Sherman. *George Eliot: A Biography.* New York: Oxford University Press, 1968.

Hall, Catherine. "Macaulay's *History of England:* A Book That Shaped Nation and Empire." In *Ten Books That Shaped the British Empire: Creating an Imperial Commons,* edited by Antoinette M. Burton and Isabel Hofmeyr, 71–89. Durham, N.C.: Duke University Press, 2014.

Hall, Matthew. *Plants as Persons: A Philosophical Botany.* Albany: State University of New York Press, 2011.

Hallam, Arthur Henry. *The Letters of Arthur Henry Hallam.* Edited by Jack Kolb. Columbus: Ohio State University Press, 1981.

———. "The Notebook of Arthur Henry Hallam," 1830. Add MS 74090 A. British Library.

———. "The Papers of Arthur Henry Hallam," 33 1816. Add MS 81296. British Library.

———. *Remains, in Verse and Prose, of Arthur Henry Hallam.* Edited by Joseph W. Overbury. London: W. Nicol, 1834.

Haraway, Donna Jeanne. *When Species Meet.* Minneapolis: University of Minnesota Press, 2010.

Hardy, Barbara Nathan. *The Novels of George Eliot: A Study in Form.* London: Athlone Press, 1994.

———. *Particularities: Readings in George Eliot.* Athens: Ohio University Press, 1983.

Harman, Graham. *Tool-Being: Heidegger and the Metaphysics of Objects.* Chicago: Open Court, 2002.

———. "The Well-Wrought Broken Hammer: Object-Oriented Literary Criticism." *New Literary History* 43, no. 2 (2012): 183–203.

Harris, Stuart. *Erasmus Darwin's Enlightenment Epic: A Study of the Evidence for Sequential Design in "The Botanic Garden" [1791] and "The Temple of Nature" [1803].* Sheffield: S. Harris, 2002.

Harris, Victor. "Allegory to Analogy in the Interpretation of Scripture." *Philological Quarterly* 45 (1966): 1–23.

Hawkings, Ann R. "Reconstructing the Boydell Shakespeare Gallery." In *Shakespeare and the Culture of Romanticism,* edited by Joseph M. Ortiz, 207–29. Farnham, U.K.: Ashgate, 2013.

Haywood, Ian. *Faking It: Art and the Politics of Forgery.* New York: St. Martin's Press, 1987.

———. *The Making of History.* Rutherford, N.J.: Fairleigh Dickinson University Press, 1986.

Hearne, Thomas. *Ductor Historicus; or, a Short System of Universal History, and an Introduction to the Study of It.* 2nd ed. 2 vols. London: Tim. Childe, 1704.

Henchman, Anna. " 'The Globe We Groan in': Astronomical Distance and Stellar Decay in *In Memoriam.*" *Victorian Poetry* 41, no. 1 (2003): 29–45.

———. *The Starry Sky Within: Astronomy and the Reach of the Mind in Victorian Literature.* New York: Oxford University Press, 2014.

Herbert, Christopher. *Culture and Anomie: Ethnographic Imagination in the Nineteenth Century.* Chicago: University of Chicago Press, 1991.

———. *Victorian Relativity: Radical Thought and Scientific Discovery.* Chicago: University of Chicago Press, 2001.

Herd, David. *Ancient and Modern Scottish Songs, Heroic Ballads, Etc. Collected from Memory, Tradition and Ancient Authors.* 2nd ed. 2 vols. Edinburgh: J. Dickson and C. Elliot, 1776.

Herdt, Jennifer. "Artificial Lives, Providential History, and the Apparent Limits of Sympathetic Understanding." In *David Hume: Historical Thinker, Historical Writer,* edited by Mark G Spencer, 37–60. University Park: Pennsylvania State University Press, 2013.

Heringman, Noah. *Sciences of Antiquity: Romantic Antiquarianism, Natural History, and Knowledge Work.* Oxford: Oxford University Press, 2013.

Herschel, William. "Catalogue of a Second Thousand of New Nebulae and Clusters of Stars; With a Few Introductory Remarks on the Construction of the Heavens." *Philosophical Transactions of the Royal Society of London* 5, no. 79 (1789): 212–55.

———. *A Treatise on Astronomy.* London: Longman, Reese, Orme, Brown, Green & Longman, 1833.

Hertz, Neil. *George Eliot's Pulse.* Stanford, Calif.: Stanford University Press, 2003.

Hill, Susan E. "Translating Feuerbach, Constructing Morality: The Theological and Literary Significance of Translation for George Eliot." *Journal of the American Academy of Religion* 65, no. 3 (Autumn 1997): 635.

Hoffman, Jesse. "Arthur Hallam's Spirit Photograph and Tennyson's Elegiac Trace." *Victorian Literature and Culture* 42, no. 4 (2014): 611–36.

Hofstadter, Douglas. *Fluid Concepts and Creative Analogies: Computer Models of the Fundamental Mechanisms of Thought.* New York: Basic Books, 1995.

Hofstadter, Douglas, and Melanie Mitchell. "The Copycat Project: A Model of Fluid Concepts and Analogy-Making." *Technical Report of the Center for Research on Concepts and Cognition.* Bloomington: Indiana University, 1991.

Hollander, Elizabeth. "Ariadne and the Rippled Nose: Portrait Likeness in *Middlemarch*." *Victorian Literature and Culture* 34, no. 1 (2006): 167–87.

Holmes, Richard. *The Age of Wonder: How the Romantic Generation Discovered the Beauty and Terror of Science.* New York: Pantheon Books, 2008.

Holyoak, Keith J., and Paul Thagard. "Analogical Mapping by Constraint Satisfaction." *Cognitive Science* 13 (1989): 295–355.

Hoselitz, Virginia. *Imagining Roman Britain: Victorian Reponses to a Roman Past.* Woodbridge, U.K.: Royal Historical Society / The Boydell Press, 2007.

Hughes, Linda K. "Constructing Fictions of Authorship in George Eliot's *Middlemarch*, 1871–1872." *Victorian Periodicals Review* 38, no. 2 (2005): 158–79.

Hume, David. *Essays and Treatises on Several Subjects.* London, 1758.

———. *The History of England: From the Invasion of Julius Caesar to the Revolution in 1688.* New ed. 8 vols. London: Cadell, 1770.

———. *The Life of David Hume, Esq. Written by Himself.* London: Strahan, 1777.

Hutcheon, Linda. *A Poetics of Postmodernism: History, Theory, Fiction.* New York: Routledge, 1988.

Hutton, James. "Theory of the Earth." *Transactions of the Royal Society of Edinburgh* 1, no. 2 (1788): 209–304.

Huxley, Aldous. *Literature and Science.* New York: Harper & Row, 1963.

Huxley, Thomas Henry. *Evolution and Ethics.* London: Macmillan, 1893.

———. *On Science and Art in Relation to Education.* Liverpool: n.p., 1883.

————. *Science and Culture*. London: Macmillan, 1881.

Hyman, Stanley Edgar. *The Tangled Bank: Darwin, Marx, Frazer and Freud as Imaginative Writers*. New York: Atheneum, 1962.

"Iconoclasm in Philosophy." *Westminster Review* 59, no. 116 (April 1853): 388–407.

Jackson, Noel. "Rhyme and Reason: Erasmus Darwin's Romanticism." *Modern Language Quarterly* 70, no. 2 (2009): 171.

Jager, Colin. *The Book of God: Secularization and Design in the Romantic Era*. Philadelphia: University of Pennsylvania Press, 2007.

James, Henry. "Daniel Deronda: A Conversation." *Atlantic Monthly* 38, no. 130 (December 1876): 684–94.

Jameson, Fredric. *Marxism and Form; Twentieth-Century Dialectical Theories of Literature*. Princeton, N.J.: Princeton University Press, 1972.

Jauss, Hans Robert. *Toward an Aesthetic of Reception*. Translated by Timothy Bahti. Minneapolis: University of Minnesota Press, 1982.

Jobert, Louis. *The Knowledge of Medals; or, Instructions for Those Who Apply Themselves to the Study of Medals Both Ancient and Modern*. London: T. Caldecott, 1715.

Johns, Adrian. *The Nature of the Book: Print and Knowledge in the Making*. Chicago: University of Chicago Press, 2008.

Joseph, Gerhard. *Tennysonian Love: The Strange Diagonal*. Minneapolis: University of Minnesota Press, 1969.

Kant, Immanuel. *Critique of Pure Reason*. Translated by Paul Guyer and Allen W. Wood. Cambridge: Cambridge University Press, 1998.

Karl, Frederick R. *George Eliot, Voice of a Century: A Biography*. New York: W. W. Norton, 1995.

Kelley, Theresa. *Clandestine Marriage: Botany and Romantic Culture*. Baltimore: Johns Hopkins University Press, 2012.

Kidd, Colin. "The Ideological Significance of the *History of Scotland*." In *William Robertson and the Expansion of Empire*, edited by Stewart J. Brown, 122–31. Cambridge: Cambridge University Press, 1997.

Kincaid, James R. *Tennyson's Major Poems: The Comic and Ironic Patterns*. New Haven: Yale University Press, 1975.

Kingsley, Charles. *Literary and General Lectures and Essays*. Vol. 20 of *The Works of Charles Kingsley*. London: Macmillan, 1880.

Klancher, Jon P. *Transfiguring the Arts and Sciences: Knowledge and Cultural Institutions in the Romantic Age*. New York: Cambridge University Press, 2013.

Koerner, Lisbet. "Goethe's Botany: Lessons of a Feminine Science." *Isis* 84, no. 3 (1993): 470–95.

Kohn, Eduardo. *How Forests Think: Toward an Anthropology beyond the Human*. Berkeley: University of California Press, 2013.

Kuhn, Thomas S. *The Structure of Scientific Revolutions*. 3rd ed. Chicago: University of Chicago Press, 1996.

Kurnick, David. *Empty Houses: Theatrical Failure and the Novel*. Princeton, N.J.: Princeton University Press, 2012.

————. "An Erotics of Detachment: *Middlemarch* and Novel-Reading as Critical Practice." *ELH* 74, no. 3 (2007): 583–608.

LaFleur, Greta L. "Precipitous Sensations: Herman Mann's *The Female Review* (1797), Botanical Sexuality, and the Challenge of Queer Historiography." *EAL: Early American Literature* 48, no. 1 (2013): 93–123.

Lamarck, Jean-Baptiste Pierre Antoine de Monet de. *Zoological Philosophy.* Translated by Hugh Samuel Roger Elliott. London: Macmillan, 1914.

Langan, Celeste, and Maureen N. McLane. "The Medium of Romantic Poetry." In *The Cambridge Companion to British Romantic Poetry,* edited by James Chandler and Maureen N. McLane, 239–62. Cambridge: Cambridge University Press, 2008.

Latour, Bruno. *Pandora's Hope: Essays on the Reality of Science Studies.* Cambridge, Mass.: Harvard University Press, 1999.

———. *Science in Action: How to Follow Scientists and Engineers through Society.* Cambridge, Mass.: Harvard University Press, 1987.

Leavis, F. R. *The Great Tradition: George Eliot, Henry James, Joseph Conrad.* New York: New York University Press, 1963.

Leeuwenhoek, Antonie. "Observationes D. Anthonii Lewenhoeck, de Natis è Semine Denitali Animaculis." *Philosophical Transactions of the Royal Society of London* 12, nos. 133–42 (1677): 1040–6.

Leifchild, John R. *The Higher Ministry of Nature: Viewed in the Light of Modern Science, and as an Aid to Advanced Christian Philosophy.* London: Hodder & Stoughton, 1872.

———. "On the Various Contrivances by Which British and Foreign Orchids Are Fertlised by Insects." *The Athenaeum,* May 24, 1862.

Levine, Carolyn. *Forms: Whole, Rhythm, Hierarchy, Network.* Princeton, N.J.: Princeton University Press, 2015.

Levine, George Lewis. *Darwin and the Novelists: Patterns of Science in Victorian Fiction.* Chicago: University of Chicago Press, 1988.

———. *Darwin Loves You: Natural Selection and the Re-enchantment of the World.* Princeton, N.J.: Princeton University Press, 2006.

———. *Darwin the Writer.* Oxford: Oxford University Press, 2011.

———. *Dying to Know: Scientific Epistemology and Narrative in Victorian England.* Chicago: University of Chicago Press, 2002.

———. *The Realistic Imagination: English Fiction from "Frankenstein" to "Lady Chatterley."* Chicago: University of Chicago Press, 1983.

———. "Reflections on Darwin and Darwinizing." *Victorian Studies* 51 (2009): 223–45.

Lewes, George Henry. "Mr Darwin's Hypotheses—Part 2." *Fortnightly Review* 3, no. 18 (June 1868): 611–28.

———. "Mr Darwin's Hypotheses—Part 3." *Fortnightly Review* 4, no. 19 (July 1868): 61–80.

———. "Mr Darwin's Hypotheses—Part 4." *Fortnightly Review* 4, no. 23 (November 1868): 492–509.

———. *Problems of Life and Mind: First Series.* 2 vols. London: Trübner, 1875.

———. *Problems of Life and Mind: Third Series.* 2 vols. London: Trübner, 1879.

Liddell, Henry George, and Robert Scott, eds. *A Greek-English Lexicon.* Oxford: Clarendon Press, 1996.

"Life and Doctrine of Geoffroy St. Hilaire." *Westminster Review* 61, no. 119 (January 1854): 160–90.

Lightman, Bernard V. *Victorian Popularizers of Science*. Chicago: University of Chicago Press, 2007.

———. *Victorian Science in Context*. Chicago: University of Chicago Press, 1997.

Linnaeus, Carl, and Botanical Society at Lichfield. *The Families of Plants, with Their Natural Characters, According to the Number, Figure, Situation, and Proportion of All Parts of Fructification*. Lichfield: Printed by J. Jackson, 1787.

Liu, Alan. "The Power of Formalism: The New Historicism." *ELH* 56, no. 4 (1989): 721–71.

Lockhart, J. G. *Memoirs of the Life of Sir Walter Scott*. Edinburgh: A. and C. Black, 1852.

Lovejoy, Arthur O. *The Great Chain of Being: A Study of the History of an Idea*. Cambridge, Mass.: Harvard University Press, 1936.

Lukács, György. *The Historical Novel*. London: Merlin Press, 1962.

Lumsden, Alison. *Walter Scott and the Limits of Language*. Edinburgh: Edinburgh University Press, 2010.

Lynch, Deidre. *The Economy of Character: Novels, Market Culture, and the Business of Inner Meaning*. Chicago: University of Chicago Press, 1998.

———. "'Wedded to Books': Bibliomania and the Romantic Essayists." *Romantic Circles*, February 1, 2004. http://www.rc.umd.edu/praxis/libraries/lynch/lynch.html.

Macaulay, Thomas Babington. *Critical, Historical and Miscellaneous Essays*. Edited by Edwin Percy Whipple. New York: Sheldon, 1860.

———. "Neale's Romance of History: England." *Edinburgh Review* 47 (May 1828): 331–67.

Macfarren, George A. *Six Lectures on Harmony*. 2nd ed. London: Longmans, Green, Reader, and Dyer, 1867.

Machiavelli, Niccolò. "Letter to Vettori." In *Selected Political Writings*, edited by David Wootton, 1–4. Indianapolis: Hackett, 1994.

Manning, Susan. "Antiquarianism, the Scottish Science of Man, and the Emergence of Modern Disciplinarity." In *Scotland and the Borders of Romanticism*, edited by Leith Davis, Ian Duncan, and Janet Sorensen, 57–76. New York: Cambridge University Press, 2010.

———. *Poetics of Character: Transatlantic Encounters, 1700–1900*. New York: Cambridge University Press, 2013.

Mansell, Darrell. "George Eliot's Conception of 'Form.'" *Studies in English Literature, 1500–1900* 5, no. 4 (1965): 651–62.

Marcus, Sharon. *Between Women: Friendship, Desire, and Marriage in Victorian England*. Princeton, N.J.: Princeton University Press, 2007.

Martin, Robert Bernard. *Tennyson, the Unquiet Heart*. Oxford: Clarendon Press, 1980.

Martineau, Harriet. *Life in the Wilds: A Tale*. London: Charles Fox, 1832.

———. *Miscellanies*. Vol. 1. Boston: Hilliard, Gray, 1836.

Mathias, Thomas James. *The Pursuits of Literature: A Satirical Poem in Dialogue*. 3rd ed. London: T. Becket, 1797.

Matsen, Patricia P., Philip B. Rollinson, and Marion Sousa, eds. *Readings from Classical Rhetoric*. Carbondale: Southern Illinois University Press, 1990.

Matz, Aaron. *Satire in an Age of Realism*. Cambridge: Cambridge University Press, 2010.

McCaw, Neil. *George Eliot and Victorian Historiography: Imagining the National Past*. New York: St. Martin's Press, 2000.

McClintock, Anne. *Imperial Leather: Race, Gender, and Sexuality in the Colonial Contest.* New York: Routledge, 1995.

McGann, Jerome J. *The Beauty of Inflections: Literary Investigations in Historical Method and Theory.* Oxford: Clarendon Press, 1985.

———. *The Romantic Ideology: A Critical Investigation.* Chicago: University of Chicago Press, 1983.

McKeon, Michael. "Theory and Practice in Historical Method." In *Rethinking Historicism from Shakespeare to Milton,* edited by Ann Baynes Coiro and Thomas Fulton, 40–64. Cambridge: Cambridge University Press, 2012.

McLane, Maureen N. *Balladeering, Minstrelsy, and the Making of British Romantic Poetry.* Cambridge: Cambridge University Press, 2008.

Meillassoux, Quentin. *After Finitude: An Essay on the Necessity of Contingency.* Translated by Ray Brassier. London: Continuum, 2009.

———. *The Number and the Siren: A Decipherment of Mallarmé's "Coup de Dés."* Translated by Robin Mackay. Falmouth, U.K.: Urbanomic, 2012.

Meinhold, W. *Mary Schweidler, the Amber Witch.* London: John Murray, 1844.

Melas, Natalie. *All the Difference in the World: Postcoloniality and the Ends of Comparison.* Stanford, Calif.: Stanford University Press, 2007.

Mellor, Anne Kostelanetz. *Blake's Human Form Divine.* Berkeley: University of California Press, 1974.

Merrell, Floyd. *Entangling Forms within Semiosic Processes.* New York: de Gruyter Mouton, 2010.

Metzger, Bruce M., and Bart D. Ehrman. *The Text of the New Testament: Its Transmission, Corruption, and Restoration.* New York: Oxford University Press, 2005.

Michelet, Jules. *On History: Introduction to World History (1831), Opening Address at the Faculty of Letters, 9 January 1834, Preface to History of France (1869).* Translated by Flora Kimmich, Lionel Gossman, and Edward K. Kaplan. Cambridge: Open Book, 2013.

———. *The People.* Translated by G. H. Smith. London, 1846.

Mill, John Stuart. "The Spirit of the Age." In *The Spirit of the Age: Victorian Essays,* edited by Gertrude Himmelfarb, 50–79. New Haven: Yale University Press, 2007.

———. *A System of Logic, Ratiocinative and Inductive: Being a Connected View of the Principles of Evidence, and Methods of Scientific Investigation.* 2 vols. London: J. W. Parker, 1843.

Miller, Andrew H. *The Burdens of Perfection: On Ethics and Reading in Nineteenth-Century British Literature.* Ithaca, N.Y.: Cornell University Press, 2008.

Miller, J. Hillis. *Reading for Our Time: "Adam Bede" and "Middlemarch" Revisited.* Edinburgh: Edinburgh University Press, 2012.

Milne, Colin, and Carl Linnaeus. *The Institutes of Botany; Containing Accurate, Compleat and Early Descriptions of All the Known Genera of Plants.* London: Sold by W. Griffin, 1771.

Mitchell, Rebecca N. "The Rosamond Plots: Alterity and the Unknown in *Jane Eyre* and *Middlemarch.*" *Nineteenth-Century Literature* 66, no. 3 (2011): 307–27.

Mitchell, Robert. *Experimental Life: Vitalism in Romantic Science and Literature.* Baltimore: Johns Hopkins University Press, 2013.

Momigliano, Arnaldo. "Ancient History and the Antiquarian." *Journal of the Warburg and Courtauld Institutes* 13 (1950): 285–315.

Moretti, Franco. *Atlas of the European Novel, 1800–1900*. London: Verso, 1998.

Morgan, Monique R. *Narrative Means, Lyric Ends: Temporality in the Nineteenth-Century British Long Poem*. Columbus: Ohio State University Press, 2009.

Mueller-Vollmer, Kurt. *The Hermeneutics Reader: Texts of the German Tradition from the Enlightenment to the Present*. New York: Continuum, 1985.

Nealon, Jeffrey T. *Plant Theory: Biopower & Vegetable Life*. Stanford, Calif.: Stanford University Press, 2016.

———. *Post-Postmodernism; or, The Logic of Just-in-Time Capitalism*. Stanford, Calif.: Stanford University Press, 2012.

Newport, Kenneth G. C. *Apocalypse and Millennium: Studies in Biblical Eisegesis*. Cambridge: Cambridge University Press, 2008.

Newton, Isaac. *Sir Isaac Newton's "Daniel and the Apocalypse."* London: J. Murray, 1922.

Nicholas, Mary A., and Cynthia Ann Ruder. "In Search of the Collective Author: Fact and Fiction from the Soviet 1930s." *Book History* 11, no. 1 (2008): 221–44.

Nicholes, Joseph. "Vertical Context in *Middlemarch*: George Eliot's Civil War of the Soul." *Nineteenth-Century Literature* 45, no. 2 (1990): 144–75.

Nicolson, Harold. *Tennyson; Aspects of His Life, Character and Poetry*. Boston: Houghton Mifflin, 1925.

"Notes." *Nature* 46, no. 1198 (October 13, 1892): 572.

Nunokawa, Jeff. "*In Memoriam* and the Extinction of the Homosexual." *ELH* 58, no. 2 (1991): 427–38.

O'Brien, Karen. "Between Enlightenment and Stadial History: William Robertson on the History of Europe." *Journal for Eighteenth-Century Studies* 16, no. 1 (March 1993): 53–64.

———. *Narratives of Enlightenment: Cosmopolitan History from Voltaire to Gibbon*. Cambridge: Cambridge University Press, 1997.

Oliver, Susan. "Sir Walter Scott's Transatlantic Ecology." *Wordsworth Circle* 44, nos. 2–3 (2013): 115–20.

Olsen, Mark J., Darren Stephens, Devin Griffiths, Patrick Daugherty, George Georgiou, and Brent L. Iverson. "Function-Based Isolation of Novel Enzymes from a Large Library." *Nature Biotechnology* 18, no. 10 (2000): 1071–4.

"On the Various Contrivances by Which British and Foreign Orchids . . . " *British Quarterly Review* 36, no. 71 (July 1862): 243–4.

Ovsienko, Valentin, and Serge Tabachnikov. *Projective Differential Geometry Old and New: From the Schwarzian Derivative to the Cohomology of Diffeomorphism Groups*. Cambridge: Cambridge University Press, 2005.

Owen, Richard. "Darwin 'On the Origin of Species.'" *Edinburgh Review* 77, no. 111 (April 1860): 487–532.

———. *The Life of Richard Owen*. 2 vols. London: J. Murray, 1894.

———. *On the Archetype and Homologies of the Vertebrate Skeleton*. London: J. Van Voorst, 1848.

———. *Report on the Archetype and Homologies of the Vertebrate Skeleton*. London: Printed by Richard and John E. Taylor, 1847.

Packham, Catherine. *Eighteenth-Century Vitalism: Bodies, Culture, Politics*. Basingstoke: Palgrave Macmillan, 2012.

Pattison, Robert. *Tennyson and Tradition*. Cambridge, Mass.: Harvard University Press, 1979.

Peacham, Henry. *The Garden of Eloquence, Conteining the Most Excellent Ornaments, Exornations, Lightes, Flowers, and Formes of Speech, Commonly Called the Figures of Rhetorike.* London: Printed by R[ichard] F[ield] for H. Iackson dvvelling in Fleetstrete, 1593.

———. *The Garden of Eloquence, Conteyning the Figures of Grammar and Rhetorick.* London: H. Jackson, 1577.

Pearsall, Cornelia D. J. *Tennyson's Rapture: Transformation in the Victorian Dramatic Monologue.* New York: Oxford University Press, 2008.

Peirce, Charles S. *Collected Papers of Charles Sanders Peirce.* Edited by Charles Hartshorne and Paul Weiss. 8 Vols. Cambridge, Mass.: Belknap Press of Harvard University Press, 1960–6.

Phillips, Mark. *On Historical Distance.* New Haven: Yale University Press, 2013.

———. *Society and Sentiment: Genres of Historical Writing in Britain, 1740–1820.* Princeton, N.J.: Princeton University Press, 2000.

Pinkerton, John. *An Essay on Medals; or, An Introduction to the Knowledge of Ancient and Modern Coins and Medals; Especially Those of Greece, Rome, and Britain.* 3rd ed. 2 vols. London: T. Cadell and W. Davies, 1808.

Pinna, Mário C. C. "Concepts and Tests of Homology in the Cladistic Paradigm." *CLA Cladistics* 7, no. 4 (1991): 367–94.

Pinney, Thomas. "The Authority of the Past in George Eliot's Novels." *Nineteenth-Century Fiction* 21, no. 2 (1966): 131–47.

Piper, Andrew. *Dreaming in Books: The Making of the Bibliographic Imagination in the Romantic Age.* Chicago: University of Chicago Press, 2013.

Pittock, Murray. *The Reception of Sir Walter Scott in Europe.* London: Continuum, 2006.

Plutarchus, and Thomas Wyatt. *T. Wyatis Translatyon of Plutarckes Boke Of the Quyete of Mynde.* London: Pynson, 1528.

"The Poetry of Sorrow." *The Times,* November 28, 1851.

Poncelet, Jean Victor. *Traité des propriétés projectives des figures: ouvrage utile à ceux qui s'occupent des applications de la géométrie descriptive et d'opérations géométriques sur le terrain.* Paris: Gauthier-Villars, 1822.

Potkay, Adam. *The Passion for Happiness: Samuel Johnson and David Hume.* Ithaca, N.Y.: Cornell University Press, 2000.

Priestley, Joseph. *The History and Present State of Electricity with Original Experiments.* 3rd ed. 2 vols. London: C. Bathurst, and T. Lowndes [etc.], 1775.

———. *A History of the Corruptions of Christianity* . . . 2 vols. Birmingham: J. Johnson, 1782.

Priestman, Martin. Introduction to *The Collected Writings of Erasmus Darwin,* vol. 1. Bristol: Thoemmes Continuum, 2004.

———. *Romantic Atheism: Poetry and Freethought, 1780–1830.* Cambridge: Cambridge University Press, 2000.

Prout, William. *Chemistry, Meteorology, and the Function of Digestion Considered with Reference to Natural Theology.* Edited by John William Griffith. 4th ed. London: H. G. Bohn, 1855.

Quintilian. *Quintilian's Institutes of Oratory; or, Education of an Orator.* Translated by John Selby Watson. 12 vols. London: G. Bell and Sons, 1892.

Raby, Peter. *Bright Paradise: Victorian Scientific Travellers*. Princeton, N.J.: Princeton University Press, 1997.

Reill, Peter Hanns. *Vitalizing Nature in the Enlightenment*. Berkeley: University of California Press, 2005.

Reinikka, Merle A. *A History of the Orchid*. Coral Gables, Fla.: University of Miami Press, 1972.

Reiss, John O. "Natural Selection and the Conditions for Existence: Representational vs. Conditional Teleology in Biological Explanation." *History and Philosophy of the Life Sciences* 27, no. 2 (2005): 249–80.

"Results of the Census of 1851." *Westminster Review* 61, no. 120 (April 1854): 323.

Richards, I. A. *The Philosophy of Rhetoric*. New York: Oxford University Press, 1936.

Richards, Robert J. *The Romantic Conception of Life: Science and Philosophy in the Age of Goethe*. Chicago: University of Chicago Press, 2002.

Richardson, Alan. *British Romanticism and the Science of the Mind*. Cambridge: Cambridge University Press, 2005.

Ricks, Christopher. *Tennyson*. 2nd ed. Berkeley: University of California Press, 1989.

———. "Tennyson's 'Hail, Briton!' and 'Tithon': Some Corrections." *Review of English Studies* 15, no. 57 (1964): 53–5.

Ricœur, Paul. *Rule of Metaphor: The Creation of Meaning in Language*. 2nd ed. London: Routledge, 2003.

———. *Time and Narrative*. 3 vols. Chicago: University of Chicago Press, 1984.

Riehl, Wilhelm Heinrich. *The Natural History of the German People*. Translated by David J. Diephouse. Lewiston, N.Y.: Edwin Mellen, 1990.

Rigney, Ann. *The Afterlives of Walter Scott: Memory on the Move*. Oxford: Oxford University Press, 2012.

Rischin, Abigail S. "Beside the Reclining Statue: Ekphrasis, Narrative, and Desire in *Middlemarch*." *PMLA* 111, no. 5 (1996): 1121–32.

Ritson, Joseph. "Observations on the Ancient English Minstrels." In *Ancient Songs and Ballads: From the Reign of King Henry the Second to the Revolution*, edited by Joseph Frank, vol. 1. London: Printed for Payne and Foss, by T. Davison, 1829.

Roger, Jacques, and L. Pearce Williams. *Buffon: A Life in Natural History*. Ithaca, N.Y.: Cornell University Press, 1997.

Rowlinson, Matthew Charles. *Real Money and Romanticism*. Cambridge: Cambridge University Press, 2010.

Rudwick, M. J. S. *Bursting the Limits of Time: The Reconstruction of Geohistory in the Age of Revolution*. Chicago: University of Chicago Press, 2005.

Rudwick, M. J. S., and Georges Cuvier. *Georges Cuvier, Fossil Bones, and Geological Catastrophes: New Translations & Interpretations of the Primary Texts*. Chicago: University of Chicago Press, 1997.

Rudy, Jason R. *Electric Meters: Victorian Physiological Poetics*. Athens: Ohio University Press, 2009.

Rupke, Nicolaas A. *Richard Owen: Biology without Darwin*. Chicago: University of Chicago Press, 2008.

Ruse, Michael. *Darwinism and Its Discontents*. Cambridge: Cambridge University Press, 2006.

Ruskin, John. *Modern Painters*. 3rd ed. London: Smith, Elder, 1851.

Russett, Margaret. *Fictions and Fakes: Forging Romantic Authenticity, 1760–1845.* Cambridge: Cambridge University Press, 2006.

Russett, Margaret, and Joseph A. Dane. "'Everlastinge to Posterytie': Chatterton's Spirited Youth." *Modern Language Quarterly* 63, no. 2 (2002): 141–65.

"Ruth and Villette." *Westminster Review* 59, no. 116 (April 1853): 474–91.

Ruthven, Kenneth Knowles. *Faking Literature.* Cambridge: Cambridge University Press, 2001.

Ryan, Vanessa Lyndal. *Thinking without Thinking in the Victorian Novel.* Baltimore: Johns Hopkins University Press, 2012.

Sabine, George H. "Descriptive and Normative Sciences." *Philosophical Review* 21, no. 4 (1912): 433–50.

Sachs, Jonathan. *Romantic Antiquity: Rome in the British Imagination, 1789–1832.* New York: Oxford University Press, 2010.

Said, Edward W. *Beginnings: Intention and Method.* New York: Basic Books, 1975.

———. *Culture and Imperialism.* New York: Knopf, distributed by Random House, 1993.

———. *Orientalism.* New York: Vintage Books, 1979.

Saint-Amour, Paul K. *The Copywrights: Intellectual Property and the Literary Imagination.* Ithaca, N.Y.: Cornell University Press, 2003.

Saint-Hilaire, Étienne Geoffroy. "Mémoires sur la structure et les usages de l'appareil olfactif dans les poissons." *Annales des Sciences Naturelles* 7 (1825): 322–54.

Saussure, Ferdinand de. *Course in General Linguistics.* Edited by Charles Bally, Albert Sechehaye, and Albert Riedlinger. Translated by Wade Baskin. New York: McGraw-Hill, 1966.

Saussy, Haun. "Exquisite Cadavers Stitched from Fresh Nightmares: Of Memes, Hives and Selfish Genes." In *Comparative Literature in an Age of Globalization,* edited by Haun Saussy, 3–42. Baltimore: Johns Hopkins University Press, 2006.

Schleiermacher, Friedrich. "On the Different Modes of Translating." In *Translation, History and Culture,* translated by André Lefevere and edited by Susan Bassnett and André Lefevere, 141–65. London: Pinter Publishers, 1990.

Schor, Hilary M. *Curious Subjects: Women and the Trials of Realism.* Oxford: Oxford University Press, 2013.

"Science." *Westminster Review* 61, no. 120 (April 1854): 580–95.

Scott, Walter. *The Antiquary.* Edited by David Hewitt. Edinburgh: Edinburgh University Press, 1995.

———. "Ballads and Songs," n.d. National Library of Scotland.

———. "Biographical Memoir of Dr. Leyden." In *The Edinburgh Annual Register, for 1811,* vol. 4, pt. 2, xl–lxviii. Edinburgh: John Ballantyne, 1813.

———. *Edinburgh Annual Register.* Vol. 4, pt. 1. Edinburgh: John Ballantyne, 1813.

———. "Essay on Imitation of the Ancient Ballad." In *Minstrelsy of the Scottish Border,* vol. 4. Edinburgh: A. and C. Black, 1807.

———. *Guy Mannering.* Edited by Peter Garside. Edinburgh: University Press, 1999.

———. *Introductions, and Notes and Illustrations to the Novels, Tales, and Romances, of the Author of Waverley.* Edinburgh: Cadell, 1833.

———. *Ivanhoe.* Edited by Graham Tulloch. Edinburgh: Edinburgh University Press, 1998.

——. *The Letters of Sir Walter Scott, 1787–1807.* Edited by Davidson Cook and W. M. Parker. 4 vols. London: Constable, 1932.

——. *Minstrelsy of the Scottish Border: Consisting of Historical and Romantic Ballads; with a Few of Modern Date, Founded upon Local Tradition.* 2nd ed. 3 vols. Edinburgh: Ballantyne, 1803.

——. "Tales of My Landlord." *Quarterly Review* 16, no. 32 (January 1817): 430–80.

——. *Waverley.* Edited by Peter Garside. Edinburgh: Edinburgh University Press, 2007.

——. *Waverley; or, 'Tis Sixty Years Since.* London: Penguin, 1984.

Secord, James A. "Darwin and the Breeders: A Social History." In *The Darwinian Heritage,* edited by David Kohn, 519–42. Princeton, N.J.: Princeton University Press, 1985.

——. *Victorian Sensation.* Chicago: University of Chicago Press, 2000.

Sedgwick, Eve Kosofsky. *Between Men: English Literature and Male Homosocial Desire.* New York: Columbia University Press, 1985.

Sendry, Joseph. "*In Memoriam*: The Minor Manuscripts." *Harvard Library Bulletin* 27 (1979): 36–65.

Seward, Anna. *Memoirs of the Life of Dr. Darwin: Chiefly during His Residence at Lichfield, with Anecdotes of His Friends and Criticisms on His Writings.* London: Johnson, 1804.

Sewell, William H. "Marc Bloch and the Logic of Comparative History." *History and Theory* 6, no. 2 (1967): 208–18.

Shapin, Steven, and Simon Schaffer. *"Leviathan" and the Air-Pump: Hobbes, Boyle, and the Experimental Life.* Princeton, N.J.: Princeton University Press, 1985.

Shatto, Susan. "Tennyson's Revisions of *In Memoriam*." *Victorian Poetry* 16, no. 4 (1978): 341–56.

Shaw, Harry E. *Narrating Reality: Austen, Scott, Eliot.* Ithaca, N.Y.: Cornell University Press, 1999.

Shelley, Percy Bysshe. *Essays, Letters from Abroad, Translations and Fragments.* Edited by Mary Wollstonecraft Shelley. London: Edward Moxon, 1840.

Shih, Shu-mei. "Comparison as Relation." In *Comparison: Theories, Approaches, Uses,* edited by Rita Felski and Susan Stanford Friedman, 79–98. Baltimore: Johns Hopkins University Press, 2013.

Shuttleworth, Sally. *George Eliot and Nineteenth-Century Science: The Make-Believe of a Beginning.* Cambridge: Cambridge University Press, 1984.

——. *The Mind of the Child: Child Development in Literature, Science, and Medicine, 1840–1900.* Oxford: Oxford University Press, 2010.

Siegel, Jonah. *Haunted Museum: Longing, Travel, and the Art-Romance Tradition.* Princeton, N.J.: Princeton University Press, 2005.

Silverman, Kaja. *The Miracle of Analogy; or, The History of Photography, Part 1.* Stanford, Calif.: Stanford University Press, 2015.

Simmel, Georg. *The Problems of the Philosophy of History.* New York: Free Press, 1977.

Smith, Adam. *An Inquiry into the Nature and Causes of the Wealth of Nations.* 3 vols. Dublin: Whitestone, Chamberlaine, and W. Watson, 1776.

Smith, Jonathan. *Charles Darwin and Victorian Visual Culture.* Cambridge: Cambridge University Press, 2006.

Smith, William. *Stratigraphical System of Organized Fossils.* London: Printed for E. Williams, 1817.

Snow, C. P. *The Two Cultures and the Scientific Revolution*. Cambridge: Cambridge University Press, 1959.

Snyder, Laura J. *Eye of the Beholder: Johannes Vermeer, Antoni van Leeuwenhoek, and the Reinvention of Seeing*. New York: W. W. Norton, 2015.

———. *Reforming Philosophy: A Victorian Debate on Science and Society*. Chicago: University of Chicago press, 2006.

Spencer, Herbert. *The Principles of Psychology*. 3rd ed. London: Longman, 1855.

Stafford, Barbara. *Visual Analogy: Consciousness as the Art of Connecting*. Cambridge, Mass.: MIT Press, 2001.

St. Clair, William. *The Reading Nation in the Romantic Period*. Cambridge: Cambridge University Press, 2004.

Steiner, T. R. *English Translation Theory, 1650–1800*. Assen: Van Gorcum, 1975.

Stevens, L. Robert. "Darwin's Humane Reading: The Anaesthetic Man Reconsidered." *Victorian Studies* 26, no. 1 (1982): 51–63.

Stocking, George W. *Victorian Anthropology*. New York: Free Press, 1987.

Strauss, David Friedrich, and George Eliot. *The Life of Jesus, Critically Examined*. 2 vols. New York: C. Blanchard, 1856.

Sutherland, Kathryn. "Fictional Economies: Adam Smith, Walter Scott and the Nineteenth-Century Novel." *ELH* 54, no. 1 (Spring 1987): 97–127.

Swift, Jonathan. *Gulliver's Travels*. 2 vols. London: Jones, 1826.

Taylor, Charles. *A Secular Age*. Cambridge, Mass.: Belknap Press of Harvard University Press, 2007.

Taylor, Jesse Oak. "The Novel as Climate Model: Realism and the Greenhouse Effect in *Bleak House*." *Novel* 46, no. 1 (2013): 1–25.

———. *The Sky of Our Manufacture: The London Fog in British Fiction from Dickens to Woolf*. Charlottesville: University of Virginia Press, 2016.

Tennyson, Alfred. "Darwin's Gemmule." HM 19493, Bixby Locker-Lampson Collection, The Huntington Library, San Marino, Calif.

———. *In Memoriam*. Edited by Susan Shatto and Marion Shaw. Oxford: Clarendon Press, 1982.

———. *In Memoriam*. Edited by Erik Irving Gray. 2nd ed. New York: W. W. Norton, 2004.

Tennyson, Hallam. *Alfred Lord Tennyson; A Memoir by His Son*. 2 vols. New York: Macmillan, 1897.

"Thackeray's Works." *Westminster Review* 59, no. 116 (April 1853): 363–88.

Thompson, E. P. "History from Below." *Times Literary Supplement*, April 7, 1966.

Thomson, Patricia. "Sir Thomas Wyatt: Classical Philosophy and English Humanism." *Huntington Library Quarterly* 25, no. 2 (1962): 79–96.

Tosh, John. *The Pursuit of History: Aims, Methods, and New Directions in the Study of Modern History*. London: Longman, 1984.

Trevor-Roper, Hugh. "Sir Walter Scott and History." *The Listener* 86 (August 1971): 225–34.

Trigger, Bruce G. *A History of Archaeological Thought*. Cambridge: Cambridge University Press, 1989.

Trollope, Anthony. *An Autobiography*. 2 vols. Edinburgh: William Blackwood and Sons, 1888.

Trumpener, Katie. *Bardic Nationalism: The Romantic Novel and the British Empire*. Princeton, N.J.: Princeton University Press, 1997.

Tucker, Herbert F. *Epic: Britain's Heroic Muse, 1790–1910*. Oxford: Oxford University Press, 2008.

———. "The Fix of Form: An Open Letter." *Victorian Literature and Culture* 27, no. 2 (1999): 531–5.

———. *Tennyson and the Doom of Romanticism*. Cambridge, Mass.: Harvard University Press, 1988.

Turner, Frank M. *Between Science and Religion: The Reaction to Scientific Naturalism in Late Victorian England*. New Haven: Yale University Press, 1974.

Tylor, Edward Burnett. *Primitive Culture*. 2 vols. London: Murray, 1871.

Tytler, Alexander Fraser. *Essay on the Principles of Translation*. London: T. Cadell and W. Davies, 1797.

Underwood, Gary N. "Linguistic Realism in *Roderick Random*." *Journal of English and Germanic Philology* 69, no. 1 (1970): 32–40.

"The Universal Postulate." *Westminster Review* 60, no. 118 (October 1853): 513–50.

US EPA, Climate Change Division. "Future Climate Change." Overviews & Factsheets. Accessed November 25, 2015. http://www3.epa.gov/climatechange/science/future.html#Temperature.

Vance, Norman. *The Victorians and Ancient Rome*. Oxford: Blackwell, 1997.

Vickers, Neil. "Coleridge and the Idea of 'Psychological' Criticism." *Journal for Eighteenth-Century Studies* 30, no. 2 (June 2007): 261–78.

Voskuil, Lynn M. "Victorian Orchids and the Forms of Ecological Society." In *Strange Science: Investigating the Limits of Knowledge in the Victorian Age*, edited by Shalyn Claggett and Lara Karpenko. Ann Arbor: University of Michigan Press, 2016.

Walker, Alison, Arthur MacGregor, and Michael Hunter. *From Books to Bezoars: Sir Hans Sloane and His Collections*. London: British Library, 2012.

Wallace, Alfred Russel. "On the Tendency of Species to Form Varieties." *Zoological Journal of the Linnean Society* 3 (1859): 53–62.

Wasserman, Earl R. *The Subtler Language: Critical Readings of Neoclassic and Romantic Poems*. Baltimore: Johns Hopkins Press, 1959.

Whewell, William. *Astronomy and General Physics Considered with Reference to Natural Theology*. Bridgewater Treatises on the Power, Wisdom and Goodness of God as Manifested in the Creation, Treatise III. London: W. Pickering, 1833.

———. *The Philosophy of the Inductive Sciences, Founded upon Their History*. London, 1840.

White, Hayden V. *Metahistory: The Historical Imagination in Nineteenth-Century Europe*. Baltimore: Johns Hopkins University Press, 1973.

Wiesenfarth, Joseph. "*Middlemarch*: The Language of Art." *PMLA* 97, no. 3 (1982): 363–77.

Wilberforce, Samuel. "Darwin's 'On the Origin of Species.'" *London Quarterly Review* 115 (July 1860): 225–64.

Williams, John W., and Stephen T. Jackson. "Novel Climates, No-Analog Communities, and Ecological Surprises." *Frontiers in Ecology and the Environment* 5, no. 9 (October 1, 2007): 475–82.

Williams, Raymond. "The Knowable Community in George Eliot's Novels." *NOVEL: A Forum on Fiction* 2, no. 3 (1969): 255–68.

Winchester, Simon. *The Map That Changed the World: William Smith and the Birth of Modern Geology*. New York: HarperCollins, 2001.

Wippel, John F. "Metaphysics." In *The Cambridge Companion to Aquinas*, edited by Norman Kretzmann and Eleonore Stump, 85–127. Cambridge: Cambridge University Press, 1993.

Wohlfarth, Marc E. "*Daniel Deronda* and the Politics of Nationalism." *Nineteenth-Century Literature* 53, no. 2 (1998): 188–210.

Wolfson, Susan J. *Formal Charges: The Shaping of Poetry in British Romanticism*. Stanford, Calif.: Stanford University Press, 1997.

———. "Reading for Form." *Modern Language Quarterly* 61, no. 1 (2000): 1–16.

Woloch, Alex. *The One vs. the Many: Minor Characters and the Space of the Protagonist in the Novel*. Princeton, N.J.: Princeton University Press, 2003.

Woodward, John. *An Essay towards a Natural History of the Earth: And Terrestrial Bodies Especially Minerals: As Also of the Sea, Rivers, and Springs*. 2nd ed. London: Printed for T. W. for Richard Wilkin, 1702.

Woolf, D. R. *The Social Circulation of the Past: English Historical Culture, 1500–1730*. Oxford: Oxford University Press, 2003.

Wordsworth, William. *The Prelude; or, Growth of a Poet's Mind*. Edited by Ernest De Selincourt. Oxford: Clarendon Press, 1959.

Wylie, Alison. "The Reaction against Analogy." *Advances in Archaeological Method and Theory* 8 (1985): 63–111.

Yeazell, Ruth Bernard. *Art of the Everyday: Dutch Painting and the Realist Novel*. Princeton, N.J.: Princeton University Press, 2008.

Young, G. M. "Scott and the Historians." In *Sir Walter Scott Lectures, 1940–1948*, edited by Herbert John Clifford Grierson, 78–107. Edinburgh: University Press, 1950.

Zemka, Sue. *Time and the Moment in Victorian Literature and Society*. New York: Cambridge University, 2011.

Zinken, Jörg. "Discourse Metaphors: The Link between Figurative Language and Habitual Analogies." *Cognitive Linguistics* 18, no. 3 (2007): 445–66.

Zimmerman, Virginia. *Excavating Victorians*. Albany: State University of New York Press, 2008.